T0306103

Introduction to MIMO Communications

This accessible, self-contained guide contains everything you need to get up to speed on the theory and implementation of MIMO techniques.

In-depth coverage of topics such as RF propagation, space-time coding, spatial multiplexing, OFDM in MIMO for broadband applications, the theoretical MIMO capacity formula, and channel estimation, will give you a deep understanding of how the results are obtained, while detailed descriptions of how MIMO is implemented in commercial WiFi and LTE networks will help you apply the theory to practical wireless systems.

Key concepts in matrix mathematics and information theory are introduced and developed as you need them, and key results are derived step by step, with no details omitted. Including numerous worked examples, and end-of-chapter exercises to reinforce and solidify your understanding, this is the perfect introduction to MIMO for anyone new to the field.

Jerry R. Hampton is a research engineer with over 30 years' experience in communications systems engineering. He is a member of the principal professional staff in the Applied Physics Laboratory, and an Adjunct Professor in the Whiting School of Engineering, at The Johns Hopkins University, where he teaches a graduate course in MIMO wireless communications.

"This is a well-organized comprehensive treatise on MIMO principles, methods, and applications. While many concepts are introduced in intuitively pleasing ways; the integration of detailed step-by-step mathematical developments of MIMO principles, propagation models, channel characterizations, and applications of MIMO in commercial systems adds tremendous depth and understanding to the concepts. After studying this text, if readers have interests in topics not covered, they will very likely be able to understand or author for themselves advanced MIMO literature on such topics."

David Nicholson, Communications consultant

Introduction to MIMO Communications

JERRY R. HAMPTON
The Johns Hopkins University

Shaftesbury Road, Cambridge CB2 8EA, United Kingdom

One Liberty Plaza, 20th Floor, New York, NY 10006, USA

477 Williamstown Road, Port Melbourne, VIC 3207, Australia

314–321, 3rd Floor, Plot 3, Splendor Forum, Jasola District Centre, New Delhi – 110025, India

103 Penang Road, #05–06/07, Visioncrest Commercial, Singapore 238467

Cambridge University Press is part of Cambridge University Press & Assessment, a department of the University of Cambridge.

We share the University's mission to contribute to society through the pursuit of education, learning and research at the highest international levels of excellence.

www.cambridge.org
Information on this title: www.cambridge.org/9781107042834

First published 2014

A catalogue record for this publication is available from the British Library

ISBN 978-1-107-04283-4 Hardback

Additional resources for this publication at www.cambridge.org/hampton

Contents

Preface

This book is an outgrowth of a graduate course I have taught for the past four years on MIMO Wireless Communications in the Engineering for Professionals (EP) Program within the Whiting School of Engineering at The Johns Hopkins University. When I began to develop the course in the spring of 2006, I initially thought I would simply choose a textbook from the collection of numerous books that had been written on MIMO communications at that time. As I began studying these books, however, I found that, although they were each excellent in various ways, none of them was as accessible to the average practicing communications engineer or early level electrical engineering graduate student as I had hoped. Many of these books were written by experts in the field, researchers who had made seminal contributions in the area of MIMO communications, but the prerequisites needed to follow and understand the details in their presentations were often above the level of expertise of those being introduced to MIMO for the first time.

This book is my attempt to remedy this problem. In developing the course and in writing this book, I have tried to make the concepts and techniques associated with MIMO communications accessible to an average communications engineer with an undergraduate degree in electrical engineering. I assume that readers are familiar with digital communication techniques and that they have had a formal course (or its equivalent) in digital signal processing; however, I do not assume readers are familiar with information theory or are proficient in advanced matrix mathematics, areas of expertise that are normally assumed in the MIMO literature and in many of the books that have been published on this topic. When knowledge in these areas is required to understand MIMO concepts, I have attempted to include the necessary information on those topics in the book so that it is not necessary to consult external resources. In this sense, the book has been designed to be as self-contained as possible.

As its name suggests, this book is intended to provide an introduction to the field of MIMO communications, and is, therefore, by design not encyclopedic. My goal has been to provide readers new to MIMO communications with an understanding of the basic concepts and methods, thereby laying a foundation for further study and providing them with the ability to understand the vast literature on this subject.

Although my goal has been to make the concepts of MIMO understandable to the average communications engineer, I have tried to remain rigorous at the same time. One of my initial frustrations when I began searching for a textbook was that there were often large steps or gaps in derivations that were not explained, so I have attempted to fill in the details of as many gaps as possible in my book, in some cases relegating the details to appendices to avoid interrupting the flow of the text.

A third feature of this book that I hope will be useful to readers is that it contains descriptions of how MIMO concepts are implemented in practical systems. MIMO techniques have now become as commonplace in wireless communications systems as modulation and error correction coding, so there is no shortage of examples of systems that use MIMO methods. In this book, I focus on WiFi (IEEE 802.11n) and LTE and explain how these two popular wireless standards implement MIMO concepts in practice.

Chapter 1 provides an overview of MIMO communication concepts and includes a section on key matrix properties and identities that are used throughout the book. This initial chapter explains the different types of MIMO schemes, defines fundamental concepts such as spatial diversity and spatial multiplexing, and presents measured performance results that demonstrate the performance benefits of MIMO.

Chapter 2 is devoted to derivation of the MIMO capacity formula, which predicts the maximum error-free data rate that can be supported by a MIMO communication system. This formula is used later in Chapter 3 to provide useful conceptual insights into how multiple antennas enable increased spectral efficiency. Although the MIMO capacity formula is derived using concepts from information theory, the chapter introduces concepts as necessary to derive the final result and does not assume the reader has a background in that subject.

Chapter 3 explores the implications of the MIMO capacity formula and uses it to compute the communications capacities of MIMO systems under various assumptions. The concepts of eigenmodes and channel rank are examined, and the spatial multiplexing technique called eigenbeamforming is derived and explained in this chapter.

Chapter 4 discusses RF propagation in general and develops the terminology and concepts used in characterizing multipath propagation in particular.

Chapter 5 presents several theoretical MIMO propagation models that have been developed based on theory and empirical results. Expressions for the channel model when both Rayleigh fading and line-of-sight propagation exist are also presented. These models are used to derive expressions for the dependence of the MIMO capacity on antenna correlation as well as on the amount of scattering in the channel.

Chapter 6 describes Alamouti coding, which is an important practical MIMO technique used to achieve transmit diversity. This chapter begins by examining the performance of ideal maximal ratio receive combining and then shows how Alamouti coding achieves diversity gain equal to a maximal ratio receive combiner.

Chapter 7 broadens the discussion begun in Chapter 6 to consider other types of coding techniques, called space-time codes, that can be used to achieve transmit spatial diversity. This chapter focuses on space-time block codes, but also introduces the reader to space-time trellis coding concepts. The chapter describes how to perform decoding, concluding with a presentation of representative performance results.

Chapter 8 addresses spatial multiplexing, which comprises the second major class of MIMO techniques. These techniques, which exploit multipath, enable MIMO systems to transmit higher data rates than can be achieved with conventional communication systems.

Chapter 9 discusses MIMO over broadband channels. Up to this point in the book, the assumption is that the bandwidth of the transmitted signal is smaller than the coherence bandwidth of the channel; however, in modern wireless communication systems this is seldom the case. In practice, broadband systems operate by employing OFDM signaling, so this chapter reviews OFDM and then shows how OFDM is used with the narrowband MIMO techniques developed earlier to support broadband service.

Chapter 10 discusses an important practical aspect of MIMO communications – the estimation of the properties of the communications channel. Since most MIMO techniques require that either the transmitter or the receiver (or both) have knowledge of the channel, channel estimation techniques are an essential aspect of any MIMO communication system. This chapter discusses the fundamental concepts used in MIMO channel estimation and describes how practical MIMO systems perform this function.

The book concludes with Chapter 11, which describes how MIMO is implemented in WiFi and LTE wireless communication systems.

I would like to conclude by acknowledging and thanking some key people that helped make this book possible. First, I want to thank the various students who have taken my course on MIMO Wireless Communications at Johns Hopkins over the past several years. Their penetrating questions have helped me improve both the course as well as this book. To the extent that this book succeeds in helping others understand MIMO concepts, I am indebted to these students.

In addition to my students in the EP program at Johns Hopkins, I would like to acknowledge Dennis Ryan at The Johns Hopkins University Applied Physics Laboratory who chairs the Janney Publication Program, which funds, on a competitive basis, sabbaticals for employees to write books and journal papers. I would like to thank Dennis and the Janney committee for granting me a sabbatical during the summer of 2012 to finish writing this book. Thanks also go to Rob Nichols for his encouragement and willingness to accommodate my absence from normal work duties during this sabbatical.

I would also like to express my gratitude to two colleagues who have provided invaluable support during the preparation of this book. Eric Yang shared his extensive knowledge of cellular wireless standards and guided me through the labyrinth of LTE and IEEE 802.11n standards documents that I used to write Chapter 11. Thanks Eric! I would also like to offer special thanks to Feng Ouyang, another colleague, who served as a sounding board for my interminable discussions on many aspects of MIMO theory during the lengthy gestation period of this book. Feng was incredibly patient and generous with his time while sharing his insights and mathematical expertise with me. Thanks Feng! This book would not have come about without Feng's and Eric's help.

Finally, I would like to thank my wife, Dorothy, for her support and patience during this project, which has consumed far too many of my weekends and nights over the past several years. I dedicate this book to her, to our two wonderful children, Jessica and Joshua, and finally, and ultimately, to God. *Soli Deo Gloria*

Jerry R. Hampton

1 Overview of MIMO communications

This chapter lays the foundations for the remainder of the book by presenting an overview of MIMO communications. Fundamental concepts and key terminology are introduced, and a summary of important matrix properties is provided, which will be referred to throughout the book. Some experimental results showing the benefits of MIMO are also presented.

1.1 What is MIMO?

Multiple Input Multiple Output communications, abbreviated MIMO, and normally pronounced like "My-Moe," refers to a collection of signal processing techniques that have been developed to enhance the performance of wireless communication systems using multiple antennas at the transmitter, receiver, or both. MIMO techniques improve communications performance by either *combating* or *exploiting* multipath scattering in the communications channel between a transmitter and receiver. MIMO techniques in the first category combat multipath by creating what is called *spatial diversity*, and those techniques that exploit multipath do so by performing *spatial multiplexing*. These two concepts are introduced in this chapter, and we will have much more to say about them throughout the remainder of the book. The subject of MIMO communications is the study of spatial diversity and spatial multiplexing techniques.

Figures 1.1 and 1.2 show block diagrams of generic MIMO communication systems. As indicated, the characteristics of the system depend on whether the focus of the MIMO processing is on creating spatial diversity, which improves reliability by combating fading, or if the purpose is to maximize throughput by performing spatial multiplexing. If the focus is on spatial diversity, information bits are normally encoded and modulated using conventional error correction coding and modulation techniques prior to undergoing some form of *space-time coding* (STC). At the receiver, space-time decoding is performed followed by demodulation and error decoding. If the focus is on spatial multiplexing, as illustrated in Figure 1.2, the information error encoded bits are passed through a serial-to-parallel converter and the individual output streams are modulated before being transmitted over separate antennas. At the receiver, each antenna receives a signal that consists of the sum of the signals from all of the transmit antennas; therefore, it is necessary to strip off each of the transmitted streams $\{s_i\}$ before demodulating them. The block that strips off each of the transmitted streams is often referred to as

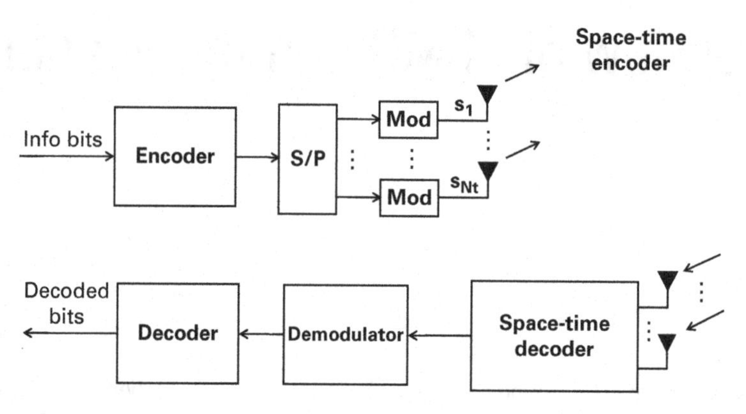

Figure 1.1 A MIMO system for *spatial diversity*.

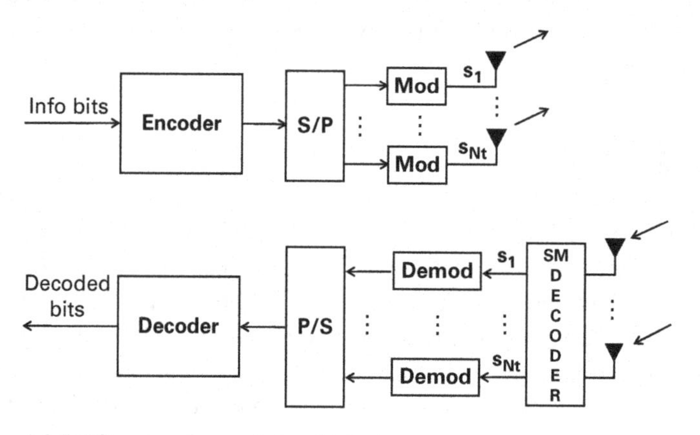

Figure 1.2 A MIMO system for *spatial multiplexing*.

an *SM decoder* or demultiplexer. There are different types of spatial demultiplexing schemes based on zero-forcing or linear minimum mean square error based methods, which we discuss in detail in Chapter 8.

Table 1.1 summarizes the relationship between a variety of different terms that are used in the MIMO literature. At this point, there are two main concepts to be clear about. The first concept is that *spatial diversity refers to techniques that are used to improve the reliability* on a communications link by combating fading and that space-time coding is the means by which this is accomplished. The second concept is that *spatial multiplexing refers to techniques that are used to increase throughput* without increasing the required bandwidth by exploiting multipath. This is done by transmitting separate data streams on each of the transmit antennas and by separating those streams at the receiver using some form of spatial demultiplexing. The details of space-time coding are the subject of Chapters 6 and 7, and spatial multiplexing is covered in Chapters 3 and 8.

Strictly speaking, MIMO refers to communication systems that have multiple antennas at both the transmitter and receiver; however, the nomenclature can be a bit

Table 1.1 Relationships of key MIMO concepts.

MIMO technique	Purpose	Approach	Method
Spatial diversity	improve *reliability*	*combat* fading	space-time coding
Spatial multiplexing	increase *throughput*	*exploit* fading	spatial demultiplexing

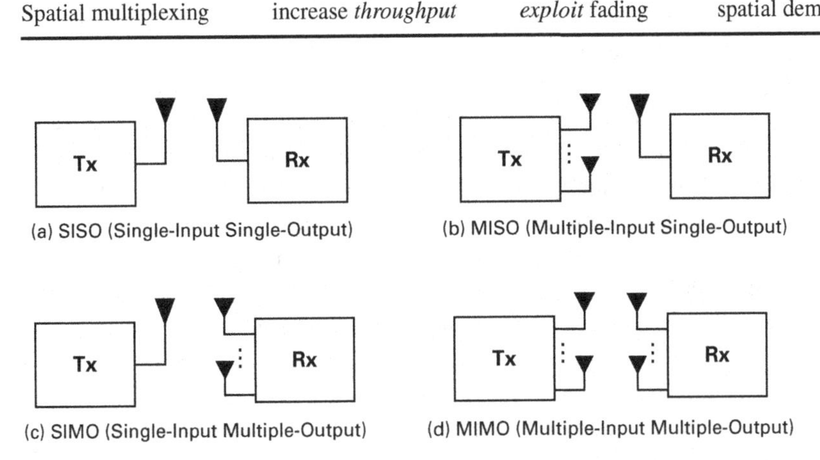

Figure 1.3 Antenna configurations and their nomenclatures used in this book.

confusing on this point and there is not always agreement on the use of terminology. In this book, we use the term MIMO in two different ways: in a broad sense to refer to a communication system that has multiple antennas at either the transmitter, the receiver, or both, and in a particular way when referring to systems that have multiple antennas at both ends of the link. When there are multiple antennas at the transmitter and only one receiver, as may occur, for example, on a cellular forward link between the base station and a single mobile user, we call that type of system a Multiple Input Single Output (MISO) system. When the opposite is true and there are multiple receive antennas but only one transmit antenna, that system is called a Single Input Multiple Output (SIMO) system. When using the term in the broad sense, we often refer to MISO and SIMO systems as particular types of MIMO configurations. Conventional communication systems that only have a single transmit antenna and a single receive antenna are called Single Input Single Output (SISO) communication systems. Figure 1.3 illustrates the four types of antenna configurations and the nomenclature used in this book.

MIMO systems with N_t transmit antennas and N_r receive antennas are referred to as $N_t \times N_r$ MIMO systems. Thus, for example, a 2×4 MIMO system implies that there are two transmit antennas and four receive antennas.

1.2 History of MIMO

The phrase "Multiple Input Multiple Output" has an interesting history. Although now it is used to describe the communication techniques that are the subject of this book, it was originally used in electric circuit and filter theory as far back as the 1950s [23].

In that original context, "MIMO" referred to circuits that had multiple input and multiple output ports. In the 1990s, however, information theorists and communication system researchers adopted this term to refer to new signal processing techniques that they were developing for communication systems having multiple antennas. In this newer use of the term, the communications channel was the reference point, and the term *multiple input* referred to the signals from multiple transmit antennas that were"entering" or "being input" to the communications channel. Similarly, the term *multiple output* referred to signals arriving at multiple receiver antennas, which were viewed as "exiting" or "being output" from the channel. The first reference to the term MIMO in this newer communications sense was in a paper by Peter Driessen and Gerry Foschini in 1999 where they published an analysis on the theoretical communications capacity of a communication system with multiple transmit and multiple receive antennas [20].

Although MIMO communications requires the use of multiple antennas, it is not the first multi-antenna technique to be developed. So what's new or unique about MIMO? To help answer that question, it is useful to place MIMO in its proper historical context. We begin by recognizing that the idea of using multiple antennas to improve aspects of communications and radar performance goes back to the beginning of the 1900s. The first use of multiple antennas was for the purpose of creating phased array antennas, which were first proposed and then demonstrated in 1905 by Karl Braun [12]. During WW II, phased array technology was used to enable rapidly-steerable radar [7], and later, phased arrays were used in AM broadcast radio to switch from groundwave propagation during the day to skywave propagation at night. This was accomplished by switching the phase and power levels supplied to the individual antenna elements daily at sunrise and sunset so that the elevation angle of the radiation pattern was towards the horizon during daylight hours and pointed slightly upward at night. This had the obvious advantage of enabling the transmitter to change the direction that it emitted energy without having to mechanically point the antenna, a challenging feat with large antennas such as those used in AM radio. Phased array technology has also long been used to perform adaptive nulling for interference and jamming avoidance.

In addition to phased array applications, multi-antenna technology has been used for more than 70 years to reduce the impact of fading on communication systems through the use of receive diversity. An early paper on the concept of receive diversity was published by H. Beverage and H. Peterson [11] in 1931. In the 1950s, receive diversity combining found extensive application on troposcatter links for military applications in which radio waves are scattered within the troposphere layer of the atmosphere [2], [84]. The scattering that occurs on troposcatter links enables communications beyond the horizon, which, other than HF, was the only way to communicate beyond the horizon prior to the advent of satellite communications. Troposcatter links were found to suffer from significant fading effects, so multiple antennas at the receiver were used to create receive diversity, which was helpful in reducing the impact of the fading.

Beginning in the 1990s, two new types of multi-antenna techniques were developed, which are the subject of this book. One of these techniques uses multiple antennas to achieve *transmit diversity*, which, like receive diversity, reduces the effect of fading. Two early papers on this technique were published in 1991 and 1993 by A. Wittneben [81] and N. Seshadri, C. Sundberg, and V. Weerackody [68], respectively. Later, Alamouti [6] published a landmark paper that described another way to achieve transmit diversity that required less processing at the receiver. Alamouti's technique has since become one of the most popular MIMO schemes in use today by nearly all wireless systems. His paper described a simple space-time coding technique for achieving transmit diversity and spurred research into other space-time coding techniques.

At about the same time that research was being conducted on transmit diversity, another class of multi-antenna techniques was being developed. Unlike those who were researching ways to use multiple antennas to combat the effects of fading, this second group of researchers was interested in developing ways of exploiting fading to support increased throughput capacity. In 1996, Gerry Foschini at AT&T Research Labs published his landmark paper on layered space-time communications, which described the underlying concept for the class of spatial multiplexing techniques that would eventually be called the Bell-Labs Layered Space-Time (BLAST) schemes [30]. In 1998 Foschini and a team from AT&T Research Labs were the first to demonstrate a laboratory prototype system that implemented a particular type of BLAST technique called vertical BLAST (i.e., V-BLAST) [31].

Since these initial breakthroughs in spatial diversity and spatial multiplexing in the late 1990s, a large body of a research has been conducted, and MIMO techniques using the spatial diversity and spatial multiplexing methods emerging from this research have been adopted in an increasing number of commercial wireless standards. The first commercial MIMO technology was introduced by Iospan Wireless Inc. in 2001. Since 2005, when the WiMAX standard first included MIMO technology, most wireless standards now include MIMO.

Figure 1.4 shows a time line of some key breakthroughs in multi-antenna technology over the past century. This diagram and the discussion above indicate that MIMO can be viewed as the latest in a long line of advances in multi-antenna technology.

1.3 Smart antennas vs MIMO

In recent years, another multi-antenna term, *smart antennas*, has become popular in the literature. What are smart antennas and what is the difference between MIMO and smart antenna technology? There is not unanimous agreement on the answer to this question. One of the first researchers to use the term smart antennas was Jack Winters at AT&T Labs [80]. In his 1998 paper, he focuses on describing ways to dynamically generate beams at a cellular base station that point in the desired direction of mobile users, and ways to create nulls that point in directions of interference. In another paper by Angeliki Alexiou and Martin Haardt in 2004 [4], however, the term smart antennas is used in a much broader sense to include not only dynamic beamforming and antenna nulling, but also spatial multiplexing and spatial diversity techniques such as Alamouti's scheme. In their use of the term, MIMO is a subset of smart antenna technology.

In this book, we use the older original concept to delineate between MIMO and smart antennas. For our purposes, smart antennas are defined as systems that employ

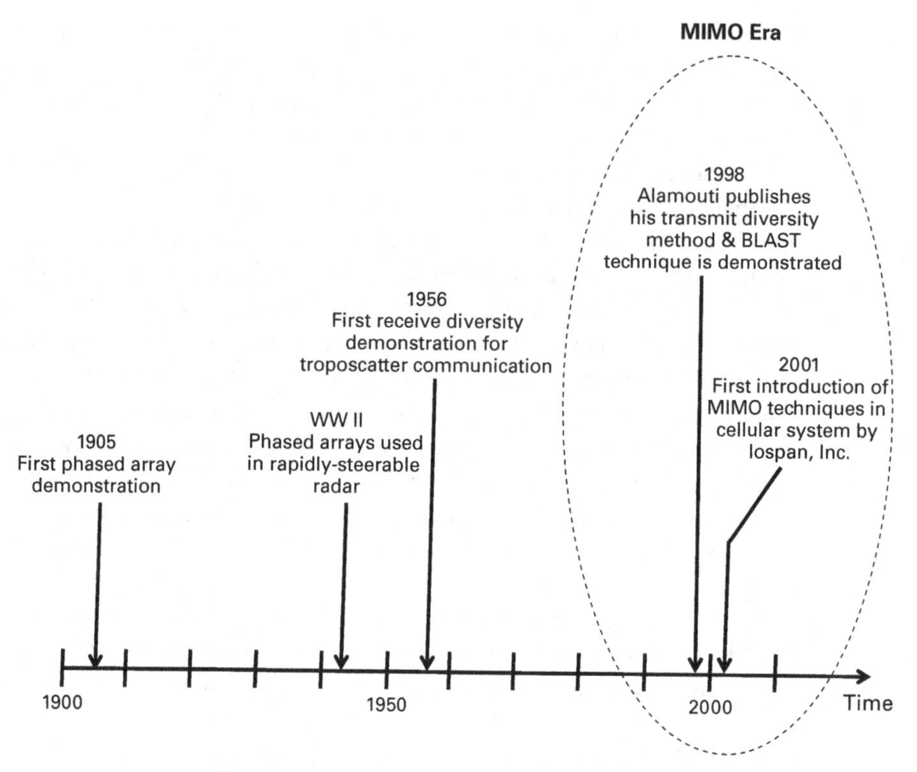

Figure 1.4 Time line of key multi-antenna advances.

techniques that are primarily designed to form beams and nulls in desired directions based on feedback from the environment. MIMO techniques, in contrast, are defined as communication systems that involve baseband signal processing techniques such as space-time coding and spatial multiplexing schemes that are not focused on pointing beams or creating nulls in space. In summary, we distinguish between smart antennas and MIMO as follows:

Smart antennas focus on:

- Conventional beamforming – directing energy in a desired physical direction;
- Adaptive nulling – creating nulls in desired directions to reduce interference.

MIMO focuses on:

- Spatial diversity – combating fading effects by creating spatial diversity through the use of baseband space-time coding techniques;
- Spatial multiplexing – using spatial multiplexing techniques to exploit multipath in order to achieve higher data rates than are possible with conventional systems having the same bandwidth.

1.4 Single-user and multi-user MIMO

Before proceeding further, a few comments should be made regarding the terms *single-user* MIMO and *multi-user* MIMO, which have been coined to describe two classes of

MIMO communications that are used in wireless systems, such as LTE and WiMAX. Single-user MIMO (SU-MIMO) refers to conventional MIMO where there is a one transmitting node and one receiving node, and the transmitter node has multiple antennas, as illustrated in Figure 1.3 (b) and (d).

In multi-user MIMO (MU-MIMO), mobile cellular users, each with a single antenna, transmit to a base station, and the base station processes the signals from each of the individual mobiles as if they were coming from multiple transmit antennas on a single node. In this case, the base station performs the same operations as the receiver in Figure 1.2, so multiple mobile users can transmit data over the same bandwidth, and the base station is able to decouple the individual data streams using spatial decoding techniques. In MU-MIMO, the individual users will not experience increased throughput; however, the overall system will. That is, MU-MIMO allows more cellular users to transmit simultaneously on the uplink path over the *same bandwidth* than would otherwise be possible.

The focus of this book is on SU-MIMO; however, with the exception of eigenbeamforming, which is a spatial multiplexing technique described in Chapter 3, the spatial multiplexing techniques we describe can be used with both SU- and MU-MIMO.

1.5 Introduction to spatial diversity

As we have just explained, one of the key purposes of MIMO communications is to improve communications reliability by combating multipath fading, which is achieved through the creation of spatial diversity. In this section, we review the concept of diversity, describe the difference between receive and transmit spatial diversity, and define three important performance metrics: *diversity order*, *diversity gain*, and *array gain*.

1.5.1 The concept of diversity

In most environments where wireless communication systems operate, the strength of the received signal varies with time, which is called *fading*. Unfortunately, fading significantly degrades communications performance by causing the probability of bit error to increase compared to what it would be if only white noise were present. Figure 1.5 shows the probability of bit error as a function of bit-energy-to-noise power spectral density, E_b/N_0, for different types of modulation in both fading and non-fading environments. The results in this figure demonstrate two important characteristics. The first is simply that fading causes the error probability to increase dramatically for a given value of E_b/N_0. The second observation is that for Rayleigh fading, which is the type of fading assumed in this figure and that often occurs in practice, the error probability decreases linearly when plotted on a logarithmic scale against E_b/N_0 plotted in dB. This is an important observation, and we use it later in this chapter.

In order to reduce the impact of fading, the concept of diversity is often employed. Diversity refers to transmitting replicas of the same signal over a fading channel in such a way that each replica fades independently of the others. When this happens, each replica tends to fade at a different time, so the probability that all the replicas

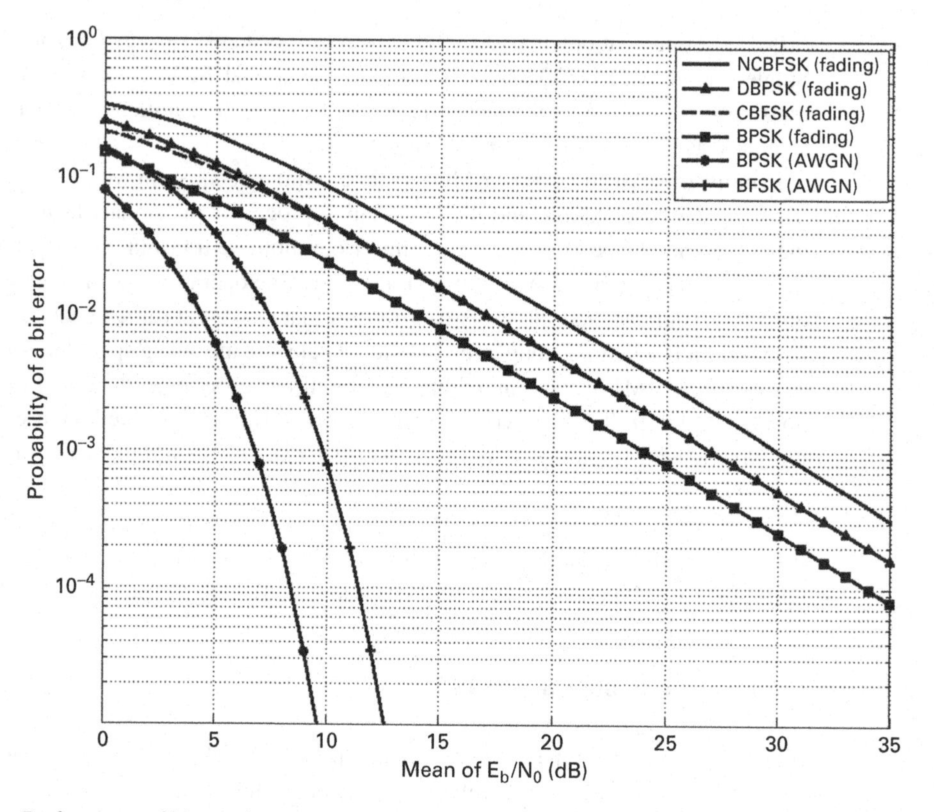

Figure 1.5 Performance of binary signaling on a Rayleigh fading channel.

fade simultaneously decreases as the number of replicas gets larger. By combining the replicas, however, the depths of the fades, and, so too, their adverse effects, can be significantly reduced because the fades do not tend to occur at the same time.

Reducing the impact of fading through diversity, therefore, involves two steps:

a) creating independent replicas of the signal; and
b) combining the replicas.

There are various ways to generate replicas of a signal for diversity purposes. One is to transmit the signal on different RF frequencies that are spaced far enough apart that the fading occurs independently on each carrier. This is called *frequency diversity*. Another diversity technique, called *time diversity*, involves transmitting the same signal at different times. In a multipath environment, this occurs naturally because the same signal arrives at the receiver by traveling over multiple physical paths, which tend to experience independent fading. Rake receivers are used to process such signals. A third way to create diversity is to transmit the same information on signals having different polarizations, called *polarization diversity*. Normally, fading is independent of signals having different polarizations. A fourth type of diversity is called *spatial diversity*, which refers to transmitting the same information over different physical paths between the transmitter and receiver. One way to create spatial diversity is to transmit a signal from one transmit antenna and receive it using multiple receive antennas. If the receive antennas are far enough apart, the fading on each path will be independent. This is the type of diversity that was originally used in the 1950s to reduce the impact of fading on troposcatter links discussed earlier.

Just as there are multiple ways to generate independent replicas of a signal, there are also different ways to combine the replicas at the receiver. The simplest type of combining is called *selective combining*, which involves comparing the replicas at each sample time and choosing the largest value for the output of the combiner. A second combining technique, called *equal gain combining*, involves adding the replicas together. The third, and most common type of combining scheme, is called *maximal ratio combining* (MRC). In MRC, the replicas are added together in the same way as they are in equal gain combining, but prior to being added they are first scaled in proportion to the signal-to-noise ratio of each replica. In Chapter 6, we discuss MRC in greater detail.

Figure 1.6 illustrates the benefits of diversity combining by plotting the output amplitude of a selective combiner in the presence of Rayleigh fading for two cases: when there is no combining (i.e., the number of signals being combined is 1), and when there are five replicas being combined. The curve associated with no combining has the deepest fades and the curve associated with five combined signals has noticeably less fading. Similar improvements occur with the other combining techniques.

1.5.2 Receive and transmit diversity

As we discussed earlier, troposcatter was one of the first types of communications techniques to use diversity combining. From that time until the 1990s, diversity techniques involved transmitting a single version of a signal and extracting replicas of the transmitted signal at the receiver and then combining those replicas. Diversity of this type is called *receive diversity* because extraction of the replicas is performed at the receiver. Figure 1.7 illustrates the architecture of a communication system that implements spatial receive diversity. As shown in this figure, the transmitted signal is denoted by s, and the communications channel has the effect of multiplying the transmitted signal by a complex value, which we call the channel response, and denote by $h_i, i = 1, \ldots, N_r$, where N_r represents the number of receive antennas. The inputs to the combiner, therefore, consist of the set of signals $\{r_i = h_i s\}$. If the receive antennas are spaced far enough apart, the random variables $\{h_i\}$ are independent, so the receiver is able to reduce the effect of fading by combining multiple independently fading signals.

In the late 1980s and early 1990s with the growing use of cellular communications, a desire for a different type of diversity architecture arose, called *transmit diversity*. The motivation for developing transmit diversity was the fact that the mobile unit in most cellular systems is small and, as a result, is often not capable of having multiple antennas. As a result, receive diversity on the forward link of cellular systems may not be possible. This led to the desire to find a spatially-based method of creating replicas of the transmitted signal at a receiver having only one antenna.

A little thought shows that this is not trivial. For example, if the base station is assumed to have multiple antennas, and if a signal, s, is simply transmitted from each of

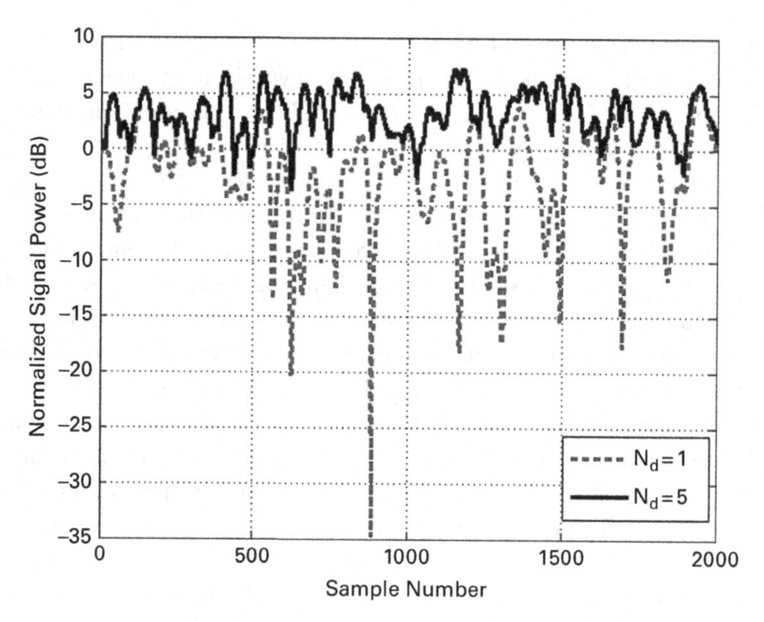

Figure 1.6 Simulated output from an equal gain combiner in Rayleigh fading for different numbers of combined signals.

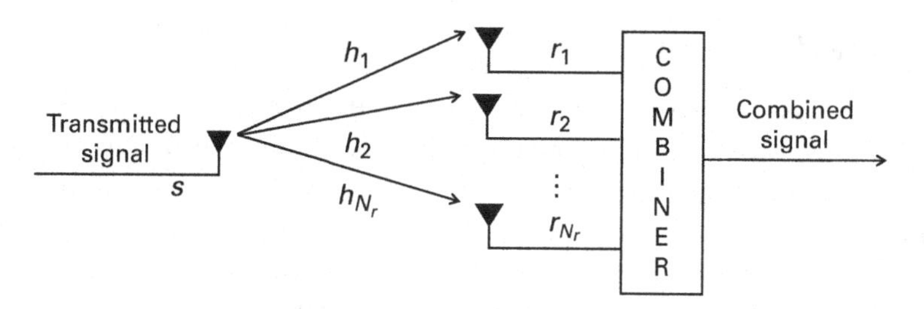

Figure 1.7 Architecture of a communication system with receive diversity combining.

these antennas, then the received signal, r, at the single receive antenna on the mobile unit is given by

$$r = s\left[\sum_{i=1}^{N_r} h_i\right]. \tag{1.1}$$

Unfortunately, this shows that the received signal is simply a scaled version of the transmitted signal, so the combiner at the mobile unit would not have access to multiple replicas of the transmitted signal if this technique were used. To correct this problem, it is necessary to perform some type of space-time coding at the transmitter. The architecture of a system with transmit diversity is depicted in Figure 1.8. One simple space-time code is the Alamouti code, which is used in most MIMO systems today. We discuss the Alamouti and other space-time codes in Chapters 6 and 7 and show how these schemes

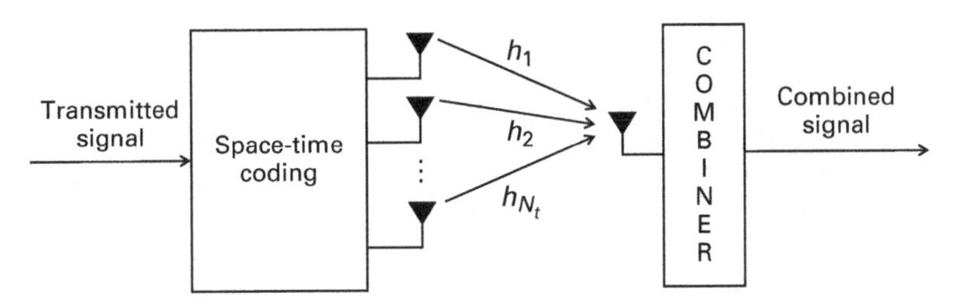

Figure 1.8 Architecture of a communication system with transmit diversity combining.

enable diversity combining on the forward link of a cellular system. For now, it is sufficient to introduce the concept and terminology of transmit diversity and to indicate that space-time coding is the means by which it is made possible.

1.5.3 Common diversity performance metrics

There are two common metrics that are used to characterize the amount of spatial diversity in a MIMO system. They are: *diversity order* and *diversity gain*. Diversity order, which we will denote by N_d, is simply the number of independent replicas of a transmitted signal that are available at the receiver for combining. Since an $N_t \times N_r$ MIMO system has up to N_tN_r independent paths between the transmitter and the receiver, it follows that spatial diversity is capable of achieving

$$\max\{N_d\} = N_tN_r. \tag{1.2}$$

Intuitively, we would expect the performance of a communication system to improve as the diversity order increases. To confirm this, Figure 1.9 shows the theoretical probability of bit error plotted as a function of average E_b/N_0 for three different types of binary modulation in Rayleigh fading, where $N_d = 1, 2, \text{and } 4$. These results assume the use of maximal ratio combining. As anticipated, for a given signal-to-noise ratio, the probability of bit error decreases as N_d increases. Furthermore, as we saw earlier in Figure 1.5, this plot also demonstrates that as E_b/N_0 becomes large, the curves approach straight lines, and that the slopes of these lines increase as the diversity order gets larger.

A common means of quantifying the benefits of diversity is to use the slope of the curve obtained by plotting bit error probability on a logarithmic scale versus mean E_b/N_0 in dB, when E_b/N_0 gets large (i.e., in the region where the curve is linear). The resulting slope is defined as the *diversity gain* of the system, which we denote by G_d. It follows that in the linear region, we can express the bit error probability mathematically as follows:

$$P_b = \zeta\left[G_c(\overline{E_b/N_0})\right]^{-G_d}, \tag{1.3}$$

where ζ is a constant that depends on the type of modulation, G_c is a constant, sometimes called the *coding gain* of the system, and the bar over E_b/N_0 denotes the mean.

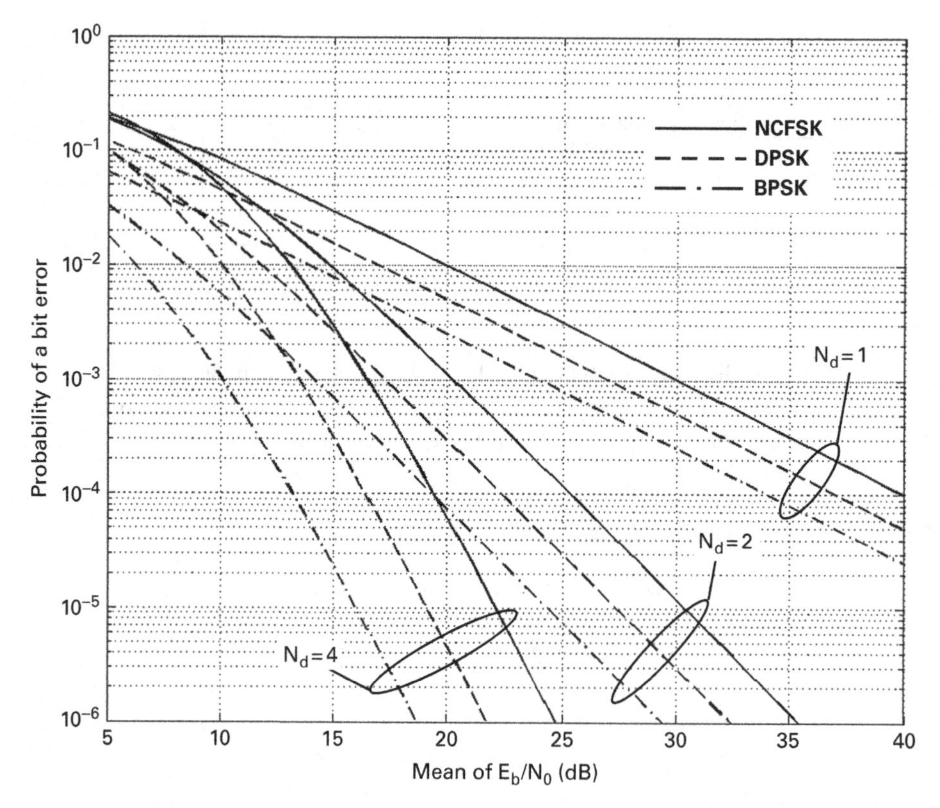

Figure 1.9 Performance of binary signals in Rayleigh fading with maximal ratio receive combining for three different diversity order values.

Example 1.1 **How to compute the diversity gain:** Assume that the bit error probability is 10^{-4} at $\overline{E_b/N_0} = 30$ dB and that it is 3.5×10^{-4} at $\overline{E_b/N_0} = 27.5$ dB. Assume that we are in the linear region of the curve. What is the diversity gain of this system?

Answer Start by taking $10\log_{10}$ of both sides of Eq. 1.3. This results in $P_b[\text{dB}] = 10\log_{10}(\zeta) - G_dG_c[\text{db}] - G_d(\bar{\rho}[\text{dB}])$, where we have used $\bar{\rho}$ to represent $\overline{E_b/N_0}$ for notational convenience. It follows that

$$P_{b_1} - P_{b_2} = -G_d(\bar{\rho}_1 - \bar{\rho}_2). \tag{1.4}$$

For the values in this example, $P_{b_1} = -40$, $P_{b_2} = 10\log_{10}(3.5 \times 10^{-4}) = -34.6$, $\bar{\rho}_1 = 30$, and $\bar{\rho}_2 = 27.5$. Therefore, $G_d = -(-40 + 34.6)/(30 - 27.5) = 2.1$.

1.5.4 Relationship between diversity order and diversity gain

The two parameters that we have defined for characterizing spatial diversity may appear fundamentally different – one refers to the number of signal replicas being combined and the other to the slope of the bit error probability versus E_b/N_0 curve – but it can be

Table 1.2 Parameter values for selected modulation schemes.

Modulation	α	β
BPSK	1/2	1
Binary, coherent orthogonal	1/2	1/2
QPSK with Gray coding	1/2	1

shown that they are normally numerically equivalent in a Rayleigh fading environment. Although it is difficult to prove this in the most general case, it can be shown to be true for specific modulation types when maximal ratio combining (MRC) is used. This section provides such a proof.

We begin with a well-known property of MRC, which states that the signal-to-noise ratio at the combiner output, ρ, is equal to the sum of the SNRs associated with the individual diversity channels, $\rho_i, i = 1, \ldots, N_d$ [14]. That is,

$$\rho = \sum_{i=1}^{N_d} \rho_i. \tag{1.5}$$

For the purpose of this proof, we consider modulation schemes where the bit error probability (BEP) has the following form:

$$P_b = \alpha \operatorname{erfc}\left(\sqrt{\beta\rho}\right); \tag{1.6}$$

where, α and β are constants and erfc is the complementary error function given by $\operatorname{erfc}(x) \triangleq (2/\sqrt{\pi}) \int_x^\infty \exp(-t^2)\,dt$. The values of α and β for three common types of modulation are listed in Table 1.2.

We next express the complementary error function in the following alternative form [16]:

$$\operatorname{erfc}(x) = \frac{2}{\pi} \int_0^{\pi/2} \exp\left(-x^2/\sin^2\theta\right) d\theta. \tag{1.7}$$

It follows that

$$\alpha \operatorname{erfc}\left(\sqrt{\beta\rho}\right) = \frac{2\alpha}{\pi} \int_0^{\pi/2} \exp\left(-\beta\rho/\sin^2\theta\right) d\theta. \tag{1.8}$$

Since $\rho = \sum_{i=1}^{N_d} \rho_i$ with MRC, it follows that the average BEP is given by

$$\begin{aligned}
\bar{P}_b &= \mathbb{E}_\rho\left\{P_b(\rho)\right\} \\
&= \frac{2\alpha}{\pi} \int_0^{\pi/2} \mathbb{E}_\rho\left\{\exp\left(-\beta \sum_{i=1}^{N_d} \rho_i/\sin^2\theta\right)\right\} d\theta \\
&= \frac{2\alpha}{\pi} \int_0^{\pi/2} \mathbb{E}_\rho\left\{\prod_{i-1}^{N_d} \exp^{-\beta\rho_i/\sin^2\theta}\right\} d\theta.
\end{aligned} \tag{1.9}$$

Assuming Rayleigh fading, it can be shown that the SNR on each diversity channel is exponentially distributed. Therefore, the probability density function (pdf) of ρ_i is given by

$$f(\rho_i) = \frac{1}{\bar{\rho}_i} \mathrm{e}^{-\rho_i/\bar{\rho}_i}, \tag{1.10}$$

where $\bar{\rho}_i$ is the mean value of the SNR on the ith diversity channel. If the diversity channels are independent, it follows that

$$\begin{aligned}
\bar{P}_b &= \frac{2\alpha}{\pi} \int_0^{\pi/2} \prod_{i=1}^{N_d} \mathbb{E}_\rho \left\{ \exp^{-\beta\rho_i/\sin^2\theta} \right\} d\theta \\
&= \frac{2\alpha}{\pi} \int_0^{\pi/2} \prod_{i=1}^{N_d} \left[\int_0^{\infty} \frac{1}{\bar{\rho}_i} e^{-\rho_i/\bar{\rho}_i} e^{-\beta\rho_i/\sin^2\theta} \, d\rho_i \right] d\theta \\
&= \frac{2\alpha}{\pi} \int_0^{\pi/2} \prod_{i=1}^{N_d} \left[\int_0^{\infty} \frac{1}{\bar{\rho}_i} e^{-\rho_i[1/\bar{\rho}_i + \beta/\sin^2\theta]} \, d\rho_i \right] d\theta \\
&= \frac{2\alpha}{\pi} \int_0^{\pi/2} \prod_{i=1}^{N_d} \frac{\sin^2\theta}{\sin^2\theta + \beta\bar{\rho}_i} \, d\theta,
\end{aligned} \tag{1.11}$$

where the last step follows from performing simple integration.

Since G_d is defined in the limit as the signal-to-noise ratio becomes large, we are interested in examining the expression for $\bar{P}_b$ when the average SNR values on each of the diversity paths approach infinity (i.e., as $\bar{\rho}_i \to \infty$, $i = 1, \ldots, N_d$). For simplicity, we assume that the average signal-to-noise ratio on each diversity channel is the same and denote $\bar{\rho}_i = \rho \; \forall i$. Therefore,

$$\begin{aligned}
\lim_{\bar{\rho}\to\infty} \bar{P}_b(\bar{\rho}) &= \lim_{\bar{\rho}\to\infty} \frac{2\alpha}{\pi} \int_0^{\pi/2} \left[\frac{\sin^2\theta}{\sin^2\theta + \beta\bar{\rho}} \right]^{N_d} d\theta \\
&= \left[\frac{2\alpha}{\pi\beta^{N_d}} \int_0^{\pi/2} \sin^{2N_d}\theta \, d\theta \right] \bar{\rho}^{-N_d}.
\end{aligned} \tag{1.12}$$

We note that the equation above has the same form as Eq. 1.3 for $\zeta = 1$, where G_c and G_d have the following forms:

$$G_c = \left[\frac{2\alpha}{\pi\beta^{N_d}} \int_0^{\pi/2} \sin^{2N_d}\theta \, d\theta \right]^{-1/N_d} \tag{1.13}$$

and

$$G_d = N_d. \tag{1.14}$$

Equation 1.14 demonstrates what we set out to show, which is that diversity gain is equal to diversity order. This is a general property that holds for many types of modulation schemes in Rayleigh fading with maximal ratio combining. A MIMO system is said to achieve *full diversity* when $N_d = G_d = N_t N_r$. MIMO systems achieve diversity through the use of space-time coding, which we describe in later chapters.

1.6 Introduction to spatial multiplexing

The second class of MIMO techniques that we study in this book is called spatial multiplexing. This section provides a brief introduction to that topic.

1.6.1 The concept of spatial multiplexing

Spatial multiplexing (SM) refers to transmitting multiple data streams over a multipath channel by exploiting multipath. By so doing, multiple data channels are able to be transmitted simultaneously over the same frequency band, enabling potentially large numbers of bits per second to be transmitted per Hertz of spectrum. Spatial multiplexing is analogous to other more common types of multiplexing schemes such as frequency-division multiplexing (FDM) and time-division multiplexing (TDM). In those schemes, multiple signals are assigned to either frequency slots in the case of FDM or time slots in TDM. In SM, multiple signals are assigned to different spatial channels instead of time or frequency slots, so the signals are transmitted at the same time over the same bandwidth. As a result, SM does not suffer from bandwidth expansion the way that TDM and FDM do.

Figure 1.10 shows a high-level block diagram of a SM MIMO system. As illustrated, there are three main components to an SM system. The first component is referred to in the diagram as a *precoder*. Its purpose is to map the multiple input streams of data that are to be transmitted onto the set of transmit antennas. The simplest form of precoder simply maps each data stream to a single unique antenna. This, of course, can only occur when the number of data streams is equal to the number of antennas. In general, the number of data streams may be less than or equal to the number of transmit antennas. In the general case, the role of the precoder may be more complex. One type of precoding that we study in Chapter 3 is called eigenbeamforming. In that scheme, the operations of the precoder depend on the characteristics of the communications channel.

The second component of an SM system is the *postcoder*, which processes the signals from the receive antennas and generates estimates of the original input data streams that originally went into the precoder at the transmitter. Since the signal at each receive antenna consists of the sum of the signals from each of the transmit antennas, the postcoder must be able to, in essence, strip off each data stream from the composite received signal. There are various ways to do this, which are discussed in detail in Chapter 8.

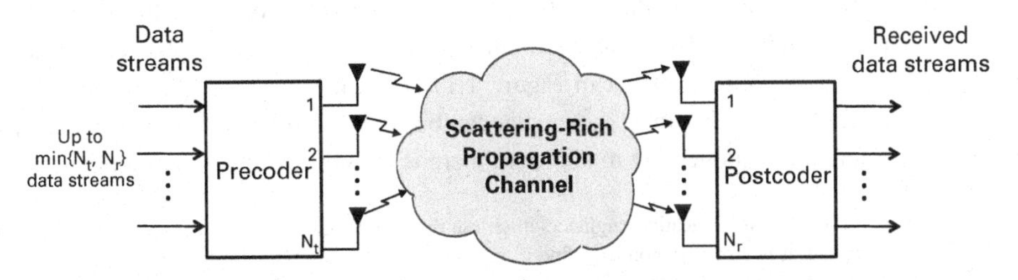

Figure 1.10 Generic diagram of a MIMO communication system that uses spatial multiplexing.

The third component of an SM system is the communications channel itself. In order for spatial multiplexing to work, the channel must have a significant amount of multipath scattering. This may seem odd since multipath is normally regarded as the enemy by communications engineers since it degrades the performance of conventional communication systems. However, since spatial multiplexing exploits multipath, its presence is necessary for SM techniques to work. In Chapter 3, we discuss the concept of *channel rank* and show why it is used to characterize the scattering richness of a multipath channel.

The spatial multiplexing literature uses a variety of different terms to describe the data signals that are able to be transmitted at the same time over an SM MIMO system. Terms such as *data streams*, *data pipes*, and *spatial channels* are common terms used to describe these data signals, and they mean the same thing. In addition, when eigenbeamforming is used, the data signals that are transmitted in parallel are often called *eigen-channels*. It can be shown that the maximum number of data streams, N_{stream}, that can be supported by a MIMO system using spatial multiplexing is given by

$$N_{\text{stream}} = \min(N_t, N_r)\,. \tag{1.15}$$

This equation shows that for an $N \times N$ MIMO system, the throughput increases linearly with the number of antennas. Since the increase in the number of data streams does not require a wider bandwidth, this equation shows that the spectral efficiency of a MIMO system also increases linearly with the number of antennas. Figure 1.11 shows the theoretical average communications capacity (in bits-per-second-per-Hertz)[1] in Rayleigh fading plotted as a function of the number of antennas (when $N_t = N_r$) for different signal-to-noise ratios, ρ. This plot illustrates the linear dependence of the average capacity on the number of antennas, which is one of the key properties that has led the wireless industry to enthusiastically adopt MIMO techniques in its modern wireless standards. These results show that even at modest SNR values, it is theoretically possible to achieve spectral efficiencies of 15 to 30 bps/Hz. It is evident from this plot that the capacity is proportional to the number of antennas, i.e., $C(N) = a(\rho)N$; where, a is the slope of the curve, which is a function of ρ. It follows that the capacity of an $N \times N$ MIMO system, $C_{\text{MIMO}}(N)$, is N times the capacity of a SISO system. That is,

$$C_{\text{MIMO}}(N) = NC_{\text{SISO}}. \tag{1.16}$$

It should be emphasized that these results are theoretical and assume ideal scattering in the channel. In practice, maximum spectral efficiencies are significantly lower than these values due to the combination of implementation limitations and the presence of ill-conditioned channels (which we define in Chapter 3). Nevertheless, the utility of the theoretical results in Figure 1.11, like any theoretical communication capacity predictions, is that they show what can be achieved in the ideal limit and they provide a reference to judge how much room there is for improvement in practical systems.

[1] Although communication engineers often use the term *capacity* in such a way that it has units of bits-per-second, it is also common to define it to have units of bits-per-second-per-Hertz. In that case, theoretical capacity is equivalent to theoretical spectral efficiency. For this discussion, we have made this assumption and use the terms interchangeably. These distinctions are clarified in Chapter 2.

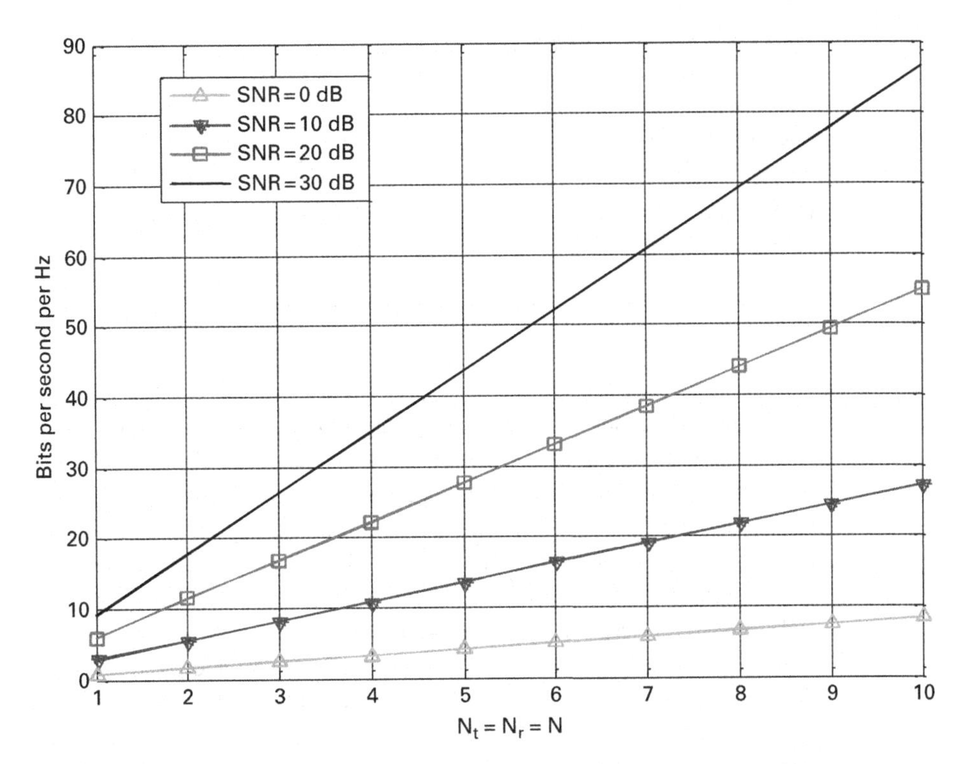

Figure 1.11 Theoretical capacity of an $N \times N$ MIMO communication system in Rayleigh fading.

1.7 Open- and closed-loop MIMO

As we will see, MIMO techniques normally require that either the transmitter or the receiver have knowledge of the characteristics of the communications channel. As such, MIMO techniques are often classified as either *open-loop* or *closed-loop*, depending on whether the transmitter or the receiver uses knowledge of the communications channel. MIMO techniques that require the transmitter to have knowledge of the channel are called closed-loop because they require the receiver to estimate the channel and to send that information back to the transmitter – hence, requiring a "closed loop". MIMO techniques that only require the receiver to have knowledge of the channel are called open loop. These terms are used throughout the MIMO literature and in the wireless standards.

Table 1.3 compares the types of MIMO techniques commonly associated with open- and closed-loop MIMO. The first thing to note is that the two basic classes of MIMO: transmit diversity and spatial multiplexing, each have separate open- and closed-loop versions. For example, when operating in an open loop manner, transmit diversity is implemented using space-time coding (STC), of which Alamouti coding is the most common. Similarly, an example of open loop spatial multiplexing is the BLAST technique, of which there are several varieties. In contrast, when operating in a closed-loop

Table 1.3 Comparison of open loop and closed MIMO.

Open loop	**Examples**
Transmit diversity	STC (e.g., Alamouti)
Spatial mutiplexing	BLAST
Closed loop	
Transmit diversity	TSD
Spatial multiplexing	eigenbeamforming

configuration, transmit diversity can be implemented using transmit selection diversity (TSD) and spatial multiplexing can be implemented using eigenbeamforming.

TSD is the simplest form of closed-loop transmit diversity where the transmitter selects one out of N_t of its available transmit antennas to transmit information at any given instant based on channel information feedback from the receiver. In general, the transmitter will use the antenna associated with the best channel response between it and the receiver. TSD was first proposed by J. Winters in 1983 [78].

At this point, it sufficient for the reader to have an understanding of the basic distinction between closed and open loop MIMO and to have a growing familiarity with the names of the various MIMO techniques. We will have much more to say about the details of these methods in later chapters. In this book, we will focus on space-time coding, BLAST, and eigenbeamforming.

1.8 The practical use of MIMO

Having introduced some of the fundamental concepts and terminology in MIMO communications, we now briefly consider its implementation in practical wireless systems.

1.8.1 Commercial MIMO implementations

The first use of MIMO techniques in a cellular system was by Iospan, Inc. in 2001. Since then, increasing numbers of wireless commercial standards have adopted the use of MIMO techniques. Table 1.4 lists prominent commercial standards that support MIMO and the maximum antenna configuration that each supports (i.e., all combinations of antennas less than those listed are supported by the standard). In general, the standards listed in this table support the following types of multi-antenna techniques:

- Alamouti space-time coding for transmit diversity;
- Eigenbeamforming spatial multiplexing;
- BLAST spatial multiplexing architectures;
- Conventional beam and null forming;
- Conventional receive diversity.

Table 1.4 Commercial wireless standards that use MIMO technology.

Wireless standard	Antenna configurations
IEEE 802.11n (WiFi)	4×4
IEEE 802.16e (WiMAX)	4×4
$HSPA^+$ (Enhanced HSPA)	2×2
LTE (3.9G)	4×4
LTE-Advanced (4G)	8×8
802.11ac (Enhanced 802.11n)	8×8

Although the details of how each of these standards implements MIMO vary, the underlying principles are the same. In this book, we focus on the fundamental concepts of MIMO, which will give the reader the necessary background to read and understand the details of the standards. Chapter 11 describes how MIMO is implemented in the IEEE 802.11n and LTE standards for the purpose of demonstrating how the concepts described in this book are implemented in practice.

1.8.2 Measured MIMO performance

This section describes performance measurements that empirically demonstrate the advantages that MIMO can provide. A particularly clear and compelling paper that demonstrates the performance benefits of MIMO was presented at the Military Communications Conference (MILCOM) in 2010 by Lai *et al.* [53]. In that paper, they present results from demonstration tests that were performed to quantify the benefits of a MIMO communication system called Mobile Networked MIMO (MNM) that was developed by the Defense Advanced Research Projects Agency (DARPA) for military applications.

The MNM MIMO radio tests were conducted at Fort Monmouth, New Jersey and in Los Angeles, California in various environments including indoor, outdoor, open-space, and in dense foliage. These tests measured the throughput gain of MIMO over SISO configurations and also measured the MIMO transmit power required to achieve the same throughput as a SISO system, thus demonstrating the ability of MIMO to extend battery life in a battery-powered transceiver. The MNM radio used in these tests consists of up to four transmit antennas and four receive antennas, operates at about 2.4 and 5 GHz, employs bandwidths ranging from 0.625 to 20 MHz, transmits data between 1.5 Mbps and 260 Mbps, and employs spatial multiplexing, space-time coding, eigenbeamforming, and receive diversity. Table 1.5 lists the parameters of the MNM radio used in this demonstration.

Two performance metrics were used in these tests to quantify the performance improvement due to MIMO. One of these metrics, called *throughput gain (TPG)*, is defined as the ratio of MIMO throughput over SISO throughput, given the same total transmit power, spectrum usage, and channel conditions. That is,

Table 1.5 MNM MIMO radio system parameter values.

RF carrier	2.4 – 2.4835 GHz and 4.9 – 5.8 GHz
Output power (total)	250 mW (2 W with external PA)
Bandwidth	0.625, 1.25, 2.5, 5, 10, 20 MHz
Data rates	1.5 Mbps – 260 Mbps
Modulation	Coded – OFDM with 64/128 FFT sizes
Antenna configuration	$m \times n \quad 1 \leq m, n \leq 4$
MIMO processing	Spatial multiplexing, Space-time coding Eigenbeamforming, Receive diversity
Number of data streams	1, 2, 3, 4

$$TPG = \frac{TP_{\text{MIMO}}}{TP_{\text{SISO}}}, \tag{1.17}$$

where TP_{MIMO} and TP_{SISO} denote the MIMO and SISO throughputs, respectively. The throughput, in turn, was computed based on the collected packet error rate (PER) and the data rate, R, as follows:

$$TP = (1 - PER)R. \tag{1.18}$$

The second performance metric used in this study was the *transmit power savings* (*TPS*), defined as the ratio of transmit power used in SISO divided by the required power for MIMO, when operating at the same throughput. This metric was used to quantify the transmit power savings of MIMO over SISO.

Table 1.6 summarizes the results from this study, separated according to whether the radios were operating within line-of-sight (LOS) of each other or were non-LOS (NLOS). The bottom row lists the average values over all cases. In these tests, the MIMO system was operated in a 4×4 configuration. The results show that the throughput is significantly higher when operating in the MIMO configuration than it is when operating in a SISO mode. The improvement in throughput is observed to range from 1.2 to 3.1 in LOS conditions and from 1.3 to 6.1 in NLOS environments. The results also show that MIMO provides significant power savings, as indicated by the values in the *TPS* columns. These power savings have potentially significant implications for battery-power transceivers.

The results from this study are interesting for two reasons. One: they show, as expected, that MIMO is capable of significantly increasing throughput. Two: they show that MIMO increases throughput by more when operating in a NLOS environment than it does in a LOS geometry. This can be explained by the fact that LOS conditions have less multipath scattering than NLOS geometries. Since spatial multiplexing uses multipath scattering to increase throughput, the lower LOS *TPG* numbers are expected. Perhaps more interesting, however, is that the throughput increase is as large as it is in LOS conditions. Conventional wisdom has assumed that MIMO would not be effective in LOS conditions; however, these results suggest the benefits of MIMO may be realizable in a broader range of conditions than has often been assumed.

Table 1.6 Summary of measured MIMO performance improvement from MNM radio field demonstrations (derived from [53]).

LOS transmission		NLOS transmission	
TPG	*TPS*	*TPG*	*TPS*
3.1	40.0	6.1	4.4
1.7	5.0	2.7	11.6
1.2	16.9	4.2	–
3.3	2.7	3.3	2.7
1.3	2.7	2.5	4.7
1.3	5.7	5.0	6,1
1.6	2.7	1.6	22.2
1.3	4.7	2.3	6.1
1.7	1.8	6.0	16.9
1.7	11.6	1.6	2.7
2.0	4.7	2.6	14.7
2.7	2.7	–	–
1.9	31.3	–	–
1.4	22.6	–	–
1.9	**10.5**	**3.3**	**9.2**

1.9 Review of matrices

We conclude this introductory chapter with a brief overview of those matrix properties that will be needed later in the book. Matrices and vectors arise naturally in the study of MIMO communications because of the use of multiple antennas. For example, if there are multiple transmit antennas, then at any instant we can represent the set of signals applied to each of the antennas as a vector $\mathbf{s} = [s_1, s_2, \ldots, s_{N_t}]^T$. Similarly, the signals at multiple receiver antennas can be represented as a receive vector $\mathbf{r} = [r_1, r_2, \ldots, r_{N_r}]^T$. Furthermore, because there are $N_t \times N_r$ combinations of transmit and receive antennas, each of which can be viewed as a separate communication channel, it is convenient to represent the overall communication channel in a MIMO system as a matrix with $N_r \times N_t$ elements. Using matrices and vectors greatly simplifies the mathematics needed to describe the behavior of MIMO systems. As a result, in order to be able to read and understand the MIMO literature, it is necessary to be conversant with matrix nomenclature and matrix properties.

This section lists those definitions and matrix identities that are needed to understand the material in this book and the majority of published papers on MIMO communications. No attempt is made to prove or elaborate on these properties; thus, this section is intended to be a reference only. The reader who wants to delve into matrix mathematics in greater detail should consult a text on that subject. Two excellent references on matrices are those by Horn and Johnson [38] and Carl Meyer [54].

1.9.1 Basic definitions

(a) **Identity matrix** An $N \times N$ square matrix with all diagonal elements equal to unity and all other elements equal to zero is called an identity matrix, and is denoted by $\mathbf{I}_N$ or just $\mathbf{I}$.

(b) **Transpose operation** The transpose of a matrix $\mathbf{A}$, which we denote by $\mathbf{A}^T$, is defined as follows: $\left[\mathbf{A}^T\right]_{ij} = [\mathbf{A}]_{ji}$. That is, the transpose of a matrix is formed by replacing its columns with its rows.

(c) **Trace** The trace of a square matrix $\mathbf{A}$ is denoted by Tr($\mathbf{A}$) and is defined by $\text{Tr}(\mathbf{A}) \triangleq \sum_i [\mathbf{A}]_{ii}$.

(d) **Orthogonal matrix** A real, square matrix $\mathbf{A}$ (i.e., a matrix with real elements) is said to be orthogonal if $\mathbf{A}\mathbf{A}^T = \mathbf{A}^T\mathbf{A} = c\mathbf{I}$, $c > 0$, where c denotes an arbitrary scalar. A matrix is said to be orthonormal if $c = 1$.

(e) **Hermitian operation** The Hermitian of a complex, square matrix $\mathbf{A}$ is denoted by $\mathbf{A}^H$. By definition $\left[\mathbf{A}^H\right]_{ij} = \left[\mathbf{A}^*\right]_{ji}$, where $\mathbf{A}^*$ denotes the matrix $\mathbf{A}$ with all elements complex conjugated. The Hermitian operation is also called the conjugate transpose of a matrix. It follows that $\left[\mathbf{A}^H\right]^H = \mathbf{A}$. A matrix $\mathbf{A}$ is said to be Hermitian if $\mathbf{A}^H = \mathbf{A}$.

(f) **Unitary matrix** A square matrix $\mathbf{U}$ is said to be unitary if $\mathbf{U}\mathbf{U}^H = \mathbf{U}^H\mathbf{U} = c\mathbf{I}$, $c > 0$, where c denotes an arbitrary real scalar. If $c = 1$, $\mathbf{U}$ is said to *normalized unitary*. In this book, as well as in most of the MIMO literature, the term unitary implies normalized unitary; hence, $c = 1$. The definition of unitary is analogous to the definition for orthogonal,the former applying to complex matrices and the latter to real matrices.

(g) **Frobenius norm** The norm of a matrix is a generalization of the concept of vector norm, which, in turn, is the length of a vector. Similarly, the norm of a matrix is the measure in some sense of the "size" of that matrix. There are different types of matrix norms; however, in most of the MIMO literature the Frobenius norm is commonly used. The Frobenius norm is defined in one of two ways. We denote the Frobenius norm of a matrix $\mathbf{A}$ using the two nomenclatures below:

$$\|\mathbf{A}\|_F \triangleq \sum_i \sum_j |A_{ij}|,$$

$$\|\mathbf{A}\|_F^2 \triangleq \sum_i \sum_j |A_{ij}|^2.$$

(h) **Linear dependence** A set of vectors $\{\mathbf{v}_k\}$ is said to be linearly dependent if there exists a set of scalars $\{a_k\}$, not all zero, such that $\sum_k a_k\mathbf{v}_k = \mathbf{0}$.

(i) **Linear independence** A set of vectors $\{\mathbf{v}_k\}$ is said to be linearly independent if the vectors are not linear dependent. That is, $\{\mathbf{v}_k\}$ are linearly independent when the only solution to $\sum_k a_k\mathbf{v}_k = \mathbf{0}$ is the trivial solution $a_k = 0\,\forall k$.

(j) **Rank** The rank of an $m \times n$ matrix $\mathbf{A}$ is defined as the largest number of columns or rows from that matrix that form a linearly independent set. The maximum number of rows is equal to the maximum number of columns. We denote the rank of $\mathbf{A}$ by $r(\mathbf{A})$. If $m \leq n$, then $1 \leq r(\mathbf{A}) \leq m$. Similarly, if $m \geq n$, then $1 \leq r(\mathbf{A}) \leq n$.

(k) **Singular and non-singular** A square matrix $\mathbf{A}$ is said to be singular if its determinant, $\det(\mathbf{A}) = 0$, and is said to be non-singular if $\det(\mathbf{A}) \neq 0$. Since the concept of determinant is not defined for a non-square matrix, the concept of singularity only applies to square matrices.

(l) **Eigenvalues and eigenvectors** Given a square matrix, $\mathbf{A}$, then the polynomial in the independent variable s formed by $\det(s\mathbf{I}-\mathbf{A})$ is called the *characteristic polynomial*, and its roots are called the eigenvalues of $\mathbf{A}$. Alternatively, the eigenvalues of $\mathbf{A}$ are the values of λ that satisfy the equation $\mathbf{Av} = \lambda\mathbf{v}$ for some non-zero vector $\mathbf{v}$. The vector $\mathbf{v}$ is said to be an eigenvector of $\mathbf{A}$ corresponding to the eigenvalues λ.

(m) **Determinant** There are multiple ways to define the determinant of a square matrix. For the purpose of this book, we define the determinant to be equal to the product of the eigenvalues of the matrix. The following additional comments apply:
   (1) The determinant is only defined for a square matrix.
   (2) The determinant of a matrix, $\mathbf{A}$, is denoted either as $\det(\mathbf{A})$ or as $|\mathbf{A}|$. Both notations are used in this book.

(n) **Null space** Given an $m \times n$ matrix $\mathbf{A}$, the null space of $\mathbf{A}$, which we denote by $N(\mathbf{A})$, is defined as the set $1 \times m$ vectors $\{\mathbf{x}|\mathbf{xA} = \mathbf{0}\}$. In other words, $N(\mathbf{A})$ is the set of all solutions $\{\mathbf{x}\}$ to $\mathbf{xA} = \mathbf{0}$.

(o) **Kronecker product** Let $\mathbf{A}$ be an $m \times n$ matrix and $\mathbf{B}$ be a $p \times q$ matrix. It follows that the Kronecker product, $\mathbf{A} \otimes \mathbf{B}$, is the $mp \times nq$ block matrix:

$$\mathbf{A} \otimes \mathbf{B} = \begin{pmatrix} a_{11}\mathbf{B} & \dots & a_{1n}\mathbf{B} \\ \vdots & \ddots & \vdots \\ a_{m1}\mathbf{B} & \dots & a_{mn}\mathbf{B} \end{pmatrix}.$$

1.9.2 Theorems and properties

(a) For any matrices $\mathbf{A}$ and $\mathbf{B}$, $(\mathbf{AB})^T = \mathbf{B}^T\mathbf{A}^T$.
(b) For any matrices $\mathbf{A}$ and $\mathbf{B}$, $(\mathbf{AB})^* = \mathbf{A}^*\mathbf{B}^*$.
(c) For any matrices $\mathbf{A}$ and $\mathbf{B}$, $(\mathbf{AB})^H = \mathbf{B}^H\mathbf{A}^H$.
(d) For any square matrices $\mathbf{A}$ and $\mathbf{B}$, $(\mathbf{AB})^{-1} = \mathbf{B}^{-1}\mathbf{A}^{-1}$.
(e) For any matrices $\mathbf{A}$ and $\mathbf{B}$, $r(\mathbf{AB}) \leq \min\{r(\mathbf{A}), r(\mathbf{B})\}$.
(f) The determinant of any identity matrix is 1.
(g) Let $\mathbf{A}$and $\mathbf{B}$ denote any two matrices having the same dimensions and a denote any scalar. Then

$$\frac{1}{a}|\mathbf{A}+\mathbf{B}| = |\mathbf{A}/a + \mathbf{B}/a|.$$

(h) The inverse of any $n \times n$ matrix, $\mathbf{A}$, exists if and only if
   (1) $r(\mathbf{A}) = n$ (i.e., $\mathbf{A}$ is full rank); and
   (2) $\det(\mathbf{A}) \neq 0$.

(i) For any square matrix, $\mathbf{A}$, it follows that

$$\left(\mathbf{A}^{-1}\right)^H = \left(\mathbf{A}^H\right)^{-1}.$$

(j) For any normalized unitary matrix $\mathbf{U}$, $\det(\mathbf{U}) = 1$.

(k) Let $\mathbf{A}$ denote an $m \times n$ matrix. Then
(1) a necessary condition for $\left(\mathbf{AA}^H\right)^{-1}$ to exist is that $n \geq m$; and
(2) a necessary condition for $\left(\mathbf{A}^H\mathbf{A}\right)^{-1}$ to exist is that $m \geq n$.

(l) For any square matrices $\mathbf{A}$ and $\mathbf{B}$, it follows that $\det(\mathbf{AB}) = \det(\mathbf{A})\det(\mathbf{B})$.

(m) For any square real matrix $\mathbf{A}$, $\det(\mathbf{A}^T) = \det(\mathbf{A})$.

(n) For any square complex matrix $\mathbf{A}$, $\det(\mathbf{A}^H) = [\det(\mathbf{A})]^*$.

(o) For any matrix $\mathbf{A}$, $\mathbf{AA}^H$ and $\mathbf{A}^H\mathbf{A}$ are Hermitian.

(p) Let $\mathbf{A}$ be an $n \times n$ matrix with eigenvalues $\{\lambda_i\}$. Then the determinant and trace of $\mathbf{A}$ can be expressed as follows:

$$\det(\mathbf{A}) = \prod_{i=1}^{n} \lambda_i,$$

$$\mathrm{Tr}(\mathbf{A}) = \sum_{i=1}^{n} \lambda_i.$$

(q) Any square matrix is singular if and only if it has one or more eigenvalues equal to zero.

(r) **Singular value decomposition** For any complex $m \times n$ matrix $\mathbf{A}$ of rank r, there are unitary matrices $\mathbf{U}_{m\times m}$ and $\mathbf{V}_{n\times n}$ and a diagonal matrix $\mathbf{D}_{r\times r} = \mathrm{diag}(\sigma_1, \sigma_2, \ldots, \sigma_r)$ such that

$$\mathbf{A} = \mathbf{U}_{m\times m} \begin{pmatrix} \mathbf{D} & \mathbf{0} \\ \mathbf{0} & \mathbf{0} \end{pmatrix} \mathbf{V}^H_{n\times n}, \text{ with } \sigma_1 \geq \sigma_2 \geq \cdots \geq \sigma_r,$$

where $\{\sigma_i\}$ are called the *singular values* of $\mathbf{A}$.

(s) Let $\mathbf{A}$ be any complex matrix and denote the singular values of $\mathbf{A}$ by $\{\sigma_i\}$ and the eigenvalues of $\mathbf{AA}^H$ (and $\mathbf{A}^H\mathbf{A}$) by $\{\lambda_i\}$. Then

$$\sigma_i = \sqrt{\lambda_i}, i = 1, \ldots, r(\mathbf{AA}^H).$$

(t) For any Hermitian matrix $\mathbf{A}$ with eigenvalues $\{\lambda_1, \lambda_2, \ldots, \lambda_r\}$, the non-zero singular values of $\mathbf{A}$, $\{\sigma_1, \sigma_2, \ldots, \sigma_r\}$, are given by

$$\sigma_i = |\lambda_i|, \; i = 1, \ldots, r(\mathbf{A}).$$

(u) **Eigenvalue decomposition** Suppose that $\mathbf{A}$ is a Hermitian matrix of dimension $m \times m$ and rank r. It follows that there is an $m \times m$ diagonal matrix $\mathbf{D}_{m\times m} = \mathrm{diag}(\lambda_1, \lambda_2, \ldots, \lambda_r, 0, \ldots, 0)$, $r \leq m$, and a normalized unitary matrix $\mathbf{U}$ dimensioned $[m \times m]$ such that

$$\mathbf{A} = \mathbf{UDU}^H,$$

where $\{\lambda_i\}$ are the eigenvalues of $\mathbf{A}$ and $r = r(\mathbf{A})$. In this case, the columns of $\mathbf{U}$ are the eigenvectors associated with the eigenvalues in $\mathbf{D}$.

(v) **QR factorization** For any $m \times n$ matrix $\mathbf{A}$, it is possible to express it as a product of two matrices as follows:

$$\mathbf{A} = \mathbf{QR},$$

where $\mathbf{Q}$ is an $m \times m$ unitary matrix and $\mathbf{R}$ is an $m \times n$ upper triangular matrix. (Note: There are various versions of this theorem. The one used above matches the Matlab convention.)

(w) **Rank plus nullity theorem** Consider an $m \times n$ matrix $\mathbf{A}$. If $m \geq n$, then

$$r(\mathbf{A}) + \dim[N(\mathbf{A})] = m.$$

If $n \geq m$, then

$$r(\mathbf{A}) + \dim[N(\mathbf{A})] = n,$$

where $\dim[N(\mathbf{A})]$, which denotes the dimension of the null space of $\mathbf{A}$, is equal to the minimum number of basis vectors needed to span that space.

(x) Let $\mathbf{A}$ be an $m \times n$ complex matrix and consider the $m \times m$ Hermitian matrix $\mathbf{AA}^H$ with eigenvalues $\{\lambda_1, \lambda_2, \ldots, \lambda_r\}$, where $r = \text{rank}(\mathbf{AA}^H)$. It follows from the definitions of trace and Frobenius norm and Section 1.9.2-(p) that

$$\text{Tr}(\mathbf{AA}^H) = \sum_{i=1}^{r} \lambda_i = \|\mathbf{A}\|_F^2.$$

(y) Let $\mathbf{A}$ and $\mathbf{B}$ denote two matrices that are dimensioned $m \times n$ and $n \times m$, respectively. It follows that

$$\frac{\partial \text{Tr}(\mathbf{AB})}{\partial \mathbf{A}} = \mathbf{B}^T \quad \text{and} \quad \frac{\partial \text{Tr}(\mathbf{AB})}{\partial \mathbf{B}} = \mathbf{A}^T.$$

It can also be shown that

$$\frac{\partial \text{Tr}\left(\mathbf{A}^H\mathbf{B}\right)}{\partial \mathbf{A}} = \mathbf{0}_{m \times n} \quad \text{and} \quad \frac{\partial \text{Tr}\left(\mathbf{AB}^H\right)}{\partial \mathbf{B}} = \mathbf{0}_{n \times m}.$$

(z) Let $\mathbf{A}$ be a square matrix dimensioned $m \times m$ and $\mathbf{I}$ be the diagonal matrix with the same dimensions as $\mathbf{A}$. Denote the eigenvalues of $\mathbf{A}$ by $\{\lambda_1, \lambda_2, \ldots, \lambda_r\}$; where, $r \leq m$ is the rank of $\mathbf{A}$. It follows that the eigenvalues of $(\mathbf{A} + \mathbf{I})$ are equal to $\{\lambda_i + 1, \ i = 1, \ldots, r\}$ and are equal to 1 for $(r + 1) \leq i \leq n$.

Problems

1.1 Consider a MIMO communication system that employs QPSK modulation and no coding, and that uses spatial diversity to mitigate the effects of Rayleigh fading. Assume that the channel experiences ideal Rayleigh fading and that the average bit error probability is measured and found to follow the behavior shown in the plot below. What is the coding gain and diversity gain of this system? How many independent spatial diversity channels exist? How many total antennas does this MIMO system have if it experiences full diversity?

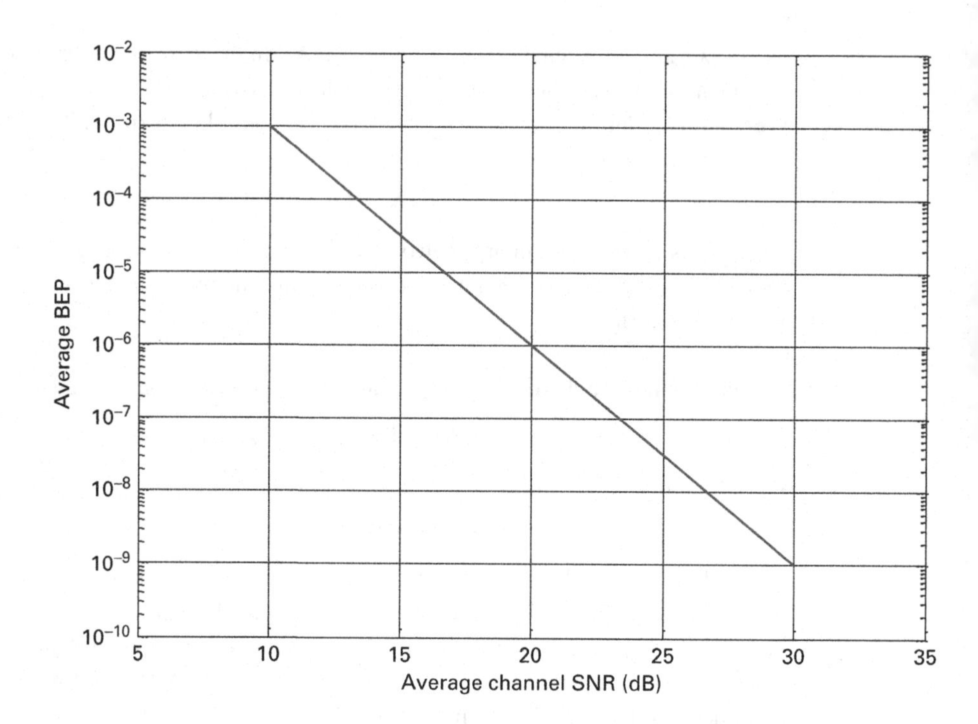

The remaining problems are designed to familiarize you with some of the matrix properties listed in Section 1.9. Most of these exercises require the use of Matlab.

1.2 (a) Create a 4×4 array, $\mathbf{A}$, where the first three rows are linearly independent vectors, and make the fourth row a multiple of the third row. Use *rank.m* in Matlab to compute the rank of $\mathbf{A}$. Since the rank of a matrix is equal to the number of linearly independent rows or columns, verify that $r(\mathbf{A}) = 3$.

(b) Repeat (a) but make the second row a multiple of the fourth row. Call this new array $\mathbf{B}$. Verify that the rank of $\mathbf{B}$ is 2 since there are only two linearly independent rows.

(c) Create a 4×4 array $\mathbf{C} = [a*\text{ones}(1,4);b*\text{ones}(1,4);c*\text{ones}(1,4);d*\text{ones}(1,4)]$, where $a, b, c,$ and d are any real or complex numbers. Verify that the $r(\mathbf{C}) = 1$.

(d) Create a 4×4 array $\mathbf{D} = \text{rand}(4,4)$ and $\mathbf{E} = \text{rand}(4,4)$. Verify that the rank of these matrices is 4. This is a general property of random matrices. In general, random matrices are full rank.

1.3 Consider the following matrices:

$$\mathbf{A} = \begin{pmatrix} a & b \\ c & d \end{pmatrix} \quad \text{and} \quad \mathbf{B} = \begin{pmatrix} e & f \\ g & h \end{pmatrix}.$$

Compute $\det(\mathbf{A})$, $\det(\mathbf{B})$, and $\det(\mathbf{AB})$. Show that $\det(\mathbf{AB}) = \det(\mathbf{A})\det(\mathbf{B})$.

1.4 Prove matrix theorem Section 1.9.2-(o).

1.5 Consider the following matrix:

$$\mathbf{A} = \begin{pmatrix} 3 & 1 \\ 1 & 3 \end{pmatrix}.$$

(a) Write down the characteristic polynomial of **A**. Use this polynomial to compute the eigenvalues of **A**.

(b) Use the expression $\mathbf{Av} = \lambda\mathbf{v}$ to find the eigenvectors $\{v_i\}$ of **A** that correspond to the eigenvalues computed in (a).

1.6 For the matrix in the previous problem, show that theorem Section 1.9.2-(p) holds.

1.7 Use Matlab to create a 4×3 array, **C**, as follows: `C = rand(4,3)`. Next, type `[U,D,V] = svd(C)`. Confirm that `U` and `V` are unitary matrices. Confirm that the singular values of `D` are in descending order and that the number of non-zero singular values is equal to the rank of `C`.

1.8 Use Matlab to create a 4×4 Hermitian matrix, $\mathbf{H} = \mathbf{AA}^H$, where **A** is an arbitrary $4 \times n$ array (i.e., pick any value for n) and choose any values for the elements of **A**. Use `svd.m` and `eig.m` to compute the singular values of **A** and **H**.

(a) Confirm that matrix theorem Section 1.9.2-(s) holds by comparing the singular values of **A** with the square root of the eigenvalues of **H**.

(b) Also confirm that matrix theorem Section 1.9.2-(t) holds by comparing the singular values of **H** with the absolute values of the eigenvalues of **H**. [Note: When performing this comparison, make sure the singular values and eigenvalues are ordered the same (i.e., compare the largest with the largest, the second largest with the second largest, etc).]

1.9 Use the Matlab routine `qr.m` to confirm the properties of QR factorization described in matrix theorem Section 1.9.2-(v). To do so, create $m \times n$ arrays by typing `A = rand(m,n)` and `C = rand(m,n) + j*rand(m,n)` for any integers m and n. Use `qr.m` to compute `Q` and `R` for `A` and `C` and confirm that the product `Q*R` yields the original matrix. Also confirm that `Q` is unitary and that `R` is upper triangular.

1.10 This problem addresses the rank plus nullity theorem.

(a) In Matlab, generate an arbitrary matrix `A = rand(m,n)` for any integers $m, n > 1$. Next, type `R = null(A)`. The columns of `R` are the basis vectors of N(**A**). Use this fact to confirm the rank plus nullity theorem.

(b) Verify that each of the basis vectors in **R** is in the null space of **A**.

(c) Demonstrate that linear combinations of these basis vectors are also in N(**A**).

2 The MIMO capacity formula

In 1948 Claude Shannon published his famous paper titled "A mathematical theory of communications," which was published in the July and October 1948 issues of the *Bell Technical Journal* [65, 66]. In that paper, he presented the fundamental concepts of what would later become the field of information theory, and derived mathematical expressions for the maximum theoretical data rate that could be transmitted over a communication system without errors. This maximum data rate, called the communications capacity, was derived for a conventional SISO system; however, the concepts he introduced in his landmark paper provided the framework for generalizing to MIMO systems. In 1999 Emre Telatar derived an expression for the theoretical capacity of a MIMO system using the concepts from information theory first developed by Shannon half a century earlier [70]. This chapter derives the MIMO capacity formula based on Telatar's arguments. The expression we develop in this chapter will be used to quantify the throughput enhancement that is possible with spatial multiplexing, as well as to provide a useful way of conceptualizing how MIMO systems work.[1]

2.1 What is information?

In its most fundamental sense, the capacity of a communication system is defined as the maximum amount of information that can be conveyed per unit of time between two points over a communications channel. In order to have a definition for capacity that is useful for engineering purposes, it is necessary to be able to quantify it – to be able to assign numerical values to the capacity of a communication system. This, in turn, requires that we be able to quantify information; thus, we need a mathematical definition of information. This section introduces the reader to the formal definition of information as it is used in information theory.

An interesting aspect of information theory is that it is based on the use of the word "information" as used in everyday speech. That is, the way we use the word "information" in everyday conversation helps motivate the way information is defined

[1] As stated in the preface, this book does not assume the reader has had a formal course in information theory; thus, basic concepts from information theory are introduced in this chapter. The reader who is already grounded in information theory should feel free to skim through the first few sections of this chapter. Readers desiring additional background are referred to [18].

mathematically in information theory. To help see this, consider the following two statements:

Statement 1 *It snowed in Anchorage, Alaska this morning.*
Statement 2 *It snowed in San Juan, Puerto Rico this morning.*

We first ask: which statement intuitively conveys more information (assuming, of course, both are true)? Most people would say that Statement 2 conveys more information because snow in San Juan, Puerto Rico is unheard of, but snow in Anchorage, Alaska is a very common event. Instinctively, we sense that we have learned more when we are told about an inherently improbable event than when we learn about something that is commonplace. This is one key property of information.

The second key property of information can be grasped by considering the following statement:

Statement 3 *It snowed in Anchorage, Alaska* ***and*** *it snowed in San Juan, Puerto Rico this morning.*

We now ask: how much information is conveyed by statement 3 compared to the information conveyed by statements 1 and 2 separately? Instinctively, we would say that the amount of information in statement 3 is equal to the amount of information in statement 1 plus the amount of information in statement 2, since statement 3 includes everything in statements 1 and 2.

From these simple observations we can deduce that information as it is commonly termed in everyday language has the following two properties:

Properties of information deduced from everyday speech

Property 1 Statements about inherently improbable events convey more information than statements about inherently likely events.
Property 2 Information is additive.

Based on these two properties it is possible to develop a mathematical definition of information. Consider an event that has an a priori probability of occurring, p. If we are told that such an event has, indeed, happened, then the amount of information, I, that is conveyed by learning of that event's occurrence is defined as follows:

$$I(p) = \log \frac{1}{p} = -\log p. \tag{2.1}$$

By inspection, we see that Eq. 2.1 satisfies Property 1 since $\log(1/p)$ increases as p decreases. Property 2 follows from the property of logarithms. To see this, consider two events with a priori probabilities of occurrence equal to p_1 and p_2. Assume, furthermore, that these two events occur independently of each other; that is, the occurrence of one of the events does not affect the likelihood that the other event will happen. Under that assumption, the probability that both events occur, $p_{1,2}$, is equal to $p_1 p_2$. If we learn that both events have occurred, then by Eq. 2.1, the amount of information that is conveyed is given by

$$I(p_{1,2}) = \log \frac{1}{p_{1,2}} = \log \frac{1}{p_1 p_2} = \log \frac{1}{p_1} + \log \frac{1}{p_2} = I(p_1) + I(p_2), \tag{2.2}$$

which demonstrates the desired additive property.

The definition of information in Eq. 2.1 is general in the sense that it does not specify the base of the logarithm. There are an infinite number of possible bases; however only base 2 and base e are used in practice. When the base 2 logarithm is used, the units of information are *bits*. When the base is e, the units of information are referred to as *nats*. In the derivation of MIMO capacity in this chapter, we occasionally alternate between these two bases depending on mathematical expediency.

2.2 Entropy

In the previous section, we introduced the mathematical definition of information associated with random events. Since random events, in turn, are characterized in terms of random variables, it is reasonable to associate information with random variables. This slight abstraction enables us to analyze the information properties of random systems using well-defined properties from probability theory.

Consider a discrete random variable, X, that can take on values $x_1, x_2, \ldots$ Each time X takes on a specific value, the instantiation of X is an event. In general, we can speak of the information associated with the event $X = x_i,\ i = 1, 2, \ldots$ and denote that information by $I(x_i)$. Using Eq. 2.1, it follows that the information associated with X taking on the value x_i is given by

$$I(x_i) = -\log p(x_i), \tag{2.3}$$

where $p(x_i) \triangleq \text{Prob}\{X = x_i\}$.

Now that we have information defined in terms of a random variable, it is natural to consider the average information associated with a given random variable. In information theory, we call the average information the *entropy*, which we denote by $H(X)$ as follows:

$$\text{Entropy} = H(X) \triangleq \mathbb{E}_X\{I(X)\}, \tag{2.4}$$

$$= \mathbb{E}_X\{-\log p(X)\}, \tag{2.5}$$

$$= -\sum_i p_X(x_i) \log p_X(x_i). \tag{2.6}$$

The entropy of a random variable also has the following interpretation: it is a measure of the uncertainty in our knowledge of the value of that random variable. To see why that is true, consider the extreme case where a random variable, X, has no uncertainty. In that case, $X = x_0$ is the value of the random variable and its distribution function is equal to 1 at $X = x_0$ and 0 at all other values of X. Applying this to Eq. 2.6 results in $H(X) = -p_X(x_0) \log p_X(x_0) = -1 \cdot \log(1) = 0$, which is consistent with our interpretation that $H(X)$ is a measure of the uncertainty in X since there is no uncertainty in this example. Although we will not show it here, it is possible to show that if $X \in \{x_1, x_2, \ldots x_N\}$, then $H(X)$ is maximized when $p_X(x_i) = 1/N \ \forall i$, which corresponds to the case where there is maximum uncertainty in the value of X – consistent with the interpretation of $H(X)$ being a measure of the uncertainty in X. One of the problems at the end of the chapter demonstrates this property for the simple case where X takes on only two values. For a more general proof, the Lagrange multiplier method is needed.

A second term that is used extensively in information theory is the *conditional entropy*, which we denote by $H(Y|X)$ and define as follows:

$$\text{Conditional entropy} = H(Y|X) = -\sum_{x,y} p(x,y) \log p_{X,Y}(y|x). \tag{2.7}$$

We use both of these terms in the subsequent steps.

2.3 Mutual information

The next term that we define is *mutual information*, which is closely related to communications capacity. Before we define mutual information, however, it is useful to consider a generic digital communication system, which is depicted in Figure 2.1. As shown, the essential elements of a communication system are an information source, an information encoder that converts the output of the information source into discrete symbols, a communications channel, and an information decoder, which converts the received symbols into a form that is compatible with the destination. Since we are concerned with digital communication systems, the information encoder converts the information, which could be analog, such as a voice signal, into a finite set of discrete encoded symbols $\{x_1, x_2, \ldots\}$. The encoded symbols are transmitted over a communications channel and are received at the destination as a set of symbols $\{y_1, y_2, \ldots\}$.

In an abstract sense, a communications channel is a system in which the output depends probabilistically on its input. That is, if the input to the channel is x_i, then the output, y_i, can take on different values depending on the properties of the channel, but the value that the received symbol takes on cannot be predicted with certainty. This leads to the following question: how much information about the occurrence of an event $X = x_i$ is provided by the occurrence of the event $Y = y_i$? That is, if the received symbol is $Y = y_i$, how much information has been learned about the event $X = x_i$? The amount

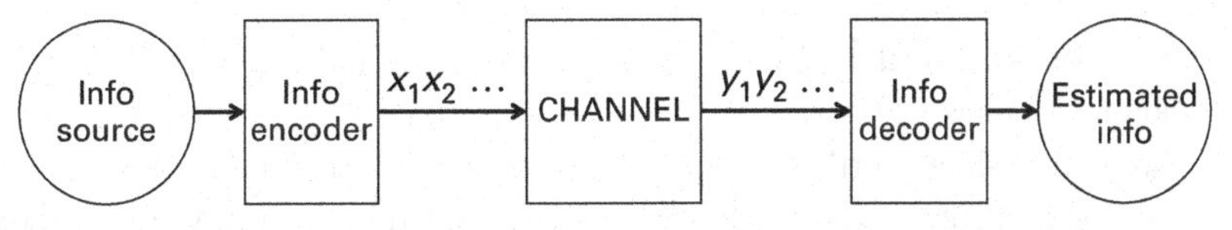

Figure 2.1 A generic digital communication system.

of information that is conveyed is called the *mutual information* (MI), which we denote by $I(x_i; y_i)$ and define as follows:

$$I(x_i; y_i) \triangleq \log\left(\frac{p(x_i|y_i)}{p(x_i)}\right). \tag{2.8}$$

Although this definition may not seem intuitive at first, it is straightforward to demonstrate that it leads to reasonable results. The following two examples examine the properties of MI.

Example 2.1 Assume X and Y are statistically independent random variables. In that case, we would not expect the received symbol to convey any information about the transmitted symbol, so the mutual information should equal 0. Under this assumption, $p(x_i|y_i) = p(x_i)$. Therefore, $I(x_i; y_i) = \log\left(\frac{p(x_i)}{p(x_i)}\right) = 0$, which is consistent with our expectations.

Example 2.2 Assume the occurrence of event $Y = y_i$ uniquely determines the event $X = x_i$. That is, we assume that if we know the received symbol, we also automatically know which symbol was transmitted. In that case, the amount of information we learn about the event $X = x_i$ from the event $Y = y_i$ is the information associated with the event $X = x_i$ itself. Under these assumptions we would expect the mutual information to be equal to $I(x_i)$.

If the event $Y = y_i$ uniquely determines the event $X = x_i$, it follows that $p(x_i|y_i) = 1$; thus, it follows from Eq. 2.8 that $I(x_i|y_i) = I(x_i)$, which, again, is consistent with our intuitive expectations.

In addition to measuring the mutual information between two specific events, such as $X = x_i$ and $Y = y_i$, it is also useful to consider the average mutual information between two random variables X and Y. We denote the average mutual information by $I(X; Y)$, which is given by

$$\begin{aligned} I(X; Y) &= \mathbb{E}_{x,y}\left\{\log\left(\frac{p(x|y)}{p(x)}\right)\right\} \\ &= \sum_{x,y} p(x, y) \log\left(\frac{p(x|y)}{p(x)}\right), \end{aligned} \tag{2.9}$$

where, for notational simplicity we have eliminated the use of the sub-indices on the random variables.

We will explain shortly how the average mutual information is related to the capacity of a communication system; however, before we do that it is useful to consider two important properties of $I(X; Y)$.

THEOREM 2.1 *The average mutual information is commutative. That is,*

$$I(X;Y) = I(Y;X). \tag{2.10}$$

Proof: $I(X;Y) = \sum_{x,y} p(x,y) \log\left(\frac{p(x|y)}{p(x)}\right)$. From Baye's theorem it follows that $p(x,y) = p(x|y)p(y) = p(y|x)p(x)$, which implies that

$$\frac{p(x|y)}{p(x)} = \frac{p(y|x)}{p(y)}.$$

It follows that

$$\begin{aligned} I(X;Y) &= \sum_{x,y} p(x,y) \log\left(\frac{p(x|y)}{p(y)}\right) \\ &= I(Y;X). \end{aligned}$$

□

THEOREM 2.2 *The average mutual information between random variables X and Y is related to the entropy of those random variables in the following way:*

$$\begin{aligned} I(X;Y) &= H(X) - H(X|Y) \\ &= H(Y) - H(Y|X), \end{aligned} \tag{2.11}$$

where $H(X|Y)$ and $H(Y|X)$ are called the conditional entropies of X and Y, and are defined as follows: $H(Y|X) \triangleq -\sum_{x,y} p(x,y)\log(x|y) = -\sum_{x,y} p(x,y)\log(y|x)$.

Proof:

$$\begin{aligned} I(X;Y) &= \sum_{x,y} p(x,y) \log\left(\frac{p(x|y)}{p(x)}\right) \\ &= -\sum_{x,y} p(x,y)\log p(x) - \left[-\sum_{x,y} p(x,y)\log(x|y)\right] \\ &= -\sum_{x} p(x)\log p(x) - \left[-\sum_{x,y} p(x,y)\log(x|y)\right] \\ &= H(X) - H(X|Y), \end{aligned}$$

where the third step follows from the fact that $\log p(x)$ is only a function of x; thus, the summation over the y component of the distribution is 1, leaving only the x dependence.

The second line in Eq. 2.11 follows directly from Theorem 2.1. □

2.4 Definition of SISO capacity

Now that we have introduced some basic concepts and nomenclature from information theory, it is possible to define the capacity of a conventional SISO communication system in a precise, mathematical way. We start by recalling that $I(X;Y)$, the average mutual information between two random variables, is the average amount of information about

X that is conveyed by knowing the value of Y. If we interpret X as the input to a communications channel and Y as its output, as illustrated in Figure 2.1, then $I(X;Y)$ can be interpreted as the average amount of transmitted information that is conveyed over the communications channel through examination of the received signal. Although the communicators do not have control over the channel, the transmitter, in principle, does have control over the distribution of the random symbols its generates, $p_X(x)$, through its design of the information encoder. Based on these considerations, the capacity of a communication system is defined to be the maximum amount of information that is conveyed by observing the received signal, where the maximization is over all possible distributions of $p_X(x)$. Mathematically, therefore, we define the capacity, C, of a SISO system as follows:

$$C \triangleq \max_{\{p_X(x)\}} I(X;Y), \tag{2.12}$$

where the maximization is performed using the following constraints:

$$0 \leq p_X(x) \leq 1 \quad \text{and} \quad \sum_x p_X(x) = 1, \tag{2.13}$$

which follow from the fact that these constraints apply to the distribution function of a random variable.

In the next section, we generalize the definition of capacity in Eq. 2.12 for the case of a MIMO communication system.

2.5 Definition of MIMO capacity

2.5.1 MIMO system model

We are now ready to apply the definition of capacity to MIMO systems. To do so, however, it is first necessary to develop a system model of a MIMO communication system. By system model, we mean a mathematical description of the relationship between the transmitted and received signals. We assume a narrowband signal, which means that the bandwidths of the communication signals are narrow compared to the coherence bandwidth of the channel, a concept that we describe precisely in Chapter 4.

We start by defining the following parameters:

$h_{ij} \triangleq$ channel response between the jth transmit antenna and the ith receive antenna,

$r_i \triangleq$ received signal at the ith receive antenna,

$s_j \triangleq$ symbol transmitted from the jth transmit antenna,

$z_i \triangleq$ noise signal at the ith receive antenna.

It follows that

$$r_i = \sum_{j=1}^{N_t} h_{ij} s_j + z_i, \qquad i = 1, \ldots, N_r. \tag{2.14}$$

In matrix form, Eq. 2.14 becomes

$$\mathbf{r} = \mathbf{H}\mathbf{s} + \mathbf{z}, \tag{2.15}$$

where

$$\mathbf{s} \triangleq \left[s_1, \ldots, s_{N_t}\right]^T, \tag{2.16}$$

$$\mathbf{z} \triangleq \left[z_1, \ldots, z_{N_t}\right]^T, \tag{2.17}$$

$$\mathbf{r} \triangleq \left[r_1, \ldots, r_{N_t}\right]^T, \text{ and} \tag{2.18}$$

$$\mathbf{H} \triangleq \begin{pmatrix} h_{1,1} & \cdots & h_{1,N_t} \\ \vdots & \ddots & \vdots \\ h_{N_r,1} & \cdots & h_{N_r,N_t} \end{pmatrix}. \tag{2.19}$$

This system model is used throughout much of the MIMO literature and is used repeatedly throughout this book. Some authors reverse the dimensions of $\mathbf{H}$ so that it is dimensioned $N_t \times N_r$. In that case Eq. 2.15 becomes $\mathbf{r} = \mathbf{s}\mathbf{H} + \mathbf{z}$ and the vectors $\mathbf{r}$, $\mathbf{s}$, and $\mathbf{z}$ become row vectors instead of column vectors. It is important for the reader to keep in mind the convention being used when reading the MIMO literature and to be able to switch between conventions.

2.5.2 Capacity

The capacity of a MIMO system is obtained by generalizing Eq. 2.12, which is done by replacing X with $\mathbf{s}$ and Y with $\mathbf{r}$. Using the relationship in Eq. 2.11, we obtain the following expression for the capacity of a MIMO system:

$$C_{\text{MIMO}} = \max_{p_{\mathbf{S}}(s_1,\ldots,s_{N_t})} \left\{H(\mathbf{r}) - H(\mathbf{r}|\mathbf{s})\right\}, \tag{2.20}$$

where the maximization is over the multivariate distribution $p_{\mathbf{S}}(\mathbf{s})$.

Equation 2.20 can be simplified using the fact that $\mathbf{r} = \mathbf{H}\mathbf{s} + \mathbf{z}$, which implies that $H(\mathbf{r}|\mathbf{s}) = H(\mathbf{H}\mathbf{s} + \mathbf{z}|\mathbf{s})$. For a given channel matrix, $\mathbf{H}$, $\mathbf{H}\mathbf{s}$ is fixed, so the only random variation in $H(\mathbf{r}|\mathbf{s})$ is due to the noise term, $\mathbf{z}$. It follows that $H(\mathbf{r}|\mathbf{s}) = H(\mathbf{z})$, which means that the MIMO capacity simplifies to the following:

$$C_{\text{MIMO}} = \max_{p_{\mathbf{S}}(s_1,\ldots,s_{N_t})} \left\{H(\mathbf{r}) - H(\mathbf{z})\right\}. \tag{2.21}$$

In order to obtain a final expression for the capacity of a MIMO system, we need to evaluate the two terms $H(\mathbf{r})$ and $H(\mathbf{z})$ in Eq. 2.21. We do that in the next two sections.

2.6 Evaluating $H(\mathbf{z})$

We evaluate $H(\mathbf{z})$ by assuming that each of the elements of $\mathbf{z}$ are independent, identically distributed (IID) real Gaussian random variables with mean μ and variance σ^2. Thus, $\mathbf{z}$ has a multivariate Gaussian distribution given by

$$p_{\mathbf{Z}}(\mathbf{z}) = \frac{1}{\left(\sqrt{2\pi}\right)^{N_r} |\mathbf{R}_{zz}|^{1/2}} \mathrm{e}^{-\frac{1}{2}(\mathbf{z}-\boldsymbol{\mu})^T \mathbf{R}_{zz}^{-1}(\mathbf{z}-\boldsymbol{\mu})}, \tag{2.22}$$

where $\boldsymbol{\mu} \triangleq \mathbb{E}\{\mathbf{z}\}$, $|\mathbf{R}_{zz}|$ denotes the determinant of $\mathbf{R}_{zz}$, and $\mathbf{R}_{zz}$ is the covariance matrix of $\mathbf{z}$ given by

$$\mathbf{R}_{zz} \triangleq \mathbb{E}\left\{(\mathbf{z}-\boldsymbol{\mu})(\mathbf{z}-\boldsymbol{\mu})^T\right\}. \tag{2.23}$$

When $\mathbf{z}$ has this property, we denote that fact using the common shorthand notation: $\mathbf{z} \sim \mathcal{N}(\boldsymbol{\mu}, \mathbf{R}_{zz})$. For the purpose of this derivation, we assume that the mean of each of the noise elements is zero; thus, $\boldsymbol{\mu} = \mathbf{0}$, so $\mathbf{z} \sim \mathcal{N}(\mathbf{0}, \mathbf{R}_{zz})$. This assumption, together with the independence of the noise terms, implies that

$$\mathbf{R}_{zz} = \sigma^2 \mathbf{I}_{N_r}, \tag{2.24}$$

where the nomenclature $\mathbf{I}_x$ denotes an identity matrix dimensioned $x \times x$.

To simplify the derivation, we begin by using the natural logarithm version of $H(\mathbf{z})$ given by

$$H(\mathbf{z}) = -\mathbb{E}\{\ln p_{\mathbf{Z}}(\mathbf{z})\}, \tag{2.25}$$

where the expectation is over the elements of $\mathbf{z}$.

Therefore,

$$\begin{aligned}
H(\mathbf{z}) &= \mathbb{E}\left\{-\frac{1}{2}(\mathbf{z}-\boldsymbol{\mu})^T \mathbf{R}_{zz}^{-1}(\mathbf{z}-\boldsymbol{\mu}) - \ln\left[\left(\sqrt{2\pi}\right)^{N_r} |\mathbf{R}_{zz}|^{1/2}\right]\right\} && \text{(a)}\\
&= \frac{1}{2}\mathbb{E}\left\{\sum_{i,j}\left[(z_i-\mu_i)\left[\mathbf{R}_{zz}^{-1}\right]_{ij}(z_j-\mu_j)\right]\right\} + \frac{1}{2}\ln\left[(2\pi)^{N_r}|\mathbf{R}_{zz}|\right] && \text{(b)}\\
&= \frac{1}{2}\sum_{i,j}\left[\mathbb{E}\{(z_j-\mu_j)(z_i-\mu_i)\}\left[\mathbf{R}_{zz}^{-1}\right]_{ij}\right] + \frac{1}{2}\ln\left[(2\pi)^{N_r}|\mathbf{R}_{zz}|\right] && \text{(c)}\\
&= \frac{1}{2}\sum_{i,j}\left[\mathbf{R}_{zz}\right]_{ji}\left[\mathbf{R}_{zz}^{-1}\right]_{ij} + \frac{1}{2}\ln\left[(2\pi)^{N_r}|\mathbf{R}_{zz}|\right] && \text{(d)}\\
&= \frac{1}{2}\sum_{i}\left[\mathbf{R}_{zz}\mathbf{R}_{zz}^{-1}\right]_{ii} + \frac{1}{2}\ln\left[(2\pi)^{N_r}|\mathbf{R}_{zz}|\right] && \text{(e)}\\
&= \frac{1}{2}\sum_{i}[\mathbf{I}]_{ii} + \frac{1}{2}\ln\left[(2\pi)^{N_r}|\mathbf{R}_{zz}|\right] && \text{(f)}
\end{aligned}$$

$$
\begin{aligned}
&= \frac{N_r}{2} + \frac{1}{2}\ln\left[(2\pi)^{N_r}|\mathbf{R}_{zz}|\right] && \text{(g)}\\
&= \frac{N_r}{2}\ln \mathrm{e} + \frac{1}{2}\ln\left[(2\pi)^{N_r}|\mathbf{R}_{zz}|\right] && \text{(h)}\\
&= \frac{1}{2}\ln\left[(2\pi\mathrm{e})^{N_r}|\mathbf{R}_{zz}|\right] \text{ nats} && \text{(i)}\\
&= \frac{1}{2}\log_2\left[(2\pi\mathrm{e})^{N_r}|\mathbf{R}_{zz}|\right] \text{ bits} && \text{(j)}\\
&= \frac{N_r}{2}\log_2(2\pi\mathrm{e}) + \frac{1}{2}\log_2|\mathbf{R}_{zz}| \text{ bits}, && \text{(k)}
\end{aligned}
\tag{2.26}
$$

where the steps between subequations (b) and (c) and between (d) and (e) follow straightforwardly from matrix properties and are addressed as problems at the end of the chapter.

Under the zero-mean, independent assumptions for the noise, which results in $\mathbf{R}_{zz}$ having the form given by Eq. 2.24, Eq. 2.26 simplifies to the following:

$$H(\mathbf{z}) = \frac{N_r}{2}\log_2(2\pi\mathrm{e}) + \frac{1}{2}\log_2|\sigma^2\mathbf{I}_{N_r}|. \tag{2.27}$$

2.7 Evaluating $H(\mathbf{r})$

We now derive an expression for $H(\mathbf{r})$ that maximizes the MIMO capacity expression in Eq. 2.21. To do so, we appeal to the following theorem that is proven in [18], which we refer to as the *entropy-maximizing theorem* (EMT).

THEOREM 2.3 (entropy-maximizing theorem) *Let* $\mathbf{x}$ *be a real-valued random vector where*

$$\mathbb{E}\{\mathbf{x}\} = \mathbf{0} \qquad \textit{and} \qquad \mathbf{R}_{xx} = \mathbb{E}\left\{\mathbf{x}\mathbf{x}^T\right\}. \tag{2.28}$$

Then $H(\mathbf{x})$ *is maximized when* $\mathbf{x} \sim \mathcal{N}(\mathbf{0}, \mathbf{R}_{xx})$, *where* $\mathcal{N}(\mathbf{0}, \mathbf{R}_{xx})$ *denotes a multivariate Gaussian distribution with zero mean and covariance matrix* $\mathbf{R}_{xx}$.

This theorem implies that $H(\mathbf{r})$ in Eq. 2.21 is maximized when $\mathbf{r} \sim \mathcal{N}(\mathbf{0}, \mathbf{R}_{rr})$; that is, when $\mathbf{r}$ is a multivariate Gaussian random variable with zero mean. Since $\mathbf{r} = \mathbf{H}\mathbf{s} + \mathbf{z}$ and $\mathbf{z} \sim \mathcal{N}(\mathbf{0}, \mathbf{R}_{zz})$, then for a given channel matrix, $\mathbf{H}$, these requirements are met if the transmitted signal vector is a multivariate Gaussian random variable with zero mean.

Under this assumption, $\mathbf{r}$ has the same distribution as $\mathbf{z}$ in Eq. 2.26 except that its covariance matrix is $\mathbf{R}_{rr}$ instead of $\mathbf{R}_{zz}$. It follows from this fact and Eq. 2.26(k) that $\max\{H(\mathbf{r})\}$ is given by

$$\max\{H(\mathbf{r})\} = \frac{N_r}{2}\log_2(2\pi\mathrm{e}) + \frac{1}{2}\log_2|\mathbf{R}_{rr}| \text{ bits}. \tag{2.29}$$

Therefore, in order to evaluate $\max\{H(\mathbf{r})\}$, it is necessary to evaluate $\mathbf{R}_{rr}$ as follows:

$$\begin{aligned}
\mathbf{R}_{rr} &\triangleq \mathbb{E}\left\{(\mathbf{r}-\boldsymbol{\mu}_r)(\mathbf{r}-\boldsymbol{\mu}_r)^T\right\} \\
&= \mathbb{E}\left\{\mathbf{r}\mathbf{r}^T\right\} \\
&= \mathbb{E}\left\{(\mathbf{Hs}+\mathbf{z})(\mathbf{Hs}+\mathbf{z})^T\right\} \\
&= \mathbb{E}\left\{(\mathbf{Hs}+\mathbf{z})\left(\mathbf{s}^T\mathbf{H}^T+\mathbf{z}^T\right)\right\} \\
&= \mathbb{E}\left\{\mathbf{Hss}^T\mathbf{H}^T+\mathbf{Hsz}^T+\mathbf{zs}^T\mathbf{H}^T+\mathbf{zz}^T\right\} \\
&= \mathbf{HR}_{ss}\mathbf{H}^T+\sigma^2\mathbf{I}_{N_r}.
\end{aligned} \tag{2.30}$$

Substituting this expression for $\mathbf{R}_{rr}$ into Eq. 2.29 yields

$$\max\{H(\mathbf{r})\} = \frac{N_r}{2}\log_2(2\pi e) + \frac{1}{2}\log_2\left|\mathbf{HR}_{ss}\mathbf{H}^T+\sigma^2\mathbf{I}_{N_r}\right| \text{ bits.} \tag{2.31}$$

2.8 Final result

2.8.1 Real signals

Since our discussion thus far has specifically focused on real signals only, we first develop the capacity expression for real signals, then generalize to complex signals. The MIMO capacity of a real signal, C_{real}, is obtained by substituting Eqs. 2.27 and 2.30 into Eq. 2.21, resulting in the following:

$$\begin{aligned}
C_{\text{real}} &= \max\{H(\mathbf{r})-H(\mathbf{z})\} \\
&= \frac{1}{2}\log_2\left|\mathbf{HR}_{ss}\mathbf{H}^T+\sigma^2\mathbf{I}\right| - \frac{1}{2}\log_2\left|\sigma^2\mathbf{I}\right| \\
&= \frac{1}{2}\log_2\left[\frac{\left|\mathbf{HR}_{ss}\mathbf{H}^T+\sigma^2\mathbf{I}\right|}{\left|\sigma^2\mathbf{I}\right|}\right] \\
&= \frac{1}{2}\log_2\left[\frac{\left|\mathbf{HR}_{ss}\mathbf{H}^T+\sigma^2\mathbf{I}\right|}{\sigma^2\left|\mathbf{I}\right|}\right].
\end{aligned} \tag{2.32}$$

We can now use the fact that $|\mathbf{I}| = 1$ (see matrix property Section 1.9.2-(f)) and matrix property 1.9.2-(g) to simplify Eq. 2.32 to the following:

$$C_{\text{real}} = \frac{1}{2}\log_2\left|\mathbf{I}+\frac{1}{\sigma^2}\mathbf{HR}_{ss}\mathbf{H}^T\right| \frac{\text{bits}}{\text{channel use}}. \tag{2.33}$$

The expression for MIMO capacity in Eq. 2.33 has units of bits/channel use. What do we mean by this? First, the fundamental unit of capacity is the bit because the base 2 of the logarithm is used. Second, the amount of information that is conveyed is the amount that the receiver learns by examining the received signal, $\mathbf{r}$, at a given instant of time; that is, as a result of one "use" of the channel. We will show shortly that the unit of bits/channel use is equivalent to bits/sec/Hz, which is what is normally referred to as spectral efficiency.

An alternative form for the capacity is to specify it in terms of bits/sec. To do this, consider a bandlimited signal with bandwidth W and duration T. The Nyquist sampling theorem states that such a signal can be completely represented with $2TW$ independent samples [59]. This means that it is possible to transmit this signal using $2TW$ samples transmitted over a period of time equal to T seconds, which is equivalent to $2W$ transmissions per second. Since each transmission is capable of conveying the amount of information specified by Eq. 2.33, it follows that the capacity in bits/sec is obtained by multiplying that expression by $2W$, which results in the following alternative expression for the MIMO capacity of a real signal:

$$C_{\text{real}} = W \log_2 \left| \mathbf{I} + \frac{1}{\sigma^2} \mathbf{H} \mathbf{R}_{ss} \mathbf{H}^T \right| \text{ bits/sec.} \tag{2.34}$$

We conclude from this discussion that since WC(bits/channel use) $=$ C(bits/sec), the units of bits/channel use is equivalent to bits/sec/W, or, equivalently, to bits/sec/Hz, as was claimed earlier.

2.8.2 Complex signals

The expression for the MIMO capacity with complex signals is similar to the capacity for real signals derived above, except for two differences. The first difference can be deduced from the fact that complex signals consist of real and imaginary components that are capable of transmitting independent streams of information; hence, we would expect the capacity for complex signals to be twice that for real signals, all else being equal. The second difference, while somewhat less obvious, can be viewed as a generalization of the transpose operation on $\mathbf{H}$ in Eqs. 2.33 and 2.34. In Telatar's paper, he shows that with complex signals the term $\mathbf{H}^T$ should be replaced with $\mathbf{H}^H$. Thus, the expressions for capacity of a MIMO system with complex signals are given by the following:

$$C_{\text{complex}} = \log_2 \left| \mathbf{I} + \frac{1}{\sigma^2} \mathbf{H} \mathbf{R}_{ss} \mathbf{H}^H \right| \frac{\text{bits}}{\text{channel use}} \tag{2.35}$$

and

$$C_{\text{complex}} = 2W \log_2 \left| \mathbf{I} + \frac{1}{\sigma^2} \mathbf{H} \mathbf{R}_{ss} \mathbf{H}^H \right| \text{ bits/sec.} \tag{2.36}$$

The approach in this section has been to use simple plausibility arguments to justify the expressions for the capacity of complex signals rather than to show rigorous proofs. This was done because the complexity of the mathematics is significantly greater when complex signals are introduced, without a concomitant increase in added insight. For a rigorous derivation when complex signals are present, the reader is referred to Telatar's original paper [70].

Although capacity formulas are presented for both real and complex signals in this chapter, in practice, most signals in actual MIMO systems are complex. For this reason, only Eqs. 2.35 and 2.36 are used in subsequent chapters. In the next chapter, we examine these capacity formulas in detail and use them to gain fundamental insights into MIMO communication system performance.

Problems

2.1 The simplest communications channel model with errors is the binary symmetric channel (BSC), which is illustrated below.

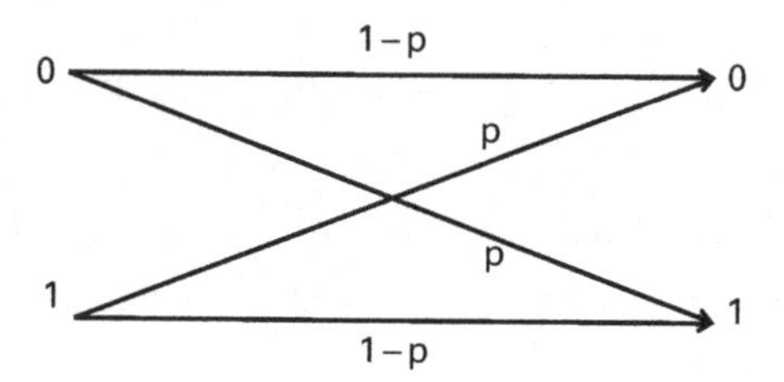

In a BSC, input to the channel is a discrete binary random variable $X \in \{0, 1\}$ and the probability of error is p. The received signals is also a discrete binary random variable, which is denoted by $Y \in \{0, 1\}$.

(a) Using Eq. 2.9, compute the average mutual information, $I(X; Y)$, as a function of p and the a priori probabilities $p_0 \triangleq \Pr(X = 0)$ and $p_1 \triangleq \Pr(X = 1)$.

(b) It can be shown that $I(X; Y)$ is maximized when $p_0 = p_1 = 0.5$. Using this fact, write down the capacity of the BSC as a function of p. [Note: the capacity is only a function of the property of the channel, which is characterized in this case entirely by p.]

(c) Plot the capacity versus p.

2.2 In the derivation of Eq. 2.26, one of the key steps was between parts (d) and (e), which assumed that $\sum_{i,j} [\mathbf{R}_{zz}]_{ji} [\mathbf{R}_{zz}^{-1}]_{ij} = \sum_i [\mathbf{R}_{zz}\mathbf{R}_{zz}^{-1}]_{ii}$, where $\mathbf{R}_{zz}$ denotes the noise covariance matrix. Prove that this relationship is true.

Hint: For any $N \times N$ matrix, $\mathbf{A}$, the i, j component of its inverse is given by

$$\left[\mathbf{A}^{-1}\right]_{i,j} = \frac{(-1)^{i+j} M_{ji}}{\det(\mathbf{A})},$$

where M_{ji} is the $(N-1) \times (N-1)$ minor obtained by deleting the jth row and the ith column of $\mathbf{A}$; although the definition of M_{ji} is not needed in the proof.

2.3 In this chapter, we derived an expression for the capacity of a MIMO system transmitting real signals and showed that it is given by:

$$C = W \log_2 \left| \mathbf{I} + \frac{1}{\sigma^2} \mathbf{H} \mathbf{R}_{ss} \mathbf{H}^T \right| \text{ bps.}$$

(a) Use this expression to derive the capacity of a SISO system in terms of the bandwidth, W, and the signal-to-noise power ratio, $\rho \triangleq \sigma_s^2/\sigma^2$; where σ_s^2 denotes the variance of the transmitted signal.

(b) Use the expression in (a) to compute the required value of E_b/N_0 as a function of ρ and the bandwidth, W. In doing so, assume that the data rate, R, equals the capacity, C. It should be noted that this expression for E_b/N_0 indicates that E_b/N_0 decreases monotonically as ρ decreases. Find the required value of E_b/N_0 in the limit as $\rho \to 0$. This value is called the Shannon limit. It is the smallest value of E_b/N_0 for which it is possible (in theory) to find a channel coding scheme that results in zero probability of error.

3 Applications of the MIMO capacity formula

In this chapter we study the MIMO capacity formula derived in the previous chapter to see what insight can be gained about MIMO communications. In order to study the MIMO capacity formula, it is necessary to make assumptions about the type of channel state information (CSI) that exists. In the MIMO literature, the term CSI refers to knowledge about the communications channel, which is equivalent to knowledge about the channel matrix, $\mathbf{H}$. In most MIMO schemes, either the transmitter, receiver, or both must know $\mathbf{H}$.[1] When the receiver has knowledge of the CSI, we call that CSIR. Similarly, when the transmitter has knowledge of the channel, we denote that by CSIT. We will find that the predictions of the capacity formula depend on the type of CSI.

We begin by examining the capacity formula under CSIR only assumptions, which leads to the concept of eigenmodes and the importance of channel rank in characterizing MIMO performance. After that, we consider systems that have both CSIT and CSIR, which leads to the concept of eigenbeamforming. Next, we present simulation results that show the predicted capacity of MIMO systems under various assumptions. Following that discussion, we derive theoretical closed-form expressions for the capacities of SIMO and MISO systems and discuss how and why SIMO and MISO systems differ. We conclude this chapter by considering the capacities of random channels, where the concepts of ergodic and outage capacity are introduced.

3.1 MIMO capacity under the CSIR assumption

Under the CSIR only assumption, the transmitter does not have knowledge about the communications channel, so there is no basis for transmitting signals in any sort of preferential way on different antennas. This fact has two implications:

1. There is no reason to transmit more energy on one antenna than another; thus, the average signal power should be the same on each transmit antenna.
2. There is no reason to introduce correlation or dependence between antennas.

In summary: under the CSIR only assumption, transmitted signals on each of the N_t transmit antennas are equi-power, independent, and uncorrelated.

[1] In practice, of course, only estimates of the channel are possible. In Chapter 10, we discuss channel estimation techniques and examine the sensitivity of MIMO performance to errors in the knowledge of $\mathbf{H}$.

To understand the implications of these assumptions, we start with the following version of the MIMO system equation:

$$\mathbf{r} = \sqrt{\rho}\,\mathbf{H}'\mathbf{s} + \mathbf{z}, \tag{3.1}$$

where $\mathbf{r}$ is the $N_r \times 1$ received signal vector, $\mathbf{H}'$ is the $N_r \times N_t$ *normalized* channel matrix, $\mathbf{s}$ is the $N_t \times 1$ transmit signal vector, $\mathbf{z}$ is the $N_r \times 1$ noise signal vector, and ρ is the signal-to-noise power ratio at the receiver. Comparison of this equation with Eq. 2.15 shows that they differ by the coefficient $\sqrt{\rho}$; that is, the product $\sqrt{\rho}\,\mathbf{H}'$ is equivalent to the physical channel matrix in Eq. 2.15. Appendix A shows that when the system equation is expressed in the form shown in Eq. 3.1, the following normalizations are implied:

$$\mathbb{E}\left\{|h'_{ij}|^2\right\} = 1, \qquad \text{(for random } \mathbf{H}'\text{)} \tag{3.2}$$

$$\|\mathbf{H}'\|_F^2 = N_r N_t, \quad \text{(for fixed } \mathbf{H}'\text{)} \tag{3.3}$$

$$\mathbb{E}\left\{|z_i|^2\right\} \triangleq \sigma^2 = 1, \tag{3.4}$$

$$\mathbb{E}\left\{|s_i|^2\right\} \triangleq \sigma_s^2 = 1/N_t, \tag{3.5}$$

where σ_s^2 denotes the signal variance applied to each transmit antenna. The form of the system equation shown in Eq. 3.1 is commonly used in MIMO literature. It has the advantage that it includes ρ explicitly, which allows the user to make theoretical capacity predictions as a function of the signal-to-noise ratio. For this reason, the form of the system equation in 3.1 is used throughout a large portion of this book.

Implications 1 and 2 above imply that the signal covariance matrix has the following form:

$$\mathbf{R}_{ss} \triangleq \mathbb{E}\left\{\mathbf{s}\mathbf{s}^H\right\} = \sigma_s^2 \mathbf{I}_{N_t}. \tag{3.6}$$

Plugging this expression for $\mathbf{R}_{ss}$ into Eq. 2.35 results in the following expression for the capacity:

$$\begin{aligned} C &= \log_2 \left| \mathbf{I}_{N_r} + \frac{1}{\sigma^2}\mathbf{H}\mathbf{R}_{ss}\mathbf{H}^H \right| \\ &= \log_2 \left| \mathbf{I}_{N_r} + \frac{\sigma_s^2}{\sigma^2}\mathbf{H}\mathbf{H}^H \right|, \end{aligned} \tag{3.7}$$

where the second equality follows from the two implications listed above for CSIR only (i.e., equal power and uncorrelated signals from different antennas).

If we now substitute $\mathbf{H} \triangleq \sqrt{\rho}\mathbf{H}'$ into Eq. 3.7 and use Eqs. 3.4 and 3.5, the capacity of the MIMO channel can be rewritten as follows:

$$C = \log_2 \left| \mathbf{I}_{N_r} + \frac{\rho}{N_t}\mathbf{H}'\mathbf{H}'^H \right|. \tag{3.8}$$

The expression for the MIMO capacity in this equation is found in many references on MIMO communications. It is important to keep in mind, however, that this expression only applies under the assumptions delineated in this section, which are only optimal under CSIR only conditions. We will see that it is possible for the capacity to exceed the values given in this expression when both CSIT and CSIR are present. Under those assumptions, the transmitter has knowledge of the communications channel matrix, so it will be found that equi-power and uncorrelated transmitted signals are not optimal and that by using the transmitter's knowledge of the channel, the channel capacity exceeds the level given by Eq. 3.8 for CSIR alone.

A note on channel nomenclature

As mentioned above, it is often convenient to normalize the channel matrix so that the signal-to-noise ratio appears explicitly in the performance equations. In Eq. 3.8, we have used the normalized version of the channel matrix and denoted it by $\mathbf{H}'$. For notational simplicity, however, we drop the prime in future discussions and simply use $\mathbf{H}$ to denote the channel matrix irrespective of whether it is normalized or not.

The nature of the channel matrix (i.e., whether it is normalized or not) is clear from the context. As a general rule, if an equation that involves $\mathbf{H}$ also explicitly contains the signal-to-noise ratio, then the channel matrix can be assumed to be normalized. For example, Eq. 3.7 uses the physical unnormalized channel matrix, and we note that it does not include an explicit term for the signal-to-noise ratio. In contrast, Eq. 3.8 does contain an explicit term for the signal-to-noise ratio (i.e., ρ) and the channel matrix is normalized. This allows us to drop the prime and write Eq. 3.8 in the following form:

$$C = \log_2 \left| \mathbf{I}_{N_r} + \frac{\rho}{N_t} \mathbf{H}\mathbf{H}^H \right|. \tag{3.9}$$

3.2 Eigen-channels and channel rank

In Chapter 1, we stated that multipath channels are capable of supporting the transmission of multiple independent data streams, which are variously referred to as data streams, data pipes, spatial channels, eigenmodes or eigen-channels. In this section, we examine this concept in greater detail and use the expression for the capacity of a MIMO channel to better understand the nature of these eigen-channels.

We start by considering matrix theorem Section 1.9.2-(u), which states that any Hermitian matrix can be expressed as a product of three matrices in an eigen-decomposition. Since $\mathbf{H}\mathbf{H}^H$ in Eq. 3.9 is Hermitian, it follows that we can decompose it as follows:

$$\mathbf{H}\mathbf{H}^H = \mathbf{U}_{N_r}\mathbf{D}_{N_r}\mathbf{U}_{N_r}^H, \tag{3.10}$$

where $\mathbf{D} = \operatorname{diag}\{\lambda_1, \lambda_2, \ldots, \lambda_r, 0, \ldots, 0\}$, $r = r(\mathbf{H}) = r(\mathbf{H}\mathbf{H}^H) =$ *channel rank*, and the subscripts explicitly indicate the dimensions of the square matrices in the decomposition.

Plugging Eq. 3.10 into Eq. 3.9 yields

$$\begin{aligned}
C &= \log_2 \left| \mathbf{I} + \frac{\rho}{N_t} \mathbf{U}\mathbf{D}\mathbf{U}^H \right| \\
&= \log_2 \left| \mathbf{U} \left(\mathbf{I} + \frac{\rho}{N_t} \mathbf{D} \right) \mathbf{U}^H \right| \\
&= \log_2 \left[\det(\mathbf{U}) \cdot \det \left(\mathbf{I} + \frac{\rho}{N_t} \mathbf{D} \right) \cdot \det \left(\mathbf{U}^H \right) \right] \\
&= \log_2 \left| \mathbf{I} + \frac{\rho}{N_t} \mathbf{D} \right| \\
&= \log_2 \left[\prod_{i=1}^{r} \left(1 + \frac{\rho}{N_t} \lambda_i \right) \right] \\
&= \sum_{i=1}^{r} \log_2 \left(1 + \frac{\rho}{N_t} \lambda_i \right),
\end{aligned} \tag{3.11}$$

where the third step follows from matrix theorem Section 1.9.2-(l), and step four follows from matrix theorem Section 1.9.2-(j).

Equation 3.11 has a simple interpretation. To see it, we need first to recall what the capacity is for a SISO channel with complex signals, which follows from Eq. 2.35 when $N_t = N_r = 1$. In that case, the capacity reduces to the following:

$$C_{\text{SISO}} = \log_2 \left(1 + \frac{|h|^2 \sigma_s^2}{\sigma^2} \right), \tag{3.12}$$

where h denotes the complex channel response between the single transmitter and the single receiver. Since the received signal $r = hs$, it follows that the received power $P_r = \mathbb{E}\left\{|r|^2\right\} = \mathbb{E}\left\{|h|^2 |s|^2\right\} = |h|^2 \sigma_s^2$. Since the received signal-to-noise power ratio $\rho = P_r/\sigma^2 = |h|^2 \sigma_s^2/\sigma^2$, it follows that

$$C_{\text{SISO}} = \log_2 (1 + \rho). \tag{3.13}$$

This equation shows that each term in Eq. 3.11 has the form of a SISO channel. We conclude the following:

The capacity of a MIMO channel with only CSIR can be interpreted as the sum of r SISO channels, each having power gain, $\lambda_i,\ i = 1, \ldots, r$, where the effective transmit power of a SISO channel is $1/N_t$ times the total actual transmit power.

This result shows that in principle it is possible to transmit up to r data streams over a MIMO channel, which demonstrates the importance of having large channel rank. Furthermore, because the gains of the SISO channels are given by the eigenvalues, this result also shows the importance of having large eigenvalues, or put negatively, the adverse effect of small eigenvalues. *Channel rank is a quantitative way to characterize the scattering richness of a MIMO channel.* Rank, however, is not the only factor that influences MIMO capacity as we will show. In the following section, we discuss the importance of the distribution of channel eigenvalues.

3.3 Optimum distribution of channel eigenvalues

Although the designer of a MIMO communications system does not have control over the characteristics of the communications channel in which the system operates, it is of interest to know what channel characteristics lead to good performance. In addition to large rank, the distribution of the eigenvalues is also important. The problem of finding the optimum eigenvalue distribution is a classic constrained optimization problem, as seen by applying matrix theorem Section 1.9.2-(x) to $\mathbf{H}\mathbf{H}^H$, which states that

$$\zeta \triangleq \sum_{i=1}^{r} \lambda_i = \|\mathbf{H}\|_F^2. \tag{3.14}$$

For a fixed channel matrix, this equation shows that the sum of the eigenvalues is a constant. It can be shown that under this constraint, the maximum capacity is obtained when the eigenvalues are all the same value. This can be proven using Lagrange multipliers, which are reviewed later in this chapter, but it can also be easily demonstrated empirically.

Figure 3.1 is a scatter plot showing the values of the capacity computed using Eq. 3.11 with 1000 different sets of eigenvalues that sum up to 1. In this figure, $N_t = r = 4$, so each set contains four eigenvalues. The values in each of the 1000 sets of four eigenvalues are chosen randomly. The straight horizontal line in the plot corresponds to the capacity when the eigenvalues are each equal to 1/4, which is the maximum observed capacity value. Although this is not a rigorous proof, it is a simple demonstration that is consistent with the claim.

Special case: $N \times N$ **full-rank channel with equal eigenvalues**

We now consider an ideal case where the channel is full rank, the eigenvalues are equal to each other, and the MIMO system is $N \times N$. Under this assumption, $N_t = N_r = N$ and $r = N$, so

$$\zeta = r\lambda = N\lambda \Rightarrow \lambda = \frac{\zeta}{N}. \tag{3.15}$$

Plugging this expression for λ into Eq. 3.11 yields

$$\begin{aligned} C &= \sum_{i=1}^{r} \log_2\left(1 + \frac{\rho}{N_t}\lambda\right) \\ &= \sum_{i=1}^{N} \log_2\left(1 + \frac{\rho\zeta}{N^2}\right) \\ &= N\log_2\left(1 + \frac{\rho\zeta}{N^2}\right). \end{aligned} \tag{3.16}$$

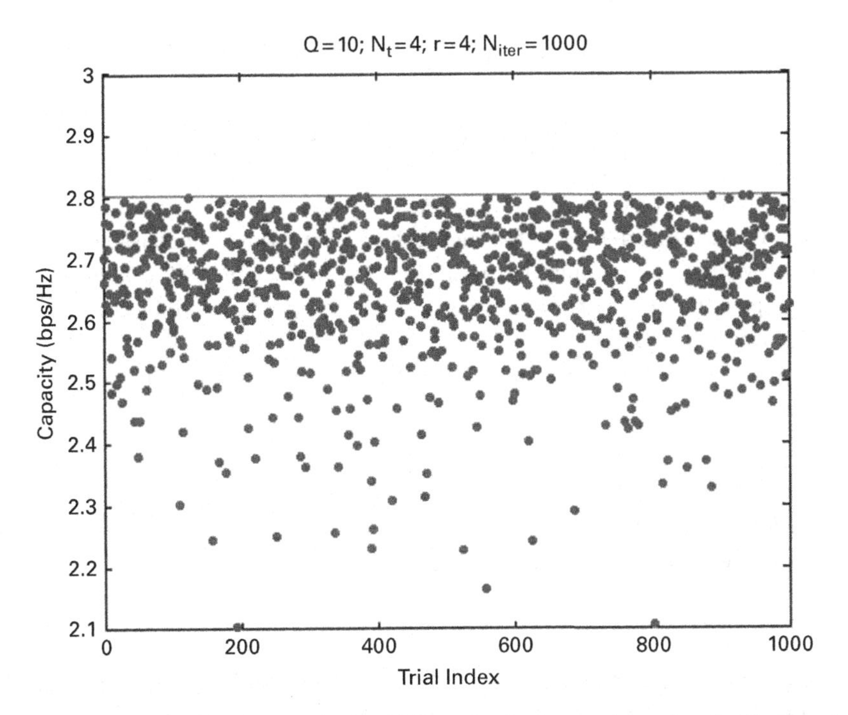

Figure 3.1 Scatter plot showing the capacity for 1000 randomly chosen eigenvalue distributions for $\rho = 10$ dB and $r = N_t = 4$. The horizontal line in the plot shows the capacity when all eigenvalues are equal.

This expression can be simplified by using the fact that for a given channel matrix, $\|\mathbf{H}\|_F^2 = N^2 = \zeta$ (see Eq. 3.3). Therefore, Eq. 3.16 becomes

$$C = N \log_2 (1 + \rho) . \tag{3.17}$$

This represents the optimum capacity that can be obtained with only CSIR.

3.4 Eigenbeamforming

We now consider the case where both the transmitter and receiver have knowledge of $\mathbf{H}$ (i.e., both CSIR and CSIT exist) and describe a practical technique that is used to implement spatial multiplexing called eigenbeamforming. We note that when the transmitter knows the channel matrix, its optimal strategy is, in general, *not* to distribute its transmit power equally among its transmit antennas, nor to make the cross-correlation between transmitted signals on different antennas zero, which, as we argued previously, are the optimum strategies under the CSIR only assumption.

We begin by writing the channel matrix in terms of the product of three matrices using singular value decomposition as described in matrix theorem Section 1.9.2-(r). When applied to the channel matrix, that theorem states that if $\mathbf{H}$ has rank r, then there are unitary matrices, $\mathbf{U}$ and $\mathbf{V}$, and a diagonal matrix $\mathbf{D} = \text{diag}\{\sigma_1, \sigma_2, \ldots, \sigma_r\}$ such that

$$\mathbf{H} = \mathbf{U}_{N_r,r}\mathbf{D}_{r,r}\mathbf{V}^H_{r,N_t}, \tag{3.18}$$

where the diagonal elements in $\mathbf{D}$ are the singular values of $\mathbf{H}$. Since it is customary to formulate the eigenbeamforming technique in terms of eigenvalues instead of singular values, we note that from matrix theorem Section 1.9.2-(s), an alternative way of expressing $\mathbf{D}$ is as $\mathbf{D} = \text{diag}\left\{\sqrt{\lambda_1}, \sqrt{\lambda_2}, \ldots, \sqrt{\lambda_r}\right\}$, where the lambdas denote the eigenvalues of $\mathbf{HH}^H$.

Assume that at some instant the transmitter has r symbols to transmit, which we denote by

$$\tilde{\mathbf{s}} \triangleq \left[\tilde{s}_1, \ldots, \tilde{s}_r\right]^T. \tag{3.19}$$

Since the transmitter is assumed to have knowledge of $\mathbf{H}$, it can compute $\mathbf{V}$ and perform the following *precoding* operation:

$$\mathbf{s} = \mathbf{V}\tilde{\mathbf{s}}, \tag{3.20}$$

which results in a transmit vector having the desired dimension $N_t \times 1$. The precoding operation can be thought of as an operation that maps $r \leq N_t$ symbols onto N_t transmit antennas. It follows that the received signal vector is given by

$$\begin{aligned} \mathbf{r} &= \sqrt{\rho}\,\mathbf{Hs} + \mathbf{z} \\ &= \sqrt{\rho}\,\mathbf{HV}\tilde{\mathbf{s}} + \mathbf{z}, \end{aligned} \tag{3.21}$$

where, as before, ρ denotes the total power received at each receive antenna from all the transmit antennas divided by the thermal noise power at each receiver.

Since the receiver is assumed to know $\mathbf{H}$, it too can perform a singular value decomposition of $\mathbf{H}$ and compute $\mathbf{U}^H$, by which it can pre-multiply the received signal. Denoting the pre-multiplied received signal by $\tilde{\mathbf{r}}$, it follows that

$$\begin{aligned} \tilde{\mathbf{r}} &\triangleq \mathbf{U}^H\mathbf{r} \\ &= \sqrt{\rho}\mathbf{U}^H\mathbf{HV}\tilde{\mathbf{s}} + \mathbf{U}^H\mathbf{z} \\ &= \sqrt{\rho}\mathbf{U}^H\mathbf{UDV}^H\mathbf{V}\tilde{\mathbf{s}} + \mathbf{U}^H\mathbf{z} \\ &= \sqrt{\rho}\mathbf{D}\tilde{\mathbf{s}} + \mathbf{U}^H\mathbf{z}. \end{aligned} \tag{3.22}$$

Since $\mathbf{D}$ is a diagonal matrix, Eq. 3.22 can be rewritten as the following set of scalar equations:

$$\tilde{r}_i = \sqrt{\rho\lambda_i}\,\tilde{s}_i + \tilde{z}_i \qquad i = 1, \ldots, r, \tag{3.23}$$

where $\tilde{z}_i \triangleq \left[\mathbf{U}^H\mathbf{z}\right]_i$.

The significance of Eq. 3.23 is that it shows that by appropriate pre-coding at the transmitter and pre-multiplying at the receiver it is possible to decompose a MIMO

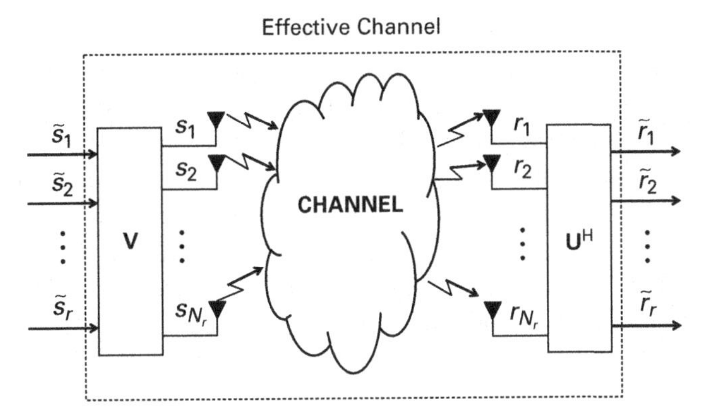

Figure 3.2 A MIMO system using eigenbeamforming.

channel into r SISO channels, each of which consists of a scaled version of the transmitted symbol plus noise. These SISO channels are sometimes referred to as eigen-channels, and the receiver can decode each of these channels using standard maximum likelihood decoding logic. The process of precoding according to Eq. 3.20 and premultiplying at the receiver according to Eq. 3.22 is called eigenbeamforming.

Figure 3.2 depicts the concept of eigenbeamforming and illustrates the effective channel that is created by the precoding and pre-multiplying operations. Because Eq. 3.23 shows that each eigen-channel has a gain equal to the square root of the eigenvalue associated with that channel, we see, again, the adverse effect of small channel eigenvalues. The signal-to-noise ratio for each eigen-channel, ρ_i, is given by

$$\rho_i = \frac{\rho \mathbb{E}\left\{|\tilde{s}_i|^2\right\}}{\sigma_{\tilde{z}_i}^2} \lambda_i \qquad i = 1, \ldots, r, \tag{3.24}$$

where $\sigma_{\tilde{z}_i}^2 \triangleq \mathbb{E}\left\{|\tilde{z}_i|^2\right\}$. Since $\tilde{z}_i \triangleq \left[\mathbf{U}^H \mathbf{z}\right]_i$, it is possible to show (see problem 3.3) that

$$\sigma_{\tilde{z}_i}^2 = \sigma^2 \qquad i = 1, \ldots, r. \tag{3.25}$$

Furthermore, the form of Eq. 3.21 implies that $\|\mathbf{H}\|_F^2 = N_t N_r$ and $\sigma^2 = 1$ (because of the normalizations given in Eqs. 3.2–3.5). It follows that

$$\rho_i = \rho \mathbb{E}\left\{|\tilde{s}_i|^2\right\} \lambda_i \qquad i = 1, \ldots, r. \tag{3.26}$$

Using the expression for the capacity of a SISO channel given in Eq. 3.13, it follows that the capacity with eigenbeamforming is given by

$$\begin{aligned} C_{\text{eig}} &= \sum_{i=1}^{r} \log_2 (1 + \rho_i) \\ &= \sum_{i=1}^{r} \log_2 \left(1 + \rho \mathbb{E}\left\{|\tilde{s}_i|^2\right\} \lambda_i\right). \end{aligned} \tag{3.27}$$

Comparison of this expression with Eq. 3.11 for CSIR-only conditions shows similarities, but also an important difference. Each of the terms in the expression for the capacity under CSIR-only conditions depends on the signal-to-noise ratio and the eigenvalue for each eigenmode. Each term in the expression for the capacity in Eq. 3.27 depends on these same quantities, except that each term also depends on the variance of each transmitted symbol; that is, on $\mathbb{E}\left\{|\tilde{s}_i|^2\right\}$, which is under the control of the transmitter. The reason for this difference is that under CSIR, the transmit power is assumed to be equally dividied between the transmit antennas. This shows that a key part of the eigenbeamforming technique is choosing appropriate powers for each of the r transmitted symbols. In the next section, we discuss how to allocate power optimally among the symbols when using eigenbeamforming.

3.5 Optimal allocation of power in eigenbeamforming

3.5.1 The waterfilling algorithm

For the purpose of this discussion, we denote by P_i the power associated with the ith symbol. That is,

$$P_i \triangleq \mathbb{E}\left\{|\tilde{s}_i|^2\right\} \qquad i = 1, \ldots, r. \tag{3.28}$$

Our goal is to find the optimum distribution for the set $\{P_i\}$ under the constraint that their sum is a fixed value. As we have mentioned, for the normalization used in Eq. 3.21, $\|\mathbf{H}\|_F^2 = N_t N_r$ and $\sigma^2 = 1$ are implied. It can be shown (see problem 3.4) that under such assumptions, it follows that

$$\sum_{i=1}^{r} P_i = 1, \tag{3.29}$$

which functions as our constraint in this problem.

Solving for the optimum set of $\{P_i\}$ is a standard constrained optimization problem, which can be solved using the Lagrange multiplier technique from the calculus of variations. The resulting algorithm that implements this solution is called the *waterfilling algorithm*. With a constrained optimization problem, there is a function that needs to be optimized that depends on a set of variables, and a constraint imposed on those variables. In our application, the function that we seek to optimize is the eigenbeamforming capacity in Eq. 3.27, which is a function of $\{P_i\}$, and the constraint is given by Eq. 3.29. The first step is to write down an expression called the *functional*, which is an equation that results from writing down the function that is to be optimized and then adding to it a scaled version of the constraint. In our case, the functional has the following form:

$$J(P_1, P_2, \ldots, P_r) = \sum_{i=1}^{r} \log_2 (1 + \rho P_i \lambda_i) + \beta \sum_{i=1}^{r} P_i, \tag{3.30}$$

where β is a constant called the Lagrange multiplier. Next, we take the partial derivative of J with respect to the variables that we are attempting to optimize (i.e., $\{P_i\}$) and set them equal to zero, which results in the following:

$$\frac{\partial J}{\partial P_i} = 0 = \frac{\frac{d\rho_i}{dP_i}}{(1+\rho P_i\lambda_i)} + \beta$$

$$= \frac{\rho\lambda_i}{(1+\rho P_i\lambda_i)} + \beta. \qquad i = 1,\ldots,r \tag{3.31}$$

Solving this equation for P_i yields

$$P_i^{\text{opt}} = \left[-\frac{1}{\beta} - \frac{1}{\rho\lambda_i}\right]_+$$

$$= \left[\mu - \frac{1}{\rho\lambda_i}\right]_+, \qquad i = 1,\ldots,r \tag{3.32}$$

where, $\mu \triangleq -1/\beta$ and

$$x_+ \triangleq \begin{cases} x, & x \geq 0 \\ 0. & x < 0 \end{cases} \tag{3.33}$$

Since $\sum_{i=1}^{r} P_i^{\text{opt}} = 1$, we can use that fact to solve for μ, resulting in the following expression:

$$\mu = \frac{1}{r}\left[1 + \frac{1}{\rho}\sum_{i=1}^{r}\frac{1}{\lambda_i}\right]. \tag{3.34}$$

The steps in the waterfilling algorithm are defined in Table 3.1 using the expressions for P_i^{opt} and μ to iteratively compute the set of values for $\{P_i\}$ that maximize C_{eig}. The resulting set $\left\{P_i^{\text{opt}}\right\}$ are then substituted into Eq. 3.27 to obtain the optimal capacity, C_{opt}. Thus,

$$C_{\text{opt}} = \sum_{i=1}^{r}\log_2\left(1 + \rho P_i^{\text{opt}}\lambda_i\right). \tag{3.35}$$

3.5.2 Discussion of the waterfilling algorithm

Now that we have defined the waterfilling algorithm, we will attempt to understand it in a more intuitive way. Figure 3.3 illustrates the algorithm for the case where $r = 5$. Part (a) depicts the values of $1/(\rho\lambda_i), i = 1,\ldots,5$ and μ after performing steps 1 and 2 of the algorithm when $p = 1$. As is evident in this figure, the eigenvalues are assumed to be sorted from largest to smallest, which is a key assumption that underlies the waterfilling algorithm in Table 3.1. Because of this fact, the quantities $1/(\rho\lambda_i)$ become larger as the index increases, which means the symbol power levels decrease with index since $P_i^{\text{opt}} = \mu - 1/(\rho\lambda_i)$. In this example, $1/(\rho\lambda_i)$, $i = 4$ and 5 are greater than μ, so $P_{r-p+1}^{\text{opt}} = P_5^{\text{opt}} < 0$. Therefore, according to the algorithm, the power associated with eigen-channel 5 should be set to 0 and p incremented to 2.

Part (b) depicts the situation after repeating steps 1 and 2 when $p = 2$. After incrementing p, the denominator in the coefficient of the expression for μ in step 1 increases, but since the number of terms in the summation decreases by 1, the quantity in brackets is smaller. Generally, but not always, μ increases each time p is incremented, which

Table 3.1 Waterfilling algorithm for eigenbeamforming.

1. Set the iteration count p to 1 and compute μ by solving the following equation:

$$\mu = \frac{1}{r-p+1}\left[1+\frac{1}{\rho}\sum_{i=1}^{r-p+1}\frac{1}{\lambda_i}\right].$$

2. Using the value of μ obtained above, solve for the power, P_i^{opt}, for the ith eigen-channel using the following equation:

$$P_i^{\text{opt}} = \mu - \frac{1}{\rho\lambda_i}. \qquad i = 1, \ldots, (r-p+1)$$

3. If the power is allocated to the channel with the lowest gain (i.e., if $P_{r-p+1}^{\text{opt}} < 0$), discard that channel by setting $P_{r-p+1}^{\text{opt}} = 0$ and rerun the algorithm with the iteration count p incremented by 1.
4. Repeat steps 1–3 until all channels have been allocated power.

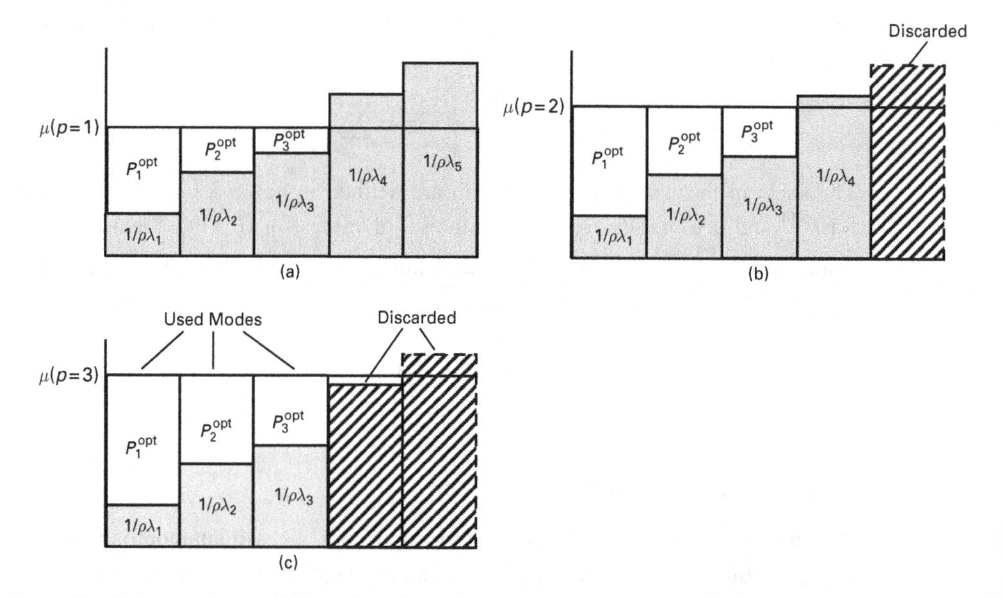

Figure 3.3 Depiction of the waterfilling algorithm where r is assumed to equal 5.

is what we have assumed in this example. Under that assumption, the values of $\{P_i\}$ increase accordingly, but despite that, $1/(\rho\lambda_4) > \mu$, so $P_4^{\text{opt}} < 0$ in this example. According to the algorithm, P_4^{opt} is set to 0 and p is incremented again.

Part (c) shows the situation after repeating steps 1 and 2 when $p = 3$. In this case, μ has increased sufficiently that $P_{r-p+1}^{\text{opt}} = P_3^{\text{opt}} > 0$. At this point, all the eigen-channel powers have been assigned non-negative values, so the algorithm is complete.

This algorithm works because it accomplishes its goal of finding the set of $\left\{P_i^{\text{opt}}\right\}$ that a) sum to 1, and b) result in the maximum capacity. We are assured of (a) even when the number of eigenvalues that are left is less than r because of the way μ is computed in step 1 of the algorithm. The reason that the capacity is maximized is because the algorithm

Table 3.2 Summary of waterfilling algorithm dependence on signal-to-noise ratio.

Signal-to-noise ratio	Waterfilling performance
$\rho \to 0$	place all power in eigenmode with largest eigenvalue
$\rho \to \infty$	apply equal power to all eigenmodes

finds the maximum number of non-zero eigen-channel power levels consistent with the constraint. Since the number of non-zero power levels is maximized, the number of non-zero terms in Eq. 3.35 is maximized, which, in turn, maximizes the capacity.

By examining Figure 3.3 we can also learn something about how the waterfilling algorithm depends on the signal-to-noise ratio. First, consider the limiting case when $\rho \to \infty$. Under that assumption, regardless of the value of p, the values of the terms $1/(\rho\lambda_i) \to 0$, which means that the grey areas in the figure shrink and all of the eigen-channel power levels approach the same value. At the opposite extreme, when $\rho \to 0$, the values of the $1/(\rho\lambda_i)$ terms become large, which means that the values of $\left\{P_i^{\text{opt}}\right\}$ become negative, resulting in the need to discard the eigen-channel powers. However, since the sum of the eigen-channel power levels must equal 1, it is necessary to retain at least one of the P_i^{opt} terms. In the limit when the signal-to-noise ratio becomes small, only P_1^{opt} is retained; thus, in the small ρ limit, the waterfilling algorithm tells the transmitter to place all its power in the eigenmode with the largest eigenvalue. Table 3.2 summarizes these observations.

3.6 Single-mode eigenbeamforming

Single-mode eigenbeamforming refers to the particular form of eigenbeamforming in which information is only transmitted over a single eigenmode – the one with the largest eigenvalue. As we have discussed, this approach is equivalent to full eigenbeamforming when the signal-to-noise ratio is low.

In single-mode eigenbeamforming, $\tilde{\mathbf{s}} = [\tilde{s}_1, 0, \ldots, 0]^T$ since only the first eigenmode is used to transmit symbols. After pre-multiplying by $\mathbf{V}$, the transmitted data vector is given by

$$\begin{aligned}
\mathbf{s} &= \mathbf{V}\tilde{\mathbf{s}} \\
&= \begin{bmatrix} v_{11} & \cdots & v_{1r} \\ v_{21} & \cdots & v_{2r} \\ \vdots & \ddots & \vdots \\ v_{N_t1} & \cdots & v_{N_tr} \end{bmatrix} \begin{bmatrix} \tilde{s}_1 \\ 0 \\ \vdots \\ 0 \end{bmatrix} \\
&= \begin{bmatrix} v_{11}\tilde{s}_1 \\ v_{21}\tilde{s}_1 \\ \vdots \\ v_{N_t1}\tilde{s}_1 \end{bmatrix}.
\end{aligned} \quad (3.36)$$

This equation shows that even though there is only one eigenmode being used to transmit data, signals are transmitted on all N_t transmit antennas. In particular, the transmit antennas are fed with scaled versions of the single data symbol being transmitted. In this sense, single-mode eigenbeamforming is similar to conventional beamforming, which applies complex weights to each antenna element in an array. In conventional beamforming, the weights are chosen to point the beam in a desired *physical* direction; however, in single-mode eigenbeamforming, the weights are chosen to match the channel matrix. Single-mode eigenbeamforming is a more abstract concept and can be thought of as a generalization of conventional beamforming.

One might ask, what benefit is there in only transmitting data using one eigenmode? The answer is that it enables higher throughput than can be achieved with a SISO system at the same signal-to-noise ratio. To see this, consider the following expression for the capacity of a MIMO system that uses single-mode eigenbeamforming:

$$\begin{aligned} C_{\text{single-mode}} &= \log_2(1 + \rho_1) \\ &= \log_2(1 + \rho P_1 \lambda_1), \end{aligned} \tag{3.37}$$

which follows from Eq. 3.27.
Since $\sum_i P_i = 1$, this implies that in the single-eigenmode case $P_1 = 1$. The capacity, therefore, reduces to

$$C_{\text{single-mode}} = \log_2(1 + \rho\lambda_1). \tag{3.38}$$

Recall that in general $\sum_i \lambda_i = \|\mathbf{H}\|_F^2$, and for the type of normalization used in Eq. 3.21, it follows that $\|\mathbf{H}\|_F^2 = N_t N_r$. Since, for a full-rank channel, $r = \min\{N_t, N_r\}$, it follows that for an $N \times N$ MIMO system, $r = N$, and if all the eigenvalues are equal, $\lambda = N$. Normally, however, the eigenvalues are not equal, so for an $N \times N$ system, it is often the case that $\lambda_1 > N$. The point is that it is easy to see that $\lambda_1 > 1$, which means that $C_{\text{single-mode}}$ in Eq. 3.38 has an effective signal-to-noise ratio of $\rho\lambda_1 > \rho$. Since the capacity of a SISO system with signal-to-noise ratio ρ is $\log_2(1 + \rho)$, it follows that the capacity of a single-mode eigenbeamforming system is greater than that of a SISO system with the same signal-to-noise ratio.

3.7 Performance comparison

We have now described three power allocation strategies: one for CSIR only, and two for CSIR combined with CSIT – full eigenbeamforming and single-mode eigenbeamforming. In this section, we compare the performance of each using results from computer simulations. In general, we find that the relative performance of these techniques depends on the signal-to-noise ratio and the relative values of N_t and N_r.

3.7.1 Results for $N_r \geq N_t$

We start by assuming that the number of receive antennas is greater than or equal to the number of transmit antennas. Under this assumption, when the channel is full rank,

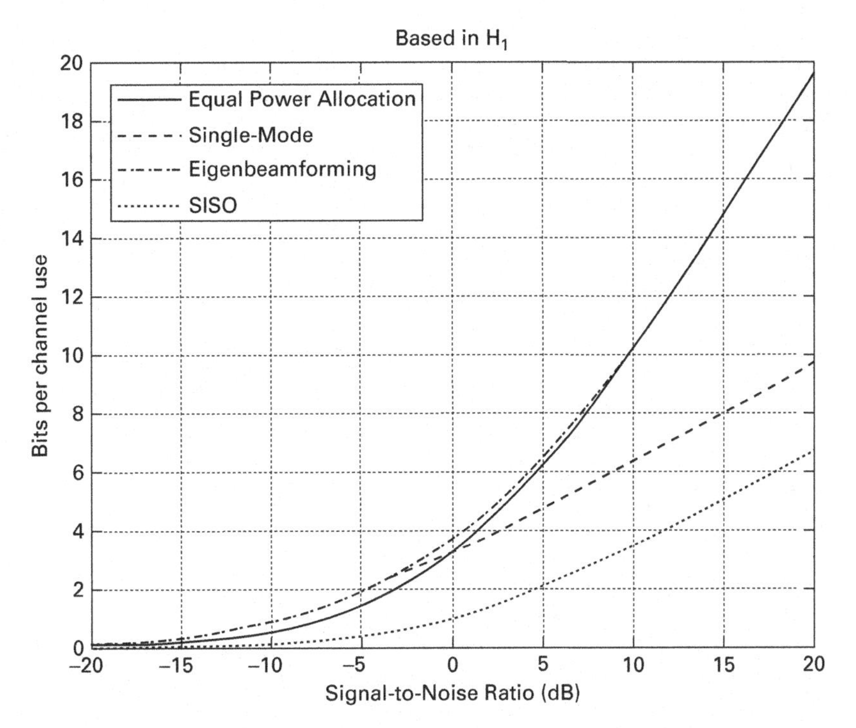

Figure 3.4 Comparison of theoretical capacity predictions for $\mathbf{H}_1$.

$r = N_t$. In that case, the channel can support up to N_t data streams. Figure 3.4 shows computer generated curves of the MIMO capacity as a function of the signal-to-noise ratio for the three different power allocation strategies described in this chapter when $N_t = 3$ and $N_r = 4$. These results are generated using the following randomly chosen channel matrix, $\mathbf{H}_1$:

$$\mathbf{H}_1 = \begin{pmatrix} 0.46 + j0.31 & -0.079 + j0.18 & 0.43 - j0.68 \\ -0.04 + j0.66 & 0.95 - j0.50 & -0.77 - j0.51 \\ 0.45 + j0.46 & 0.90 + j0.56 & 0.46 - j1.87 \\ -0.13 - j0.19 & 0.90 - j0.73 & 1.04 + j0.91 \end{pmatrix}, \tag{3.39}$$

where this matrix has the required property that, to within numerical accuracy, $\|\mathbf{H}\|_F^2 = N_t N_r = 12$.

This plot shows several noteworthy features:

1. As the signal-to-noise ratio gets large, the equal power allocation and eigenbeamforming curves become equal to each other, consistent with what was predicted in Table 3.2.
2. As the signal-to-noise ratio becomes small, the eigenbeamforming and single-mode eigenbeamforming curves become the same, which is consistent with the fact that the waterfilling algorithm used with eigenbeamforming tells the transmitter to place all its power in the largest eigenmode.

Table 3.3 Eigenmode gains as a function of ρ for the eigenbeamforming example shown in Figure 3.4.

SNR(dB)	P_1	P_2	P_3
−10	1	0	0
−9	1	0	0
−8	1	0	0
−7	1	0	0
−6	0.9058	0.0942	0
−5	0.8223	0.1777	0
−4	0.756	0.244	0
−3	0.7034	0.2966	0
−2	0.6616	0.3384	0
−1	0.6283	0.3717	0
0	0.6019	0.3981	0
1	0.581	0.419	0
2	0.5643	0.4357	0
3	0.5511	0.4489	0
4	0.5108	0.4296	0.0596
5	0.4743	0.4098	0.1159
6	0.4453	0.3941	0.1606
7	0.4223	0.3816	0.1962
8	0.404	0.3717	0.2244
9	0.3894	0.3638	0.2468
10	0.3779	0.3575	0.2646
11	0.3687	0.3525	0.2787
12	0.3615	0.3486	0.29
13	0.3557	0.3455	0.2989
14	0.3511	0.343	0.306
15	0.3474	0.341	0.3116
16	0.3445	0.3394	0.3161
17	0.3422	0.3382	0.3196
18	0.3404	0.3372	0.3224
19	0.3389	0.3364	0.3247
20	0.3378	0.3358	0.3265
21	0.3369	0.3353	0.3279
21	0.3361	0.3349	0.329
23	0.3356	0.3345	0.3299
24	0.3351	0.3343	0.3306
25	0.3347	0.3341	0.3312
26	0.3345	0.3339	0.3316
27	0.3342	0.3338	0.332
28	0.334	0.3337	0.3322
29	0.3339	0.3336	0.3325
30	0.3338	0.3336	0.3326

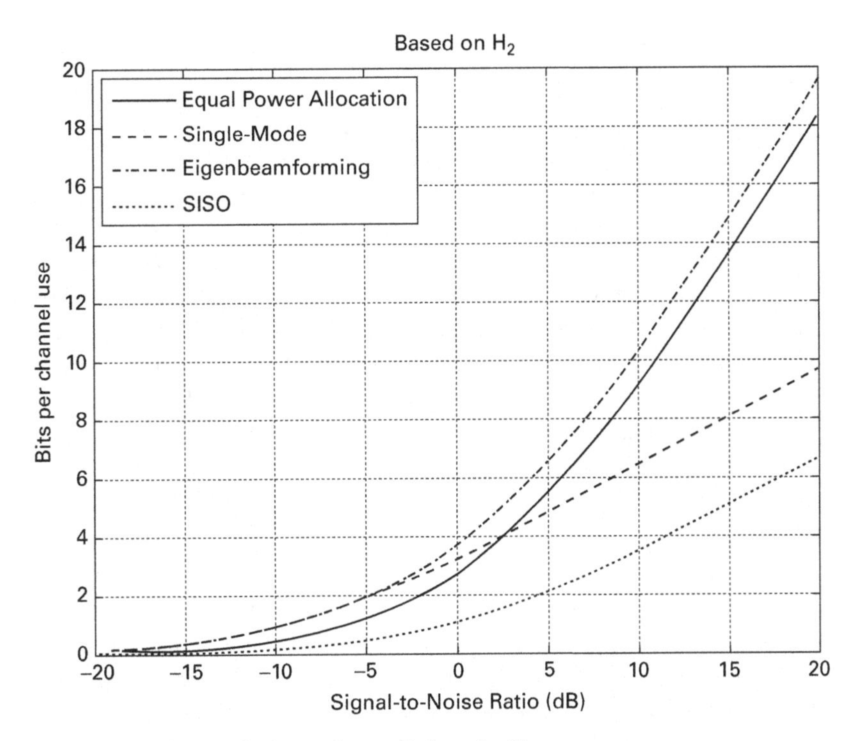

Figure 3.5 Comparison of theoretical capacity predictions for $\mathbf{H}_2$.

3. The single-mode eigenbeamforming technique has capacity that is equal to or smaller than the other MIMO techniques considered, which is to be expected since single-mode eigenbeamforming only transmits data on one eigenmode compared to $r = 3$ eigenmodes for the other schemes.
4. Finally, we have included the capacity prediction for a SISO system, which is smaller than any of the MIMO schemes, as expected. The plot confirms that even though single-mode eigenbeamforming only transmits data on a single eigenmode, its capacity is still larger than a SISO system.

It is also instructive to examine the eigenmode gains (i.e., the values of $\{P_i\}$) assigned by the waterfilling algorithm associated with the eigenbeamforming curve in Figure 3.4. These values are listed in Table 3.3 as a function of ρ. This shows that the values of $\{P_i\}$ have the properties discussed above. For example, examination verifies that the sum of the eigenmode gains is equal to 1. Secondly, we note that at low values of ρ, only eigenmode 1 is assigned a non-zero gain, consistent with Table 3.2. Thirdly, we observe that as ρ increases, the eigenmode gains become increasingly close in value, again consistent with Table 3.2.

3.7.2 Results for $N_t > N_r$

The capacity results when $N_t > N_r$ are qualitatively different in certain respects than they are when $N_r > N_t$. Figure 3.5 compares the theoretical capacity predictions for a

4×3 MIMO system using a 3×4 channel matrix, $\mathbf{H}_2$, which we derive from $\mathbf{H}_1$ as follows:

$$\mathbf{H}_2 = \mathbf{H}_1^T. \tag{3.40}$$

For a full-rank channel, such as $\mathbf{H}_2$, $N_t > N_r = r$, so there are more transmit antennas than eigenmodes that can carry independent data streams. The implication of this is that the equal power allocation strategy is wasteful because it attempts to excite more eigenmodes than the channel can support. Instead, it should place more energy in fewer data streams. At large SNR, we have seen that the waterfilling algorithm used with eigenbeamforming allocates the same power to all r eigenmodes. For this reason, we would expect equal power allocation capacity to be less than the capacity associated with eigenbeamforming in general, and that at large SNR the degradation in performance should be equal to

$$\text{Degradation} = 10\log_{10}(N_t/r). \tag{3.41}$$

Since in this example $N_t = 4$ and $r = N_r = 3$, this implies that we should find that equal power allocation performs $10\log_{10}(4/3) = 1.25$ dB worse than eigenbeamforming when ρ is large. Examination of Figure 3.5 shows that the equal gain allocation curve is, in fact, shifted to the right by 1.25 dB relative to the eigenbeamforming curve at large ρ, consistent with this prediction. At small SNR, full eigenbeamforming becomes equivalent to single-mode eigenbeamforming, resulting in the two curves merging in that regime. This phenomenon is the same regardless of whether N_t is greater or less than N_r.

3.8 Capacities of SIMO and MISO channels

The theoretical expressions for the capacities of SIMO and MISO channels have particularly simple forms that help lend useful insight when comparing these two multi-antenna configurations. We derive and discuss these expressions in this section.

3.8.1 SIMO capacity

CSIR only

We start by considering a SIMO system under the CSIR-only assumption. Recalling that in general $\|\mathbf{H}\|_F^2 = \sum_{i=1}^{r} \lambda_i$, then under the SIMO assumption

$$\|\mathbf{H}\|_F^2 = \lambda, \tag{3.42}$$

where λ is the eigenvalue of $\mathbf{H}\mathbf{H}^H$. This expression follows from the fact that $r = 1$ for a full-rank SIMO channel. Plugging Eq. 3.42 into Eq. 3.11 yields the following expression for the capacity:

$$C_{\text{SIMO,CSIR}} = \log_2\left(1 + \rho\|\mathbf{H}\|_F^2\right). \tag{3.43}$$

Since $\|\mathbf{H}\|_F^2 = N_t N_r$, it follows that for a SIMO system, $\|\mathbf{H}\|_F^2 = N_r$. So, we can rewrite (3.43) as follows:

$$C_{\mathrm{SIMO,CSIR}} = \log_2(1 + \rho N_r). \tag{3.44}$$

This expression shows that for a SIMO system with CSIR only, the capacity increases as the logarithm of N_r when ρ becomes large.

How does this expression compare with the capacity for a full-rank $N \times N$ system given in Eq. 3.17? As we saw earlier, the capacity of a full-rank $N \times N$ system with equal eigenvalues increases linearly with the number of antennas compared with logarithmically in the SIMO case. Although it is clear that the capacity of an $N \times N$ system should be larger than a SIMO system, these expressions provide a quantitative comparison.

CSIR plus CSIT

Since SIMO systems only have one transmit antenna, there is nothing the transmitter can do to exploit its knowledge of the channel. Therefore, the SIMO capacity when both CSIR and CSIT exist, $C_{\mathrm{SIMO,CSITR}}$, is the same as it is under CSIR alone. We denote this as follows:

$$C_{\mathrm{SIMO,CSITR}} = C_{\mathrm{SIMO,CSIR}}. \tag{3.45}$$

3.8.2 MISO capacity

CSIR only

In the case of MISO systems, $\mathbf{H}$ is $[1 \times N_t]$, so

$$\|\mathbf{H}\|_F^2 = \lambda, \tag{3.46}$$

the same as for the SIMO configuration. Since $\|\mathbf{H}\|_F^2 = N_t N_r$, it follows that for a MISO system, $\|\mathbf{H}\|_F^2 = N_t$. Combining this with Eq. 3.11 yields

$$C_{\mathrm{MISO,CSIR}} = \log_2(1 + \rho), \tag{3.47}$$

which is the same as the expression for a SISO system.

Why are the capacities of MISO and SISO systems the same under the CSIR-only assumption? The short answer is that having multiple transmit antennas only provides a capacity benefit if the transmitter has knowledge of the channel and can use that knowledge to adjust its transmission strategy. If the transmitter had knowledge of the channel, it could weight the signals it applies to each of its antennas to create a physical beam that pointed in the direction of the receiver. Creating such a beam would increase the signal-to-noise ratio at the receiver and allow the transmitter to increase its data rate.

Another way of answering the question is to say that with CSIR only, a MISO system cannot have array gain, which is needed to increase the throughput over a SISO system. Without such knowledge, the capacity is no better than a SISO system.

It should be emphasized that although this result says the capacities of MISO and SISO systems are the same when only CSIR exists, this statement should not be interpreted as saying that there is no benefit whatever to having multiple transmit antennas with CSIR. As we discussed in Chapter 1 and as we will discuss in more detail in later chapters, a MISO system with only CSIR can employ space-time coding to achieve transmit spatial diversity, which reduces the effect of fading and allows the transmitter to transmit higher data rates, with the same probability of bit error, than would be possible with a SISO system operating at the same average signal-to-noise ratio using the same modulation and coding scheme as the MISO system. This is a practical benefit; however, our statement about the capacities being the same means that, in principle, the maximum error free data rates of the two systems are the same. This distinction, while perhaps subtle, should be kept in mind.

CSIR plus CSIT

Although a MISO channel has at most only one eigenvalue, when the channel is known by the transmitter, single-mode eigenbeamforming can be performed, which, as we have seen, can increase the capacity even though data are being transmitted over only one eigen-channel. The capacity of a MISO system with both CSIR and CSIT is obtained by first recalling that for a MISO channel with a single eigenvalue, λ, it follows from matrix theorem Section 1.9.2-(x) that $\lambda = \|\mathbf{H}\|_F^2 = N_t$. Combining this with Eq. 3.38 yields

$$C_{\text{MISO,CSITR}} = \log_2 (1 + \rho N_t) . \tag{3.48}$$

Comparison of Eqs. 3.48 and 3.45 shows that the capacity expressions for MISO and SIMO systems when both CSIR and CSIT exist have the same form except that the SIMO equation has N_r in the expression and MISO has N_t. If we assume that both SIMO and MISO systems have the same number of antennas, the capacity expressions for both systems are identical. This is an interesting result that may seem surprising at first. In the SIMO case, the transmitter cannot employ single-mode eigenbeamforming because there is only one transmit antenna, so its capacity is the same as it is under CSIR. In the MISO case, the transmitter is assumed to employ single-mode eigenbeamforming and its capacity is different when both CSIR and CSIT exist than it is under CSIR only. Despite these differences, the SIMO and MISO capacities are the same under CSIR and CSIT conditions when the number of antennas is the same. The explanation for this somewhat unexpected result is that when both CSIR and CSIT exist, the transmitter and receiver can both weight the signals on their respective antennas (the transmitter in the case of MISO and the receiver in the case of SIMO) in response to the channel. We conclude that MISO and SIMO systems are equivalent in terms of how they can exploit

Table 3.4 Capacity formulas for SIMO and MISO communication systems.

CSI type	SIMO	MISO	Comments
CSIR	$\log_2(1+\rho N_r)$	$\log_2(1+\rho)$	MISO capacity is less than SIMO capacity because without CSIT, the transmitter cannot generate antenna array gain
CSIR & CSIT	$\log_2(1+\rho N_r)$	$\log_2(1+\rho N_t)$	MISO and SIMO systems are equivalent in terms of how they are able to exploit their knowledge of the channel

their knowledge of the channel, but only when both the transmitter and receiver have knowledge of the channel.

Table 3.4 summarizes the capacity expressions for the SIMO and MISO configurations. The first row shows that when only CSIR exists, the capacity of a SIMO system is larger than a MISO system with the same number of antennas. To see why this is so, consider a SIMO system in which a single symbol, s, is transmitted at some instant of time. Since the receiver is assumed to know the values of the channel matrix elements, the receiver can multiply the received signals at each antenna, $\{r_i,\ i=1,\ldots,N_r\}$, by the complex conjugate of the corresponding channel gain. The resulting signal, r, is given by

$$\begin{aligned} r &= r_1 h_{1,1}^* + r_2 h^* h_{2,1} + \cdots + r_{N_r} h_{N_r,1}^* \\ &= s\left(|h_{1,1}|^2 + |h_{2,1}|^2 + \cdots + |h_{N_r,1}|^2\right), \end{aligned} \tag{3.49}$$

where the term in parentheses in the second line of the equation is the array gain. In contrast, since the transmitter is not assumed to know the channel, there is no processing that it can do to create a similar array gain; thus, the MISO capacity is smaller than the SIMO capacity.

The second row of the table considers the case where both CSIR and CSIT exist. In that case, the transmitter and receiver are able to use their knowledge of the channel in the same way to enhance performance, which results in equivalent capacities for MISO and SIMO systems when the number of antennas is the same.

3.9 Capacity of random channels

In our discussions so far, we have assumed that the channel matrix is fixed. Of course, in the real world $\mathbf{H}$ is time-varying, and so, too, is the capacity. In this section we consider the capacity of randomly-varying channels and introduce the concepts of ergodic and outage capacities.

3.9.1 Definition of $\mathbf{H}_w$

The channel matrix varies over time due to the following:

a) dynamic properties of the channel;
b) movement through a spatially varying field; or
c) both of the above.

In the next chapter, we show that when there is no direct path between the transmitter and receiver, the magnitude of the elements of $\mathbf{H}$ is Rayleigh distributed. This means that $\mathfrak{Re}\{h_{ij}\}$ and $\mathfrak{Im}\{h_{ij}\} \sim N(0, \sigma_h^2)$.

DEFINITION *Let the magnitude of the elements of* $\mathbf{H}$ *be Rayleigh distributed with* $\sigma_h^2 = 1/2$*, which implies that* $\mathbb{E}\{|h_{ij}|^2\} = 1$*. Furthermore, assume that the channel is spatially white, which means that* $\mathbb{E}\{h_{ij}h_{mn}^*\} = \delta_{im}\delta_{jn}$*. When these conditions hold, the channel matrix is denoted by* $\mathbf{H}_w$*.*

This nomenclature is used extensively in the MIMO literature, so we also adopt it in this book.

3.9.2 Capacity of an $\mathbf{H}_w$ channel for large N

Although it is normally not possible to obtain a closed-form expression for the capacity of a MIMO system in randomly varying channels, there are certain cases where it is possible. One such example is when there are equal numbers of transmit and receive antennas, and the number of antennas is large. Under that condition, it is possible to use the following theorem [58]:

THEOREM 3.1 *For an* $N \times N$ *MIMO system where* $\mathbf{H} = \mathbf{H}_w$*,*

$$\lim_{N\to\infty} \mathbf{H}_w\mathbf{H}_w^H \longrightarrow N\mathbf{I}_N. \tag{3.50}$$

One of the implications of this theorem is that when N is large, the eigenvalues of $\mathbf{H}_w\mathbf{H}_w^H$ are equal to N and the rank $r = N$. It follows from Eq. 3.11 that under CSIR,

$$\begin{aligned} C &= \sum_{i=1}^{r} \log_2\left(1 + \frac{\rho}{N}\lambda_i\right) \\ &= \sum_{i=1}^{r} \log_2\left(1 + \frac{\rho}{N}N\right) \\ &= N\log_2(1+\rho). \end{aligned} \tag{3.51}$$

This shows that when the number of antennas becomes large for an $N \times N$ MIMO system:

a) The capacity becomes deterministic; and
b) The capacity increases linearly with N.

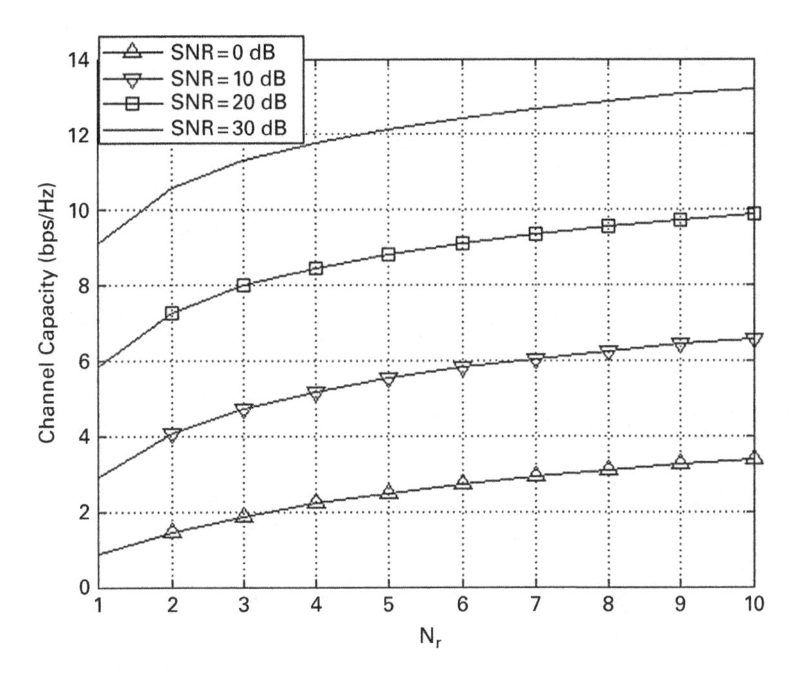

Figure 3.6 Ergodic capacity of a $[1 \times N_r]$ SIMO system in Rayleigh fading.

It should be noted that this behavior is the same as we saw earlier in Eq. 3.17 for a deterministic $N \times N$ MIMO system with a full-rank channel, and equal eigenvalues.

3.9.3 Ergodic capacity

Since the capacity, C, of a MIMO system is a function of $\mathbf{H}$, if $\mathbf{H}$ is random, so is C. Therefore, when the channel varies randomly C is a random variable that can be characterized in terms of all the various statistics that are applicable to any random variable. The simplest statistic is the mean of the capacity. In MIMO and information theory literature, the mean value of the capacity is called the *ergodic capacity*.

Although a closed-form expression for the ergodic capacity of a MIMO system in Rayleigh fading has been derived [67], because of its complexity, we have chosen to generate results using computer simulations instead. Figures 3.6–3.8 show the ergodic capacity when $\mathbf{H} = \mathbf{H}_w$ is plotted versus the number of antennas for different signal-to-noise ratio values for SIMO, MISO, and $N \times N$ MIMO antenna configurations, respectively. These results show that the behavior of the ergodic capacity is similar to the results for a deterministic channel. For example, in Table 3.4 the SIMO capacity increases as the logarithm of N_r but is independent of the number of antennas for the MISO configuration. The results in Figures 3.6 and 3.7 show similar dependence. In addition, Figure 3.8 shows that the ergodic capacity increases linearly with the number of antennas for an $N \times N$ configuration, which is also observed to occur in Eq. 3.16 for a deterministic $N \times N$ full-rank channel.

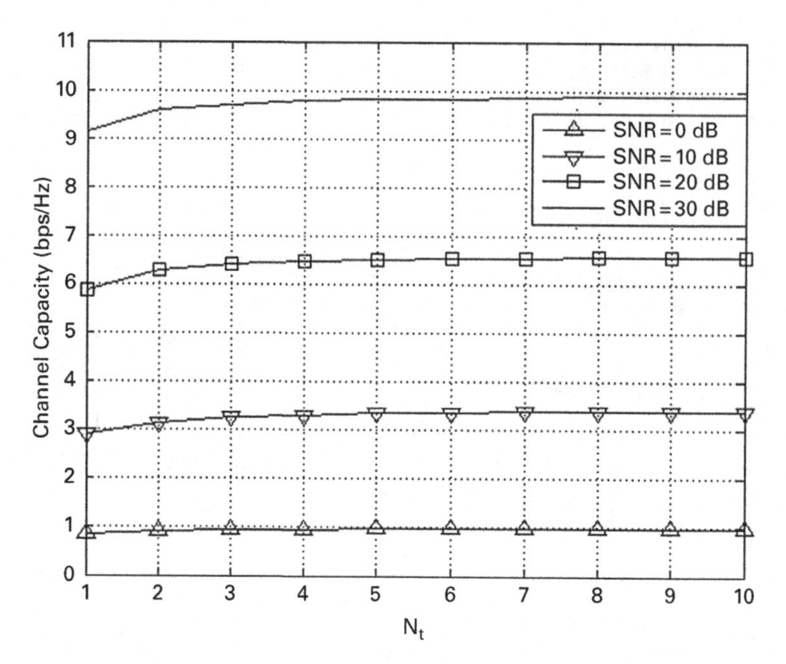

Figure 3.7 Ergodic capacity of a $[N_t \times 1]$ MISO system in Rayleigh fading.

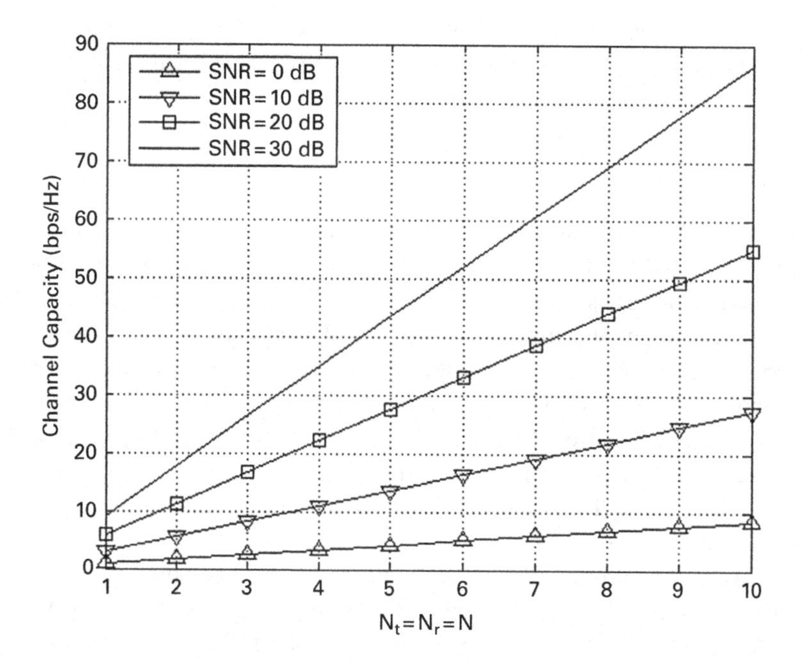

Figure 3.8 Ergodic capacity of an $[N \times N]$ MIMO system in Rayleigh fading.

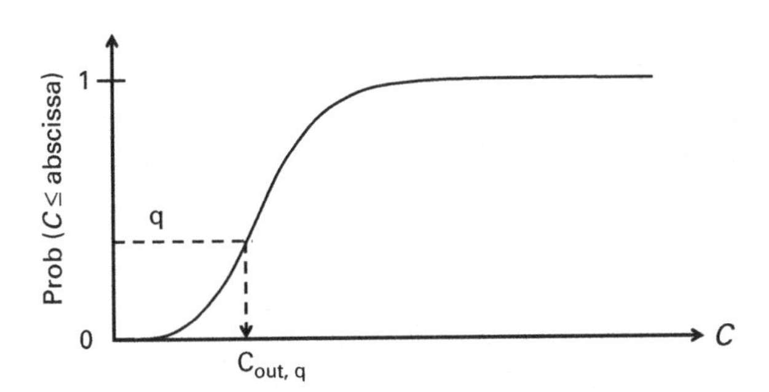

Figure 3.9 A generic cumulative distribution function of the capacity with an illustration of how the outage capacity is defined.

3.9.4 Outage capacity

Another common measure of the capacity of a MIMO channel is the *outage capacity*. The outage capacity is based on the cumulative distribution of C as illustrated in Figure 3.9. We define the outage capacity as follows:

DEFINITION *The q-percent outage capacity, $C_{\text{out},q}$, is defined as the data capacity that is guaranteed* $(1-q) \times 100\%$ *of channel realizations. That is,*

$$Pr\left\{C \leq C_{\text{out},q}\right\} = q. \tag{3.52}$$

Figures 3.10 and 3.11 show the 1-percent and 10-percent outage capacity plotted versus the number of antennas for an $N \times N$ MIMO system. These results show that outage capacity, like ergodic capacity, increases linearly with the number of antennas. A comparison of these plots with Figure 3.8 shows that $C_{\text{out},1}$ and $C_{\text{out},10}$ are less than the ergodic capacity, as expected. For example, at $\rho = 10$ dB, $C_{\text{out},1} \simeq 24$ bps/Hz and $C_{\text{out},10} \simeq 26$ bps/Hz when $N = 10$. By comparison, the ergodic capacity for the same number of antennas is approximately 28 bps/Hz. The outage capacity results in these figures were obtained using computer simulations.

In the case of SIMO and MISO systems, relatively simple closed-form solutions exist for the outage capacity that can be used in place of Monte Carlo computer simulations to make performance predictions. In the following, we derive an expression for the outage capacity of a SIMO system. It is straightforward to extend this result to a MISO system, which is left to the reader as a problem at the end of the chapter.

We begin the derivation by recalling from Eq. 3.43 that the instantaneous capacity for a SIMO channel is $C = \log_2(1 + \rho\|\mathbf{H}\|_F^2)$. Therefore, using the definition of the outage capacity, it follows that

$$\Pr\left\{\log_2(1 + \rho\|\mathbf{H}\|_F^2) \leq C_{\text{out},q}\right\} = q, \tag{3.53}$$

which implies that

$$\Pr\left\{\|\mathbf{H}\|_F^2 \leq \frac{2^{C_{\text{out},q}} - 1}{\rho}\right\} = q. \tag{3.54}$$

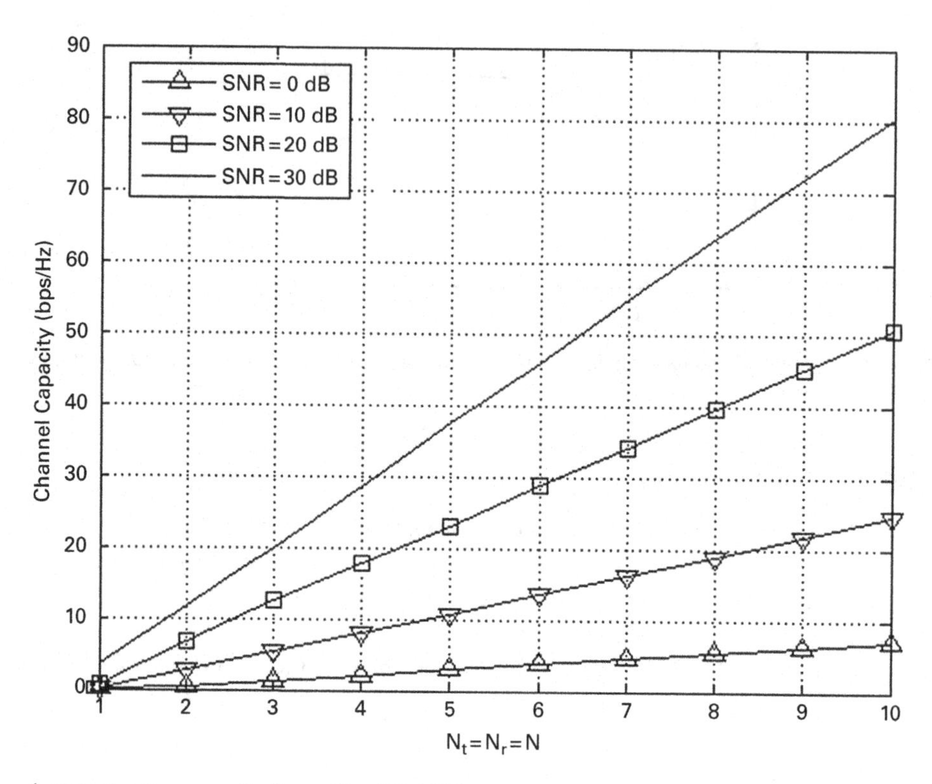

Figure 3.10 1-percent outage capacity for an $N \times N$ MIMO system.

If we assume that $\mathbf{H} = \mathbf{H}_w$, then

$$\|\mathbf{H}\|_F^2 = \sum_{i=1}^{N_r} \left(\mathfrak{Re}^2 \{h_w(i)\} + \mathfrak{Im}^2 \{h_w(i)\} \right). \tag{3.55}$$

By definition, $\mathfrak{Re}\{h_w(i)\}$ and $\mathfrak{Im}\{h_w(i)\} \sim N(0, 1/2)$. In general, the sum of the square of n zero mean Gaussian random variables, each with variance σ^2, has a Gamma distribution with the following probability density function:

$$fx(x) = \frac{1}{(2\sigma^2)^n \Gamma(n/2)} \left[x^{n/2-1} e^{-x/2\sigma^2} \right]. \tag{3.56}$$

Denoting $\|\mathbf{H}\|_F^2$ by x, it follows from Eq. 3.54 that

$$\begin{aligned} q &= \int_0^{\left(2^{C_{\text{out},q}}-1\right)/\rho} f_{\|\mathbf{H}\|_F^2}(x)\, dx \\ &= \frac{1}{\Gamma(N_r)} \int_0^{\left(2^{C_{\text{out},q}}-1\right)/\rho} x^{N_r-1} e^{-x}\, dx, \end{aligned} \tag{3.57}$$

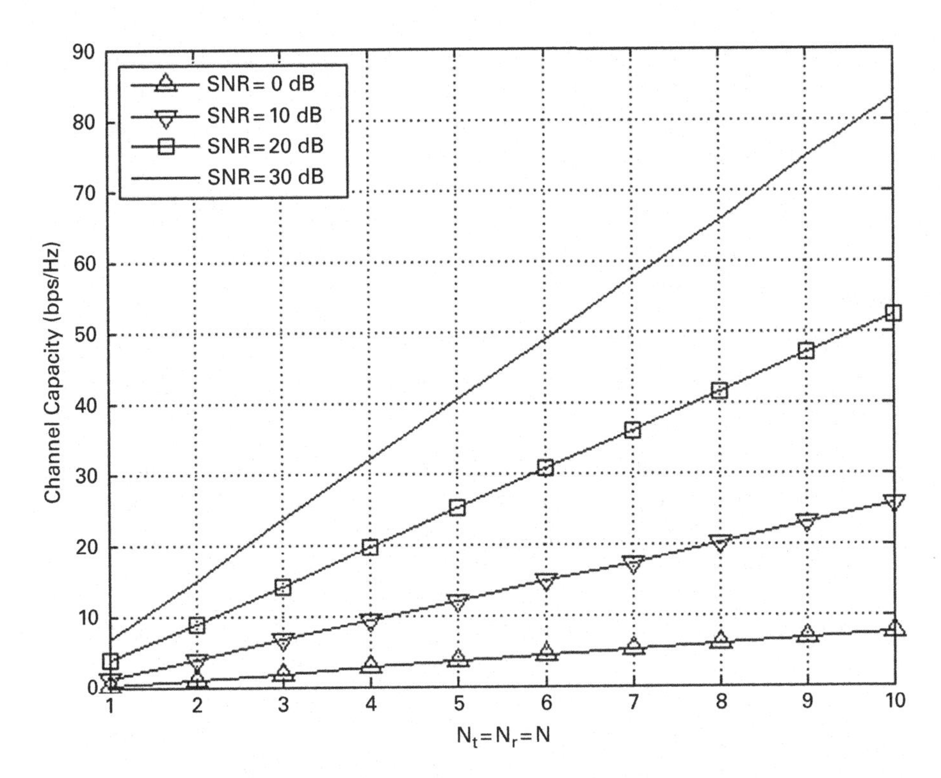

Figure 3.11 10-percent outage capacity for an $N \times N$ MIMO system.

where the integral expression on the right side of the equation is recognized as being the *incomplete gamma function*. In general, the incomplete gamma function is denoted by $P(x,a)$ and is defined as[2]

$$P(x,a) \triangleq \frac{1}{\Gamma(a)} \int_0^x t^{a-1} e^{-t}\, dt; \tag{3.58}$$

thus, we can express Eq. 3.57 as

$$q = P\left(\frac{2^{C_{\text{out},q}} - 1}{\rho}, N_r\right). \tag{3.59}$$

It follows from this equation that we can express the q-percent outage capacity for a $[1 \times N_r]$ SIMO system in the following compact form:

$$C_{\text{out},q} = \frac{\ln\left(1 + \rho P^{-1}(q, N_r)\right)}{\ln 2}, \tag{3.60}$$

where $P^{-1}(y,a)$ denotes the inverse incomplete gamma function, which is defined such that $P^{-1}\left(P\left(x,a\right),a\right) = x$.

It should be noted that Eq. 3.60 can be evaluated using the built-in Matlab function called `gammaincinv()` to solve for $P^{-1}(q, N_r)$.

[2] Some authors define the incomplete gamma function without the $1/\Gamma(a)$ coefficient. In this book we define it the same way that Matlab does for convenience with using the built-in Matlab function for evaluating the incomplete gamma function.

As a final comment regarding outage capacity, it should be noted that there is a difference between a MIMO system's outage capacity and the theoretical maximum throughput that can be achieved. The reason is that $C_{\mathrm{out},q}$ only occurs $(1-q)\%$ of the time; so, the throughput, R_q, associated with $C_{\mathrm{out},q}$ is given by

$$R_q = (1-q)C_{\mathrm{out},q}. \tag{3.61}$$

Problems

3.1 Prove that the expression on the third line of Eq. 3.11 is equal to the expression on the fourth line of that equation. [Hint: Use matrix theorems Section 1.9.2-(o) and 1.9.2-(n).]

3.2 In this chapter we stated that the capacity of a MIMO system under CSIR is maximized when the eigenvalues of the channel are equal. Although we didn't prove it, we showed evidence of this fact in Figure 3.1. Use the Lagrange multiplier technique to prove this analytically.

3.3 In Eq. 3.25, we stated that the variance of $\left[\mathbf{U}^H\mathbf{z}\right]_i$ is equal to the variance of $\mathbf{z}_i$. Prove this.

3.4 Prove Eq. 3.29.

3.5 Derive an expression for the outage capacity of a MISO system in terms of the inverse incomplete gamma function. Assume $\mathbf{H} = \mathbf{H}_w$.

3.6 Write a program called *CvsSNR_EqualPowerAllocation* that computes the capacity of an $N_t \times N_r$ MIMO system under the assumption that only the receiver has knowledge of the channel. If using Matlab, the function should have the following form:

```
C = CvsSNR_EqualPowerAllocation(SNRvec_dB, H, calcType);
```

where SNRvec_dB is a vector of SNR values in dB, H is the $N_r \times N_t$ channel matrix, calcType specifies the type of calculation method, and C is a vector of capacity values for each SNR value in SNRvec_vec. If calcType = 1, compute the capacity without using eigenvalues; if calcType = 2, compute the capacity using eigenvalue decomposition. Run this function using $\mathbf{H}_1$ in Eq. 3.39 and duplicate the equal power allocation curve in Figure 3.4.

3.7 Write a function called *CvsSNR_SingleMode* that computes the capacity of an $N_t \times N_r$ MIMO system that uses single-mode eigenbeamforming. The program should return the capacity, C, and the weights to be applied to each of the transmit antennas in a variable called Weights. If using Matlab, the function should have the following form:

```
[C, Weights] = CvsSNR_SingleMode(SNRvec_dB, H);
```

Run this function using $\mathbf{H}_1$ in Eq. 3.39 and duplicate the single-mode curve in Figure 3.4.

3.8 Write a program called *CvsSNR_Eigenbeamforming* that computes the capacity of an $N_t \times N_r$ MIMO system that employs eigenbeamforming and that models waterfilling to optimally distribute the transmitted power over the various eigenmodes. If using Matlab, the function should have the following form:

```
[C, gain] = CvsSNR_Eigenbeamforming(SNRvec_dB, H)
```

where "gain" is a [length(SNRvec_dB) x rank(H)] dimensioned array and each row contains the weights $\left\{P_i^{\text{opt}} : i = 1, \ldots, \text{rank}(\mathbf{H})\right\}$ for each SNR value. Run this function using $\mathbf{H}_1$ in Eq. 3.39 and duplicate the eigenbeamforming curve in Figure 3.4.

3.9 Write a program called *ErgodicCapacity* that computes the ergodic capacity, C, of a general $N_t \times N_r$ MIMO system with CSIR. If using Matlab, the function should have the following form:

```
C = ErgodicCapacity(SNRdB,Nt, Nr, Niter);
```

where SNRdB specifies the SNR in dB, Nt and Nr are the number of transmit and receive antennas, respectively, and Niter is the number of Monte Carlo iterations used to estimate the ergodic capacity. Use this program to duplicate the curves in Figures 3.6, 3.7, and 3.8.

3.10 Write a program called *OutageCapacity* that computes the outage capacity, C, of a general $N_t \times N_r$ MIMO system with CSIR. If using Matlab, the function should have the following form:

```
C = OutageCapacity(SNRdB, Nt, Nr, Niter, q);
```

where the parameters are the same as above, and q denotes the outage probabability in percent (0 to 100). Use this program to duplicate Figures 3.10 and 3.11.

3.11 Write a program called *OutageCapacitySIMOTheory* that computes the capacity of a SIMO system based on the theory described in this chapter. If using Matlab, the function should have the following form:

```
C = OutageCapacitySIMOTheory(Nrvec, SNRdB, q);
```

where Nrvec is a vector containing values for the number of receive antennas. Use this program to plot the capacity versus the number of receive antennas for q = 10%. Compare the predictions based on theory to the outage capacity computed based on a Monte Carlo approach generated by the *OutageCapacity* program in the previous problem for $N_t = 1$.

4 RF propagation

In the previous chapters we examined some of the implications of the MIMO capacity formula. As we have seen, the statistics of the MIMO capacity are dependent on the statistics of the channel matrix and the average signal-to-noise ratio at the receiver, which we denote by ρ. Both the statistics of $\mathbf{H}$ and the value of ρ depend on the propagation characteristics of the channel; thus, an understanding of propagation is important in order to predict and understand the performance of MIMO communication systems. Because MIMO techniques are designed to operate in a scattering environment, we focus on channel phenomena that give rise to scattering and multipath. Without scattering and multipath, the channels between the various combinations of transmit and receive antennas are correlated, which results in poor MIMO performance. In this chapter, the fundamental concepts and terminology of multipath propagation are reviewed.

4.1 Phenomenology of multipath channels

In any wireless communications path between a transmitter and a receiver, signals arrive at the receiver through various propagation mechanisms. In general, RF energy propagates between two points in one of two ways: *directly* or *indirectly*. Direct propagation refers to transmission of RF energy along a direct path between the transmitter and the receiver that does not involve any reflections, scattering, ducting, or diffractive bending. Direct propagation is called *free space* propagation and is said to undergo free space attenuation. Indirect propagation, in contrast, involves any one or a combination of the following: reflection, diffraction, scattering, or refraction. Because of these mechanisms, indirect propagation enables RF energy to propagate over multiple paths between the transmitter and the receiver, giving rise to multipath propagation. It follows that multipath propagation is capable of supporting communication over non-line-of-sight (NLOS) paths. Although most NLOS paths involve some amount of multipath, multipath can also occur on line-of-sight (LOS) paths, and this has important implications for the utility of MIMO when the transmitter and receiver are within LOS of each other. Figure 4.1 depicts the various propagation components for a general wireless channel.

The effect of the various propagation mechanisms can be seen by measuring received power versus distance from a transmitter, as illustrated notionally in Figure 4.2. In this figure, the wiggly curve depicts the received signal. Although there is only one received

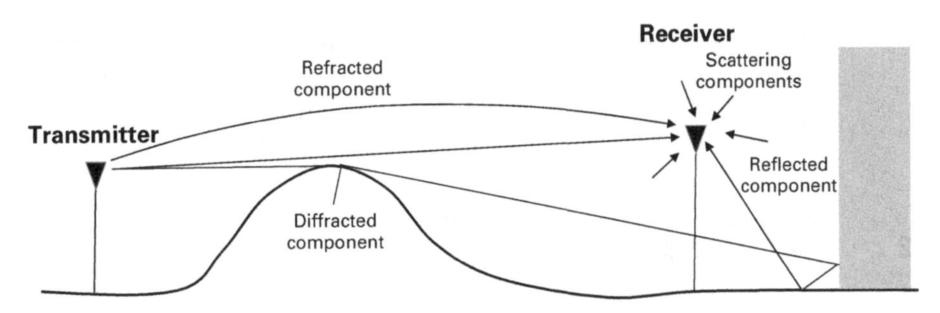

Figure 4.1 Propagation mechanisms in a wireless channel.

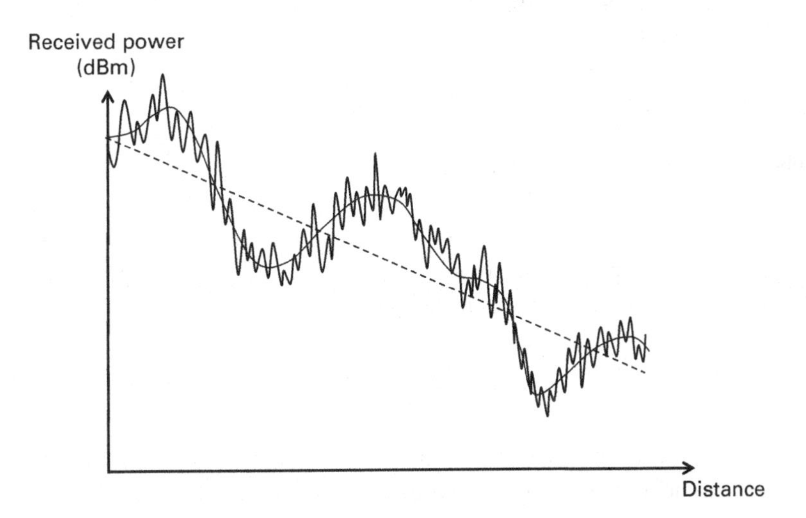

Figure 4.2 Notional illustration of the dependence of received power on distance between a transmitter and receiver over a multipath channel.

signal, we observe that there are three different propagation components, each with its own characteristic scale. The feature with the largest scale in this plot is the linear trend towards lower received power when the power and distance are plotted in dB. This component is denoted by the dashed line. This linear behavior is often observed in real channels, which indicates that the received power tends to decrease with the inverse of the distance raised to some power – so-called *power law propagation*. The second largest scale observed in this plot is depicted by the light solid curve, which is referred to as *large-scale fading*. Such fading, which is sometimes called *shadowing*, is normally due to large objects along the propagation path that block some of the transmitted RF energy. Lastly, the smallest scale feature, called *small-scale fading*, is denoted by the thick curve. The small-scale fading, which is characteristic of a NLOS multipath channel, is caused by constructive and destructive interference between signals arriving at the receiver over different paths resulting from potentially different propagation mechanisms.

In general, MIMO performance is primarily affected by the largest scale, which impacts the value of ρ, and the smallest scale, which impacts the statistics of $\mathbf{H}$. For this reason, we focus on these two propagation components in the remaining sections.

4.2 Power law propagation

Power law propagation, depicted by the dashed line in Figure 4.2, can take on various properties depending on the nature of the wireless channel. In general, the received signal power, P_r, can be computed using the Friis Equation, named after Harald T. Friis [32], which has the following form:

$$P_r = \frac{P_t G_t G_r}{L_{\text{prop}}(d) L_{\text{other}}}, \tag{4.1}$$

where P_t denotes the transmit power, G_t and G_r denote the antenna gains of the transmitter and receiver, respectively, $L_{\text{prop}}(d)$, denotes the propagation loss when the distance between the transmitter and receiver is d, and L_{other} denotes the other losses due to all sources other than propagation (e.g., cable losses, antenna pointing loss, implementation loss). The signal-to-noise power ratio, ρ, is obtained by dividing P_r by the noise power as follows:

$$\rho = \frac{P_t G_t G_r}{L_{\text{prop}}(d) L_{\text{other}} k T_0 F B}, \tag{4.2}$$

where k denotes Boltzmann's constant, $T_0 = 290$ K is the reference temperature, F is the noise figure of the receiver, and B denotes the bandwidth of the received signal.

The simplest case occurs when the mean path loss obeys *free space loss* behavior. In that case, we denote the propagation loss is by $L_{fs}(d)$, which has the following form:

$$L_{fs}(d) = \left(\frac{4\pi df}{c}\right)^2, \tag{4.3}$$

where c denotes the speed of light and f is the RF frequency. Free space loss occurs when the transmitter and receiver are in line-of-sight of each other and the impact of obstacles in their vicinity can be ignored.

Another commonly used propagation loss model is the *two-ray model*. The model, as its name suggests, models the propagation loss between a transmitter and receiver by combining two rays, a direct ray and a reflected ray, as illustrated in Figure 4.3. In [44], the received power, P_r, for the geometry in this figure is shown to have the following form:

$$P_r = P_t G_t G_r \left(\frac{\lambda}{4\pi}\right)^2 \left|\frac{1}{d_1} + \frac{1}{d_2}\Gamma e^{-jk\Delta d}\right|^2, \tag{4.4}$$

where d_1 and d_2 denote the direct and reflected path lengths, respectively, Γ is the reflection coefficient for the reflected path, and $\Delta d \triangleq d_2 - d_1$. The reflection coefficient, in turn, is given by the following expression:

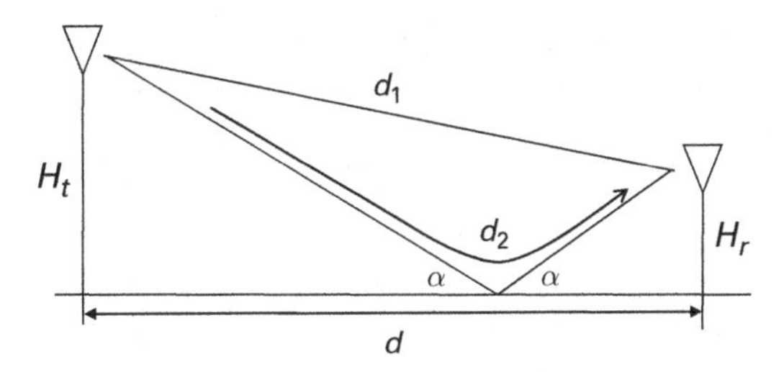

Figure 4.3 Two-ray model geometry.

$$\Gamma(\alpha) = \frac{\sin\alpha - a\sqrt{\epsilon - \cos^2\alpha}}{\sin\alpha + a\sqrt{\epsilon - \cos^2\alpha}}, \tag{4.5}$$

where α is the angle of the reflected ray relative to the surface, $a = 1/\epsilon$ or 1 for vertical and horizontal polarization, respectively, and $\epsilon = \epsilon_r - j60\sigma\lambda$, where ϵ_r is the dielectric constant of the reflecting surface relative to unity in free space and σ is the conductivity of the reflecting surface in mhos per meter.

Under most practical circumstances, the reflection angle, α, is quite small and $d \gg H_t + H_r$. In that case, $d_1 \cong d_2 \triangleq d$ and $\Gamma \cong -1$. Under these assumptions, Eq 4.4 simplifies to

$$P_r = 4P_tG_tG_r\left(\frac{\lambda}{4\pi d}\right)^2 \sin^2\left(2\pi H_tH_r/\left(\lambda d\right)\right). \tag{4.6}$$

Examination of Eq. 4.6 shows that, for small d, the envelope of the received power decreases as $1/d^2$ and that at large d, P_r decreases as $1/d^4$. The transition between these two regimes occurs when the argument in the sine term becomes smaller than about $\pi/2$. The distance at which this occurs is often referred to as the *breakpoint distance*, which is denoted by d_B. It follows that

$$d_B = \frac{4H_tH_r}{\lambda}. \tag{4.7}$$

Figure 4.4 shows the comparison between the two-ray model predictions and measured data collected by the author. The measured data in this plot were collected by measuring the propagation loss between a fixed transmitter and a mobile receiver van in a flat open field at 450 MHz. It shows measured attenuation as a function of distance between two points. The two-ray predictions are shown superimposed, denoted by "Theory" in the legend. In the measurements, the heights of the transmitter and receiver were 2.3 and 2 meters, respectively. It follows from Eq. 4.7 that the breakpoint is equal to 43.4 meters at 450 MHz. Examination of this plot confirms that the slope of the measured data changes at a distance approximately equal to that value.

Free space and two-ray propagation are, of course, only two of many possible power law models that are found to model propagation environments. In these two models, path loss increases as the square of the distance or as the distance to the fourth power. In general, large-scale path loss, while not necessarily obeying square-law or fourth-law behavior, is observed to increase as d^n, where the exponent, n, can be any positive real value. Table 4.1 lists empirically determined values of n for various environments. The

Table 4.1 Typical path loss exponents for various propagation environments.

Environment	Exponent
Free space	2
Urban area cellular	2.7 to 3.5 ([61], Table 4.2)
LOS urban	2 to 4 ([39])
NLOS urban	5 to 20 ([39])
Indoor LOS	1.6 to 1.8 ([61], Table 4.2)
Indoor obstructed	4 to 6 ([61], Table 4.2)
IEEE 802.11n	2 (before d_B); 3.5 (after d_B) ([26])

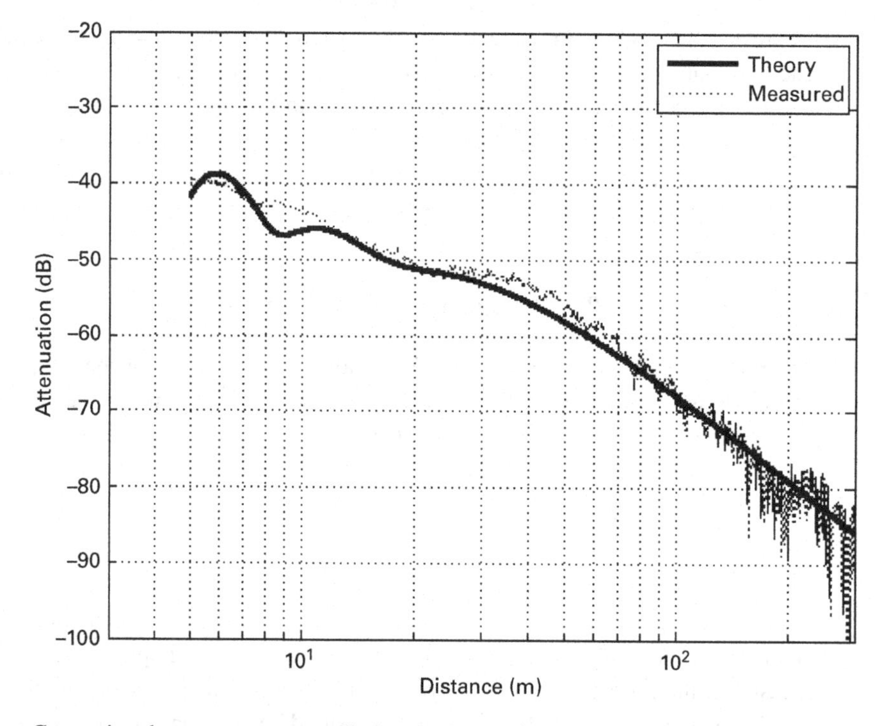

Figure 4.4 Comparison between two-ray predictions and measured path loss at 450 MHz.

last entry specifies the exponents used in the official propagation models defined as part of the IEEE 802.11n standard, which is primarily concerned with indoor propagation. We see that the exponents for the 802.11n models are similar to the theoretical two-ray values.

4.3 Impulse response of a multipath channel

This section lays the mathematical foundation for characterizing the details of multipath behavior. To do this, it is convenient to view a multipath communications channel as a filter, which allows us to use some of the concepts in filter theory to help describe multipath channels. One of the fundamental properties of a filter is its impulse response function; therefore, we consider the impulse response function of a multipath channel in this section.

We start by considering a general transmitted signal, $s(t)$, which we can write as follows:

$$s(t) = \Re\mathfrak{e}\left\{u(t)e^{j2\pi f_c t}\right\}, \tag{4.8}$$

where $u(t)$ is the lowpass equivalent (LPE) of the signal, and f_c denotes the carrier frequency. In a multipath environment, the received signal arrives over multiple paths with different gains, $\{\alpha_i(t)\}$ and delays $\{\tau_i(t)\}$. The resulting received signal is, therefore, given by

$$\begin{aligned} x(t) &= \sum_i \alpha_i(t)s(t-\tau_i(t)) \\ &= \sum_i \alpha_i(t)\mathrm{Re}\left\{u(t-\tau_i(t))e^{j2\pi f_c(t-\tau_i(t))}\right\} \\ &= \mathrm{Re}\left\{\left[\sum_i \alpha_i(t)u(t-\tau_i(t))e^{-j2\pi f_c\tau_i(t)}\right]e^{j2\pi f_c t}\right\}, \end{aligned} \tag{4.9}$$

where the term in brackets is the LPE of the received signal. We denote the LPE of the received signal by $r(t)$, which has the following form:

$$r(t) \triangleq \sum_i \alpha_i(t)u(t-\tau_i(t))e^{-j2\pi f_c\tau_i(t)}. \tag{4.10}$$

Based on these definitions, we can formulate the low-pass picture of a multipath channel as shown in Figure 4.5. This figure depicts the LPE input to the channel, $u(t)$, and the LPE output of the channel (i.e., the LPE received signal), $r(t)$. The channel itself is modeled as a filter with a linear time-variant (LTV) impulse response function, $c(t;\tau)$; thus, the received signal is equal to the convolution of the transmitted signal with the impulse response of the channel. At a given time, t, $r(t)$ and $u(t)$ are therefore related as follows:

$$\begin{aligned} r(t) &= c(t;\tau) * u(t) \\ &= \int_{-\infty}^{\infty} c(t;\tau)u(t-\tau)d\tau, \end{aligned} \tag{4.11}$$

where $*$ denotes convolution.

We will find it useful later in the chapter to express the received LPE signal, $r(t)$, in terms of the Fourier transforms of $c(t;\tau)$ and $u(t)$, which we denote by $C(t;f)$ and $U(f)$, respectively. It is straightforward to show that

$$r(t) = \int_{-\infty}^{\infty} C(t;f)U(f)e^{j2\pi ft}df, \tag{4.12}$$

where $C(t;f) = \int_{-\infty}^{\infty} c(t;\tau)e^{-j2\pi f\tau}\,d\tau$ and $U(f) = \int_{-\infty}^{\infty} u(t)e^{-j2\pi ft}\,dt$.

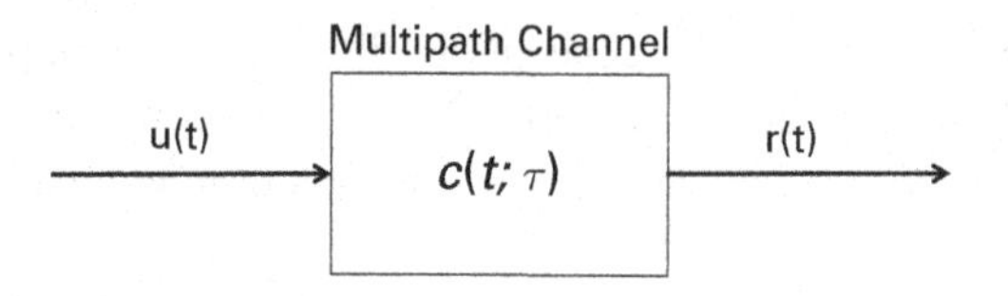

Figure 4.5 Low-pass channel model.

In order for Eq. 4.11 to give the same expression for $r(t)$ as in Eq. 4.10, the impulse response function must have the following form:

$$c(t;\tau) = \sum_i \alpha_i(t) \mathrm{e}^{-\mathrm{j}2\pi f_c \tau_i(t)} \delta(\tau - \tau_i(t)), \tag{4.13}$$

which we use later in this chapter when we consider the statistics of the amplitude of a received signal in multipath.

A key difference between a multipath channel and a simple linear time-invariant (LTI) filter is that multipath channels are time-variant. This is the reason that the impulse response of the multipath channel in Figure 4.5 has two time variables: t and τ. Understanding the distinction between these two time variables may be confusing at first.

To help explain the idea, we begin with a simple example, which is illustrated in Figure 4.6. This figure shows the simplest possible multipath channel – one that consists of a transmitter and receiver, plus one scatterer that moves, denoted by "S". The top two drawings show the geometry at two different instants of time: $t = t_1$ and $t = t_2$. The only difference between these two pictures is the location of the scatterer at the two instants of time. For simplicity, we assume that the scatterer is stationary at time t_1, then it moves instantaneously to a new point at time t_2. The scatterer is assumed to be located on a line that passes between the transmitter and receiver, and is initially 25% further from the transmitter than the distance between the transmitter and receiver. At time t_2, the scatterer is assumed to move along the same line to a point that is twice as far away from the receiver as it is at t_1. The transmitter is assumed to emit an impulse of RF energy at times t_1 and t_2.

The lower lefthand plot shows the received power arriving at the receiver as a result of the two impulse emissions by the transmitter. Because the transmitter and receiver are located a distance d_0 apart, it takes the impulse d_0/c seconds to reach the receiver over the direct path and $(d_0 + 2d_0/4)/c = (3/2)d_0/c$ seconds to reach the receiver over the scattered path when the scatterer is at location 1. At the later time, t_2, the transmitter emits a second impulse, which again takes d_0/c seconds to arrive at the receiver over the direct path and $(d_0 + 2d_0/2)/c = 2d_0/c$ seconds to arrive at the receiver over the scattered path since the scatterer is now assumed to be located $d_0/2$ from the transmitter. The received impulses arrive at the receiver at the times shown on the right side of the time line in the lower lefthand part of the figure.

Because the channel is dynamic (due to the fact that the scatterer moves), this leads to the impulse response of the channel varying with time. The lower lefthand part of Figure 4.6 shows two impulse response functions plotted on a single time axis. The

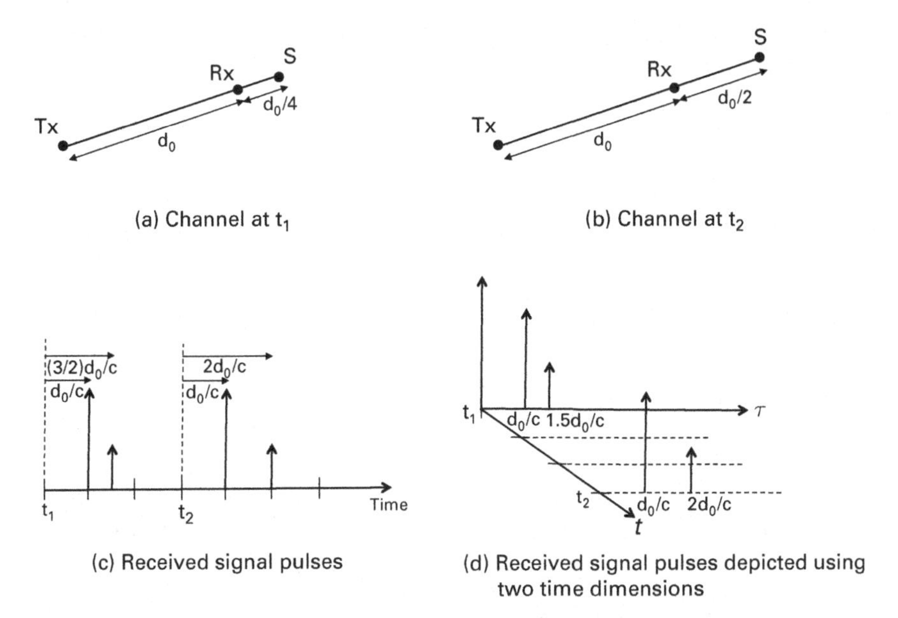

Figure 4.6 A simple multipath channel with one scatterer that moves. This figure illustrates how the impulse response of a time-variant multipath channel can be depicted using two time dimensions, although there is really only one physical time.

lower righthand part of the figure shows an alternative way of plotting the response of this channel. In this portion of the figure, the t-axis denotes the times at which the transmitter emits its impulses, and the τ axis denotes the time it takes the impulses to arrive at the receiver relative to the time at which the impulses are emitted. It is common to view multipath channel behavior using two time variables in this manner; however, it is important to keep in mind that this is done for convenience in conceptualizing and analyzing multipath channels. There is, in fact, only one physical time dimension.

Figure 4.7 depicts $c(t; \tau)$ using two time dimensions. As in the simple example, the t time dimension corresponds to the time dimension over which the channel changes, and the τ dimension corresponds to the time dimension over which the channel responds to an impulse that occurs at a given value of t. In real multipath channels there are usually many scatterers, which results in the impulse response having a continuous appearance in the τ dimension.

4.4 Intrinsic multipath channel parameters

There are four key parameters used to characterize the intrinsic properties of multipath channels. In this context, the word *intrinsic* means that the parameters refer to the properties inherent in the channel and are not related to the nature of the signal that is being transmitted through the channel. Later in this chapter we consider different ways of characterizing channels based on the relationship between these intrinsic parameters

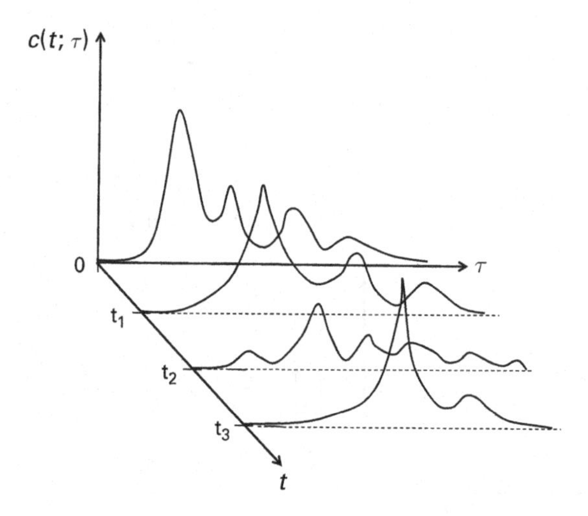

Figure 4.7 Notional depiction of a generic time-variant multipath impulse response function.

and certain features of the signals being transmitted. Two of the intrinsic parameters are related to how the multipath channel varies in the t dimension, and the other two intrinsic parameters are related to variations in the τ dimension. We begin by considering the two parameters that characterize the τ dimension.

4.4.1 Parameters related to τ

The two intrinsic multipath channel parameters that characterize variations in the τ dimension are *delay spread*, T_m, and *coherence bandwidth*, B_c. These parameters are related to the *time dispersion* properties of the channel. In general, dispersion refers to dispersing or spreading a signal out in some dimension. **Time dispersion refers to spreading a signal out in the τ time dimension due to the effects of multipath**. It is important to keep in mind that delay spread and coherence bandwidth do not provide information about the time-varying nature of the channel (i.e., the dependence of the channel on t), only on its temporal response to an impulse at some instant of time.

4.4.1.1 Delay spread (T_m)

Relation to channel impulse response

Delay spread is a measure of the range of propagation delays experienced by the multipath components of a signal when transmitted over a time-dispersive multipath channel. The theoretical definition of delay spread is in terms of the autocorrelation of the channel impulse response function, $c(t, \tau)$. For this purpose, it is typical to assume that the multipath channel is wide sense stationary (WSS) in t, which means that the autocorrelation of the channel response at two points in time (i.e., at two values of t) is only a function of the difference between the two time instants and not the instants of time themselves. Under this assumption, the autocorrelation function (ACF) is defined as follows:

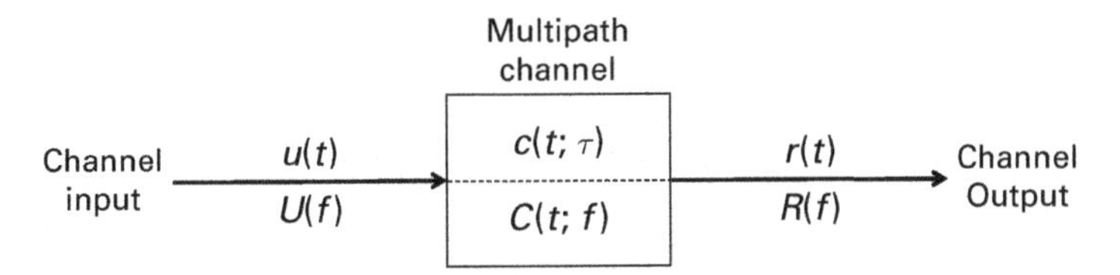

Figure 4.8 The time and frequency-domain input–output relationships of a lowpass equivalent multipath channel.

$$\phi_c(\Delta t; \tau_1, \tau_2) \triangleq \frac{1}{2}\mathbb{E}\left\{c^*(t, \tau_1)c(t + \Delta t; \tau_2)\right\}, \tag{4.14}$$

where the WSS assumption is reflected in the fact that the ACF is a function of Δt only.

In most wireless channels, the attenuation and phase shift of the channel associated with path delay τ_i are uncorrelated with the path associated with $\tau_j, i \neq j$. This is called *uncorrelated scattering* (US). When a channel is both WSS and has uncorrelated scattering, it is said to be a *WSSUS* channel. Under this assumption

$$\phi_c(\Delta t; \tau_1, \tau_2) = \phi_c(\Delta t; \tau_1)\delta(\tau_1 - \tau_2). \tag{4.15}$$

If we consider a single instant of time (t), then $\Delta t = 0$, so

$$\begin{aligned} \phi_c(\Delta t; \tau_1, \tau_2) &= \Phi_c(0; \tau_1)\delta(\tau_1 - \tau_2) \\ &= \phi_c(\tau) \\ &= \frac{1}{2}\mathbb{E}\left\{|c(\tau)|^2\right\}, \end{aligned} \tag{4.16}$$

where $\phi_c(\tau)$ is called the *multipath intensity profile*, the *delay power spectrum,* or the *power delay profile* of the channel.

What is the physical meaning of $\mathbb{E}\left\{|c(\tau)|^2\right\}$ in Eq. 4.16? To help answer this question, consider Figure 4.8, which depicts the lowpass equivalent input and outputs of a multipath channel in both time and frequency domains. As with any filter, the relationship between the frequency domain representations of the input and output is given by

$$R(f) = U(f)C(t;f), \tag{4.17}$$

which implies that

$$|R(f)|^2 = |U(f)|^2\,|C(t;f)|^2. \tag{4.18}$$

Since Parseval's theorem states that $\int_{-\infty}^{\infty} |r(t)|^2\,dt = \int_{-\infty}^{\infty} |R(f)|^2\,df$, it follows that the total received energy, $E_r = \int_{-\infty}^{\infty} |r(t)|^2\,dt$ is expressed as

$$E_r = \int_{-\infty}^{\infty} |U(f)|^2\,|C(t;f)|^2\,df. \tag{4.19}$$

If we assume that the input to the channel is a delta function, then $U(f) = 1$, so

$$E_r = \int_{-\infty}^{\infty} |C(t;f)|^2 \, df = \int_{-\infty}^{\infty} |c(t;\tau)|^2 \, d\tau = \int_{-\infty}^{\infty} |c(\tau)|^2 \, d\tau, \tag{4.20}$$

where the third equality holds because the channel is assumed to be WSS. Since energy has units of [power]×[time], this equation shows that $|c(\tau)|^2$ is the received power as a function of τ. This means that in practice the multipath intensity profile can be computed by transmitting an impulse and measuring the received power as a function of time. The resulting time-varying function is the multipath intensity profile. Figures 4.9 and 4.10 show examples of measured delay profiles in residential outdoor environments with and without a direct path component.

These examples show two typical characteristics of delay profiles. First, they show that there is a general, though non-monotonic, trend towards weaker power as the delay increases. This fall-off in power is due to the fact that larger delay values are associated with energy that has propagated over a longer distance, resulting in greater propagation loss. The second feature we see is that there are bumps in the delay profile, indicating that there is more energy associated with some delays than others. This is caused by the scatterers not being uniformly distributed in space. If, for example, there were only a small number of discrete scatterers, then the delay profile would consist of discrete spikes corresponding to the propagation delays of the paths between the transmitter and each of the scattering objects. In most realistic environments, there are a very large number of scatters distributed throughout space, so the delay profile is not discrete; however, because the scatterers are not uniformly distributed, some delays have more energy associated with them than others.

Definitions of T_m

There are two commonly used definitions for the delay spread of a channel. First, it is important to emphasize that normally when defining T_m, all delays are specified relative to the delay associated with the direct component of the transmitted signal, which is the first component of the signal to arrive at the receiver. When specifying the delay relative to the direct component, the delay is referred to as the *excess delay*. The delay spread is normally specified either by a) the mean excess delay, $\bar{\tau}$, or b) the rms excess delay, σ_τ, which are formally defined as follows:

$$\bar{\tau} = \frac{\sum_k \phi_c(\tau_k)\tau_k}{\sum_k \phi_c(\tau_k)} \tag{4.21}$$

and

$$\sigma_\tau = \sqrt{\bar{\tau^2} - (\bar{\tau})^2}, \tag{4.22}$$

where

$$\bar{\tau^2} = \frac{\sum_k \phi_c(\tau_k)\tau_k^2}{\sum_k \phi_c(\tau_k)}. \tag{4.23}$$

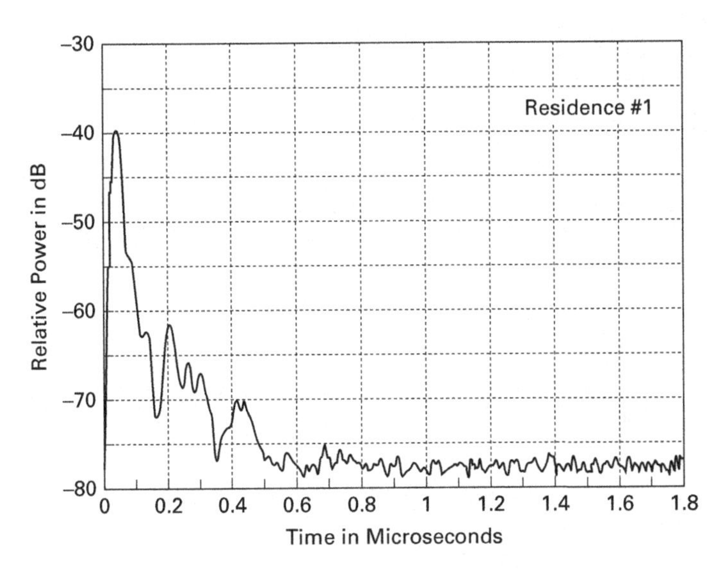

Figure 4.9 Example of a measured multipath delay profile in a residential environment when there is a direct path component combined with scattering. Frequency is 850 MHz (from [19]).

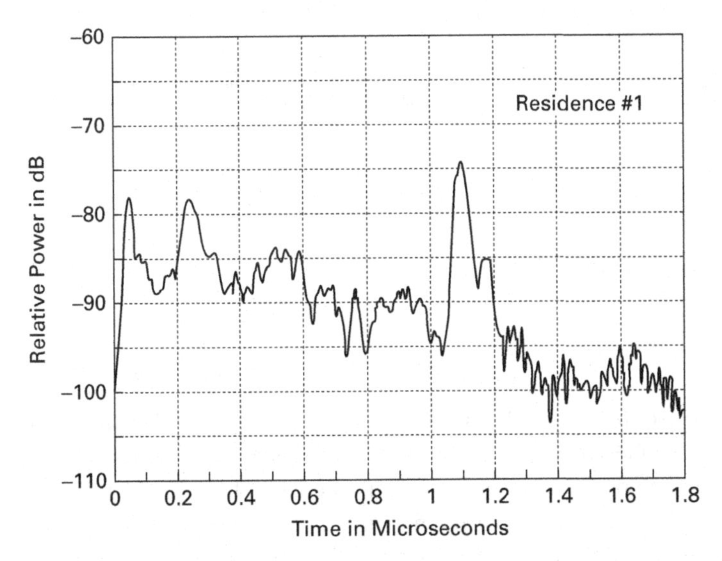

Figure 4.10 Example of a measured multipath delay profile in a residential environment in the absence of a direct path component. Frequency is 850 MHz (from [19]).

Example 4.1 What are the mean and rms excess delay spread values for the power delay profile in Figure 4.11?

Answer Start by computing $\bar{\tau}$. For convenience, denote –100 dBm by P. It follows from Eq. 4.21 that

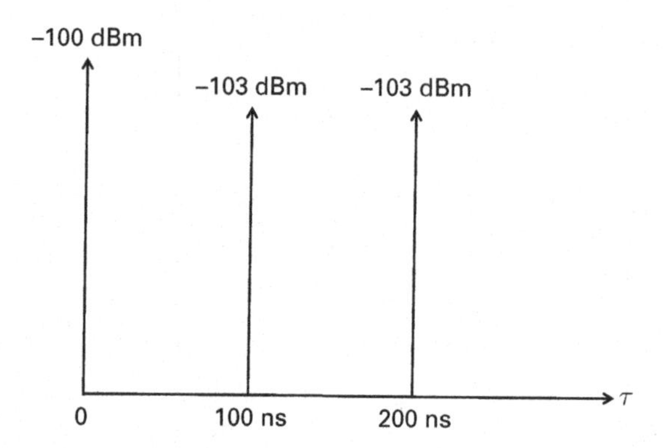

Figure 4.11 Example power delay profile.

$$\bar{\tau} = \frac{(P)(0) + (0.5P)(100) + (0.5P)(200)}{P + 0.5P + 0.5P}$$
$$= \frac{150}{2}$$
$$= 75 \text{ ns.}$$

Next, compute $\overline{\tau^2}$. It follows from Eq. 4.23 that

$$\overline{\tau^2} = \frac{(P)(0)^2 + (0.5P)(100^2) + (0.5P)(200^2)}{P + 0.5P + 0.5P}$$
$$= \frac{5 \times 10^4}{4}$$
$$= 12\,500 \text{ ns}^2.$$

Equation 4.22 implies that

$$\sigma_\tau = \sqrt{12,500 - (75)^2}$$
$$= 82.9 \text{ ns.}$$

In addition to illustrating the steps used in computing the delay spread, this example also highlights the fact that only the relative amplitude of the power delay profile is needed; absolute power levels are not. This is evident from the fact that we arbitrarily replaced the true power level of the maximum component with a variable P. This, of course, means that delay spread values can be measured without needing to compute the absolute received power levels, eliminating the needed for complex calibrations during the measurements.

Typical RMS delay spread values

The delay spread values of multipath channels depend on the nature of the scattering, but are relatively insensitive to RF frequency over the range of frequencies typically used by commercial wireless devices, which range from about 1 to 5 GHz. A summary of measured RMS delay values is given in Table 4.2.

Table 4.2 Measured RMS delay spread for selected environments (from [61]).

Environment	Frequency (MHz)	RMS delay spread (σ_τ) (ns)
Urban (New York)	910	1300
Urban (San Francisco)	892	20 000
Typical suburban	910	300
Extreme suburban	910	2000
Indoor (office bldg.)	1500	25
Indoor (office bldg.)	850	270
Indoor (office bldg.)	1900	90

4.4.1.2 Coherence bandwidth (B_c)

The coherence bandwidth, B_c, is a measure of the range of frequencies over which a communications channel passes all spectral components of a signal with equal gain and delay. It is related in a simple way to the delay spread, as we will show shortly. To do so, we begin by defining the transfer function of a channel, $C(t;f)$, as follows:

$$C(t;f) \triangleq \int_{-\infty}^{\infty} c(t;\tau)\mathrm{e}^{-\mathrm{j}2\pi f\tau}\,d\tau. \tag{4.24}$$

Next, consider the autocorrelation of $C(t;f)$, which we denote by $\Phi_C(t_1,t_2;f_1,f_2)$ and define as follows:

$$\Phi_C(t_1,t_2;f_1,f_2) \triangleq \frac{1}{2}E\left\{C^*(t_1;f_1)C(t_2;f_2)\right\}. \tag{4.25}$$

If we assume the channel is wide sense stationary, then the autocorrelation will only depend on the difference in time, Δt. In that case,

$$\Phi_C(t_1,t_2;f_1,f_2) \rightarrow \Phi_C(\Delta t;f_1,f_2) = \frac{1}{2}E\left\{C^*(t;f_1)C(t+\Delta t;f_2)\right\}. \tag{4.26}$$

Furthermore, if we assume an uncorrelated channel, then the channel will be WSSUS and it can be shown (see end of chapter problems) that Eq. 4.26 reduces to

$$\Phi_C(\Delta t;f_1,f_2) \rightarrow \Phi_C(\Delta t;\Delta f) = \int_{-\infty}^{\infty} \phi_c(\Delta t;\tau)\mathrm{e}^{-\mathrm{j}2\pi\Delta f\tau}\,d\tau, \tag{4.27}$$

where Δf is the difference between the two frequencies, f_1 and f_2.

If we consider a single instant of time, $\Delta t = 0$, so $\phi_C(\Delta t;\Delta f)$ becomes a function of Δf only, which we denote by $\Phi_C(\Delta f)$. That is, $\Phi_C(\Delta t = 0;\Delta f) \triangleq \Phi_C(\Delta f)$, and Eq. 4.27 reduces to

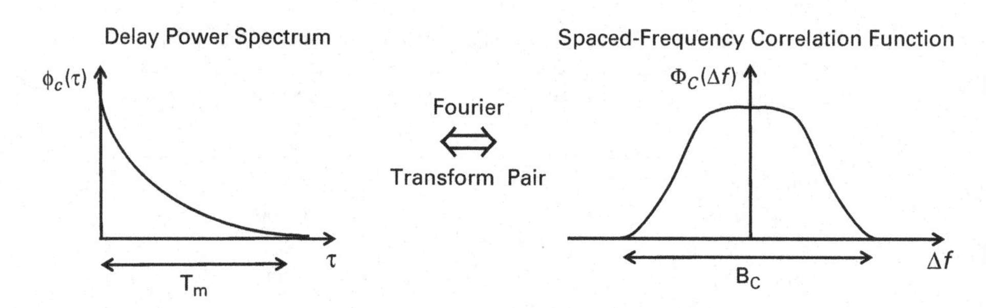

Figure 4.12 Relationship between the power delay profile and the spaced-frequency correlation function.

$$\Phi_C(\Delta f) = \int_{-\infty}^{\infty} \phi_c(\tau) e^{-j2\pi \Delta f \tau} d\tau, \tag{4.28}$$

where $\Phi_C(\Delta f)$ is called the *spaced-frequency correlation function*. For the purpose of defining the coherence bandwidth, it is useful to define the normalized spaced-frequency correlation function, $\Phi'_C(\Delta f)$, which is defined as follows:

$$\Phi'_C(\Delta f) \triangleq \Phi_C(\Delta f) / \max\{\Phi_C(\Delta f)\}. \tag{4.29}$$

It follows that Φ'_C has a maximum value equal to 1.

Equation 4.28 shows that the multipath power delay profile and the spaced-frequency correlation function, the two key functions related to τ, are a Fourier transform pair, as illustrated in Figure 4.12. B_c is typically defined in one of two ways, called the 50th and 90th percentile bandwidths, which we denote by $B_{c,50}$ and $B_{c,90}$, respectively. The nth percentile coherence bandwidth ($0 \leq n \leq 100$) is defined as the bandwidth over which the normalized spaced-frequency correlation function is greater than $n/100$. It can be shown that [61]

$$B_{c,50} \simeq \frac{1}{5\sigma_\tau} \tag{4.30}$$

and

$$B_{c,90} \simeq \frac{1}{50\sigma_\tau}. \tag{4.31}$$

Example 4.2 Compute the 50th and 90th percentile coherence bandwidths associated with the multipath channel shown in Figure 4.11.

Answer Example 4.1 showed that $\sigma_\tau = 82.9$ ns. It follows that $B_{c,50} = 1/(5 \times 82.9 \times 10^{-9}) = 2.4$ MHz, and that $B_{c,90} = 1/(50 \times 82.9 \times 10^{-9}) = 240$ kHz.

Before we complete our discussion of coherence bandwidth, it is worthwhile examining the spaced-frequency correlation function in a little more detail in order to develop a physical interpretation of $\Phi_C(\Delta f)$. To do that, consider Eq. 4.17, which specifies the frequency domain relationships between the LPEs of the transmitted signal, $U(f)$, the received signal, $R(f)$, and the LPE of the frequency response of the channel, $C(t;f)$. Assume that two pure complex sinusoidal signals (i.e., signals with the form $\cos(2\pi ft) + \mathrm{j}\sin(2\pi ft)$) at frequencies f_1 and f_2 are transmitted simultaneously at some time t. Since the Fourier transforms of these signals are equal to $\delta(f - f_1)$ and $\delta(f - f_2)$, respectively, it follows that the received signals associated with these sinusoids are given by

$$\text{Sinusoid at } f_1\text{: } R(f) = \begin{cases} C(t;f_1), & f = f_1 \\ 0, & \text{otherwise} \end{cases} \tag{4.32}$$

and

$$\text{Sinusoid at } f_2\text{: } R(f) = \begin{cases} C(t;f_2), & f = f_2 \\ 0. & \text{otherwise} \end{cases} \tag{4.33}$$

It follows from Eq. 4.26 that $\Phi_C(\Delta f) \triangleq \Phi_C(\Delta t = 0; \Delta f) = \frac{1}{2}\mathbb{E}\{C^*(t;f_1) \times C(t;f_2)\}$ is, in fact, the correlation between the channel outputs when the input sinusoids are separated in frequency by $\Delta f = f_1 - f_2$. The correlation bandwidth is equal to the frequency separation for which this correlation is large.

4.4.2 Parameters related to t

The two intrinsic multipath channel parameters that characterize variations in the t dimension of a channel are the *Doppler spread*, B_d, and the *coherence time*, T_c. These two parameters are analogous to T_m and B_c, except that they characterize the *frequency dispersion* properties of the channel. Frequency dispersion refers to the spreading out of signal in the frequency dimension.

4.4.2.1 Doppler spread (B_d)

The physical cause of Doppler spread

Doppler spread refers to the broadening of a transmitted signal's bandwidth as it propagates through a multipath channel due to relative motions between the transmitter, scatterers, and the receiver. Figure 4.13 illustrates the basic principle behind Doppler spread for a simple case where a transmitter is moving relative to two stationary scattering objects with a stationary receiver. The scatterers are depicted by the two grey circles. In this example, we assume that the transmitter emits a pure carrier with a frequency equal to f_c and that it is moving towards the scatterer on the right with speed equal to v and away from the scatterer on the left with the same speed. Because of Doppler, the frequency received by the lefthand scatterer is $f_c - f_d$ and the received frequency by the righthand scatterer $f_c + f_d$, where $f_d = vf_c/c$ denotes the Doppler shift. Some of the energy from each scatterer gets emitted in the direction of the receiver, so the signal at the receiver consists of energy from the two scatterers plus energy that arrives directly from the transmitter with a received frequency equal to f_c. The spectrum of the transmitted signal, which consists of a single component at f_c, is shown in the top right of the

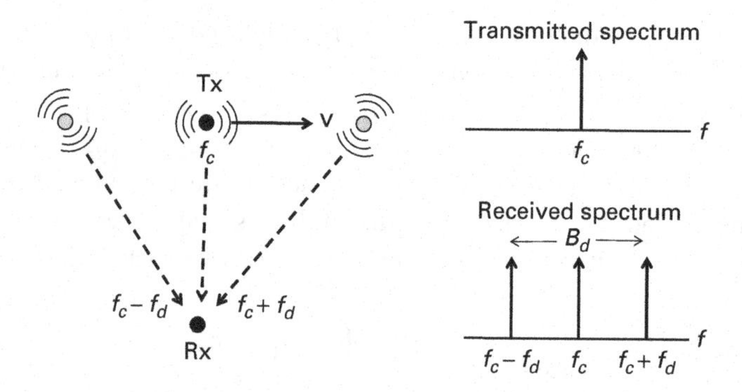

Figure 4.13 A simple illustration showing how Doppler spread occurs.

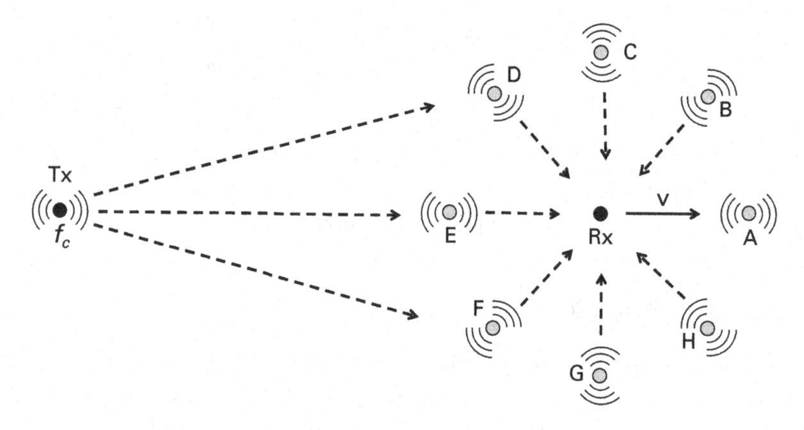

Figure 4.14 Geometry associated with Clark's Doppler spread model.

figure. The spectrum of the signal at the receiver depicted in the lower right consists of three spectral components at $f_c - f_d, f_c$, and $f_c + f_d$; thus, the total received bandwidth is $2f_d$, which demonstrates that spectral broadening has occurred. In this case, a reasonable definition for the Doppler spread is $B_d = 2f_d$.

Clark's Doppler spread model

The simple example above was provided merely to show, in principle, how relative motion in a channel can lead to a broadening of the spectrum. This is obviously only one of many possible scenarios, however. Another type of geometry in which spectral broadening occurs is shown in Figure 4.14. In this figure, it is the transmitter that is fixed and the receiver that is moving. In addition, we assume that the receiver is surrounded by a circular ring of scatterers and that there is no direct component between the transmitter and receiver. This is an idealized depiction of what happens when a vehicle is driving through an urban environment and receives a signal from a cell tower after scattering from buildings and objects surrounding the receiver.

In the geometry shown in Figure 4.14, the signal arrives at the receiver with many different frequencies. For example, energy scattered from scatterers **A** and **E** arrives at the receiver with frequencies equal to $f_c + f_d$ and $f_c - f_d$, respectively. On the other hand, energy from scatterers **C** and **G** arrives with frequency equal to f_c because the receiver is moving perpendicular to them. Energy from the other scatterers will arrive with frequencies that fall between these two extremes. If we imagine that there is an infinite number of scatterers in a circle surrounding the receiver, the notion of a frequency spectrum emerges. For the geometry shown in Figure 4.14, Gans [33] has analyzed the power spectral density of the received signal as a function of frequency. In his analysis, he assumes that the arriving electromagnetic energy comes from all directions with equal amplitudes. The resulting spectrum is referred to as *Clark's Model.* For the case of a vertical quarter-wave whip antenna, Gans shows that the received spectrum, $S(f)$, has the following form:

$$S(f) = \frac{1.5}{\pi f_d \sqrt{1 - \left(\frac{f - f_c}{f_d}\right)^2}}, \tag{4.34}$$

where $f_d = v f_c / c$ is the maximum Doppler shift. Figure 4.15 shows three example plots of the Doppler spectrum based on Eq. 4.34 for speeds equal to 20, 40, and 60 mph, assuming a carrier frequency equal to 2.4 GHz. The U-shape of these curves is characteristic of the Doppler spectrum associated with a uniform circle of scatterers. The minimum at $(f - f_c) = 0$ corresponds to energy arriving at the mobile receiver from scatterers that are located in a direction perpendicular to motion. The large values at the left and right extremes of each curve correspond to the energy arriving from scatterers in front of the mobile (positive $(f - f_c)$) and from those directly behind the mobile (negative $(f - f_c)$). The maxima of each curve occur at $|f - f_c| = f_d = v f_c / c$.

Doppler model used in IEEE 802.11n

The Clark Doppler spectrum is a convenient theoretical model for cellular wireless applications such as those encountered in the LTE, LTE-Advanced, and WiMAX wireless systems. In WLAN applications, however, the transmitter and receiver are often fixed and it is the channel that undergoes dynamics. In that case, different Doppler models may be used. In this section, we describe the Doppler model used in development and testing of 802.11n devices.

In [26], a general expression is given for the Doppler spectrum that applies to indoor environments, such as those in which WiFi devices operate. The resulting Doppler spectrum has the following form:

$$S(f) = \frac{1}{1 + A\left(\frac{f - f_c}{f_d}\right)^2}, \tag{4.35}$$

where A is a constant that is set equal to 9. It follows that $S(f_d) = 0.1$. The maximum Doppler spread, f_d, in the 802.11n model is based on an assumed environmental speed equal to 1.2 km/hr, which implies that $f_d = 6$ and 3 Hz at $f_c = 5.25$ and 2.4 GHz, respectively.

Figure 4.16 shows the Doppler spectrum based on Eq. 4.35 for 2.4 GHz and 5.25 GHz. The contrast between this indoor spectrum and the Clark spectrum is dramatic.

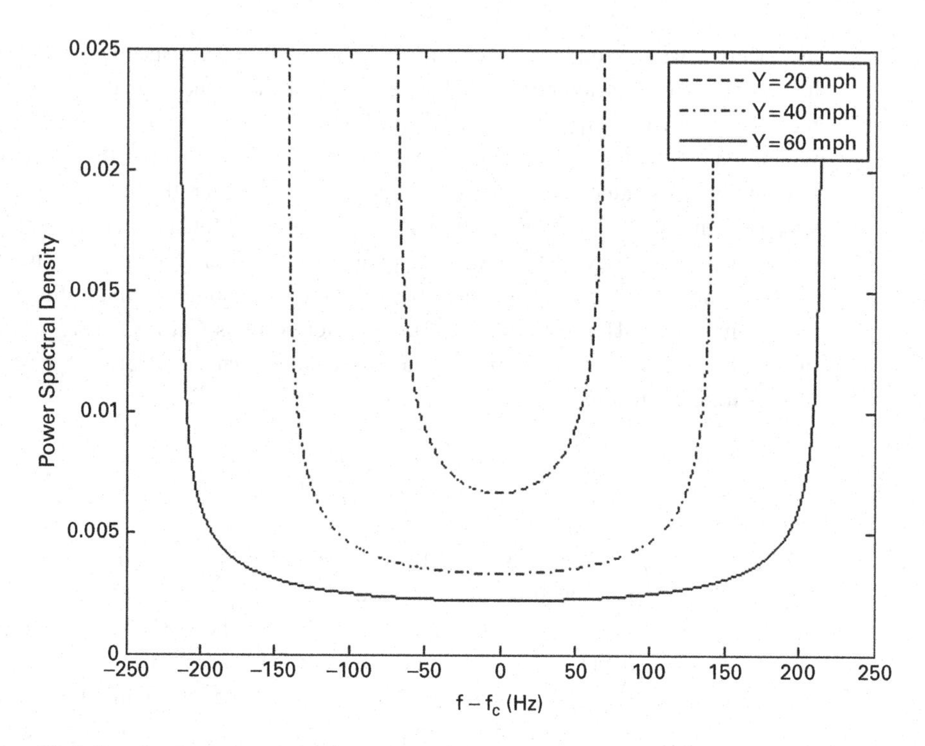

Figure 4.15 Clark Doppler power spectrum for an unmodulated CW carrier at 2.4 GHz.

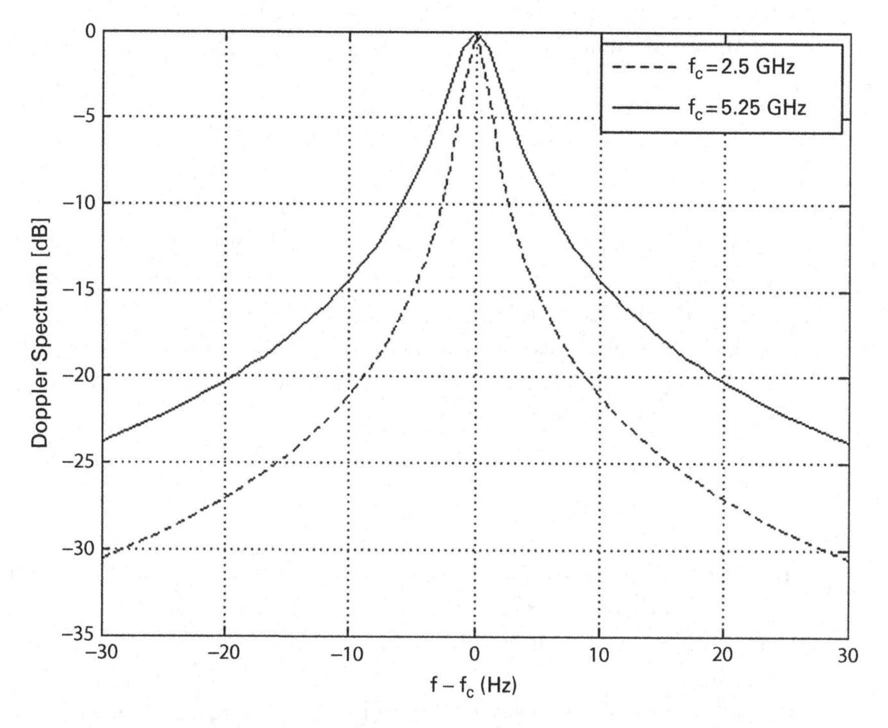

Figure 4.16 Doppler spectrum used in IEEE 802.11n.

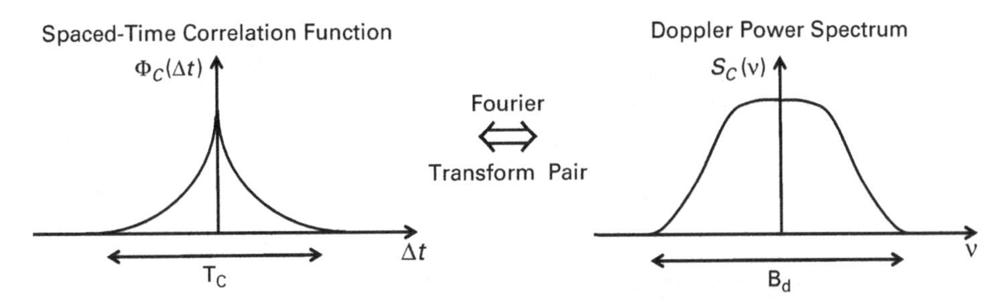

Figure 4.17 Relationship between the spaced-time correlation function and the Doppler power spectrum.

General Doppler power spectrum

In this section, we discuss the Doppler power spectrum in its most general sense and explain how it is related to $\Phi_C(\Delta t; \Delta f)$ defined in Eq. 4.27. The general definition of the Doppler power spectrum, which we denote by $S_C(\nu; \Delta f)$, is as follows:

$$S_C(\nu; \Delta f) \triangleq \int_{-\infty}^{\infty} \Phi_C(\Delta t; \Delta f) e^{-j2\pi\nu\Delta t} d\Delta t. \tag{4.36}$$

If we consider a single transmit frequency, then $\Delta f = 0$ and the Doppler spectrum simplifies to

$$S_C(\nu) \triangleq S_C(\nu; \Delta f = 0) = \int_{-\infty}^{\infty} \Phi_C(\Delta t) e^{-j2\pi\nu\Delta t} d\Delta t, \tag{4.37}$$

where ν in this equation is equivalent to $f - f_c$ in our earlier discussion. This implies that the Doppler spectrum is the Fourier transform of $\Phi_C(\Delta t)$, which is called the *spaced-time correlation function*. We depict this Fourier transform pair relationship in Figure 4.17.

To see that the relationship in Eq. 4.37 makes sense from an intuitive standpoint, consider the case where the channel is time-invariant. Under that assumption, there are no dynamics to create Doppler broadening, so we should expect the Doppler spectrum to be a delta function. For a time-invariant channel, the frequency channel response of the channel does not change with time, so $\Phi_C(\Delta t)$ is simply a constant for all values of Δt. This implies that the Doppler spectrum $S_C(\nu) = \delta(\nu)$, which then implies that $B_d = 0$, as we would expect.

4.4.2.2 Coherence time (T_c)

The coherence time is the maximum time that can elapse between transmitting two identical signals and there still being a high probability that the received signals will be highly correlated. Clearly, if the channel is varying slowly, the coherence time will be large, and vice versa. Since slowly varying channels have narrow Doppler spectra (i.e., small Doppler bandwidths), then the coherence time and Doppler spread are inversely proportional to one another. That is,

$$T_c \propto \frac{1}{B_d}. \tag{4.38}$$

A conservative measure of the coherence time is the so-called 50-percentile value, denoted by $T_{c,50}$, which is defined as the time separation for which the time correlation is equal to 0.5. It can be shown that [61]

$$T_{c,50} \simeq \frac{9}{16\pi B_d}. \tag{4.39}$$

4.5 Classes of multipath channels

In the previous section, we talked about the intrinsic properties of multipath channels. In this section, we talk about how to categorize channels based on the relationship between these intrinsic properties and the nature of the signals being transmitted through them. This results in the following four classes of multipath channels:

1. Flat fading;
2. Frequency-selective fading;
3. Fast fading; and
4. Slow fading.

We consider each of these types of channels in the following subsections.

4.5.1 Flat fading

DEFINITION *A channel is said to be* flat *if the bandwidth of the signal* $W < B_c \Longrightarrow T_s > T_m$.

Under this condition, Eq. 4.12 indicates that

$$\begin{aligned} r(t) &= \int_{-\infty}^{\infty} C(t;0)U(f)e^{j2\pi ft}df \\ &= C(t;0)u(t), \end{aligned} \tag{4.40}$$

where we have made the approximation that $C(t;f) = C(t;0)$ since flat fading implies that $W < B_c$, which means that the channel is approximately constant (and equal to $C(t,f = 0)$) over the bandwidth of the transmitted signal. Equation 4.40 shows that under flat fading, the effect of the channel is a simple multiplication of the transmit signal by a complex gain.

Next, consider time to be discrete, where t_k denotes the time at which the kth symbol, $s(k) \triangleq u(t_k)$, is transmitted, and define the channel response at that time to be $h(k) \triangleq C(t_k;0)$. It follows that $r(k) = h(k)s(k)$, where $r(k) \triangleq r(t_k)$. If we now consider the addition of noise, $z(k)$, then we can express the system model for flat fading as follows:

$$r(k) = h(k)s(k) + z(k). \tag{4.41}$$

It should be emphasized that the expression in Eq. 4.41 does not impose any constraints on the variances of the individual terms. If, however, we use the conventions defined in Eqs. 3.2–3.5 for convenience in modeling, then Eq. 4.41 becomes

$$r(k) = \sqrt{\rho}h(k)s(k) + z(k), \tag{4.42}$$

where ρ denotes the signal-to-noise ratio. The MIMO form of this equation follows directly and is equivalent to Eq. 3.1. The preceding discussion provides a justification for that equation.

4.5.2 Frequency-selective fading

DEFINITION *A channel is said to be* frequency selective *if its bandwidth* $W > B_c \Longrightarrow T_s < T_m$.

Under this condition, we seek to derive an expression for the received signal in terms of the transmitted signals, the impulse response of the channel, and the noise at the receiver. That is, we seek a system equation for a frequency-selective channel that corresponds to Eq. 4.42 for flat fading.

We begin by defining $u(t)$ to be a bandlimited LPE signal with bandwidth equal to W. It follows that such a signal can be expressed as an infinite sum [58], given by

$$u(t) = \sum_{l=-\infty}^{\infty} u\left(\frac{l}{W}\right) \frac{sin\left[\pi W(t - l/W)\right]}{\pi W(t - l/W)}. \tag{4.43}$$

This well-known result goes by a variety of names, including the *Nyquist sampling theorem, Shannon sampling theorem*, the *Nyquist–Shannon sampling theorem*, or simply the *sampling theorem*. It is straightforward to show that the Fourier transform of $u(t)$, which we denote by $U(f)$, is given by

$$U(f) = \begin{cases} \frac{1}{W}\sum_{l=-\infty}^{\infty} u(l/W)\mathrm{e}^{-\mathrm{j}2\pi fl/W}, & |f| \le W/2 \\ 0. & \text{otherwise} \end{cases} \tag{4.44}$$

Using Eq. 4.12, it follows that

$$\begin{aligned} r(t) &= \int_{-\infty}^{\infty} C(t;f)U(f)\mathrm{e}^{\mathrm{j}2\pi ft}df + z(t) \\ &= \int_{-\infty}^{\infty} C(t;f)\frac{1}{W}\sum_{l=-\infty}^{\infty} u(l/W)\mathrm{e}^{\mathrm{j}2\pi f(t-l/W)}df + z(t) \\ &= \frac{1}{W}\sum_{l=-\infty}^{\infty} u(l/W)\left[\int_{-\infty}^{\infty} C(t;f)\mathrm{e}^{\mathrm{j}2\pi f(t-l/W)}df\right] + z(t) \\ &= \frac{1}{W}\sum_{l=-\infty}^{\infty} u(l/W)c(t;(t - l/W)) + z(t) \\ &= \frac{1}{W}\sum_{l=-\infty}^{\infty} u(t - l/W)c(t;l/W) + z(t) \end{aligned}$$

$$= \sum_{l=-\infty}^{\infty} u(t - l/W) \left[\frac{1}{W} c(t; l/W) \right] + z(t), \tag{4.45}$$

where we have added a thermal noise term, $z(k)$, and where a simple change of variables was used in going from step 4 to step 5.

If we now assume that the channel impulse response duration is T_m, then the number of symbol periods in T_m, L, is given by $L \triangleq T_m/T_s$. It follows that $c \neq 0$ only for $l = 0, 1, \ldots L-1$, so Eq. 4.45 becomes

$$r(t) = \sum_{l=0}^{L-1} u(t - l/W) \left[\frac{1}{W} c(t; l/W) \right] + z(t). \tag{4.46}$$

We now consider discrete time instants $t_k = k/W$, $k = 0, 1, \ldots,$ and establish the following definitions:

$$s(k-l) \triangleq u\left((k-l)/W\right), \tag{4.47}$$

$$h^{(l)}(k) \triangleq \frac{1}{W} c(k/W; l/W), \tag{4.48}$$

and

$$z(k) \triangleq z(k/W). \tag{4.49}$$

Substituting these definitions into Eq. 4.46 results in

$$r(k) = \sum_{l=0}^{L-1} h^{(l)}(k) s(k-l) + z(k). \tag{4.50}$$

This equation is the system equation for a SISO system operating in a frequency-selective fading environment. We note that, unlike Eq. 4.41 for flat fading, in frequency-selective fading the effect of the channel is not a simple complex gain applied to each symbol.

The expression in Eq. 4.50 does not impose any constraints on the variances of the individual components. For simulation purposes, however, it is often convenient to assume the following normalizations:

$$\sum_{l=0}^{L-1} |h^{(l)}(k)|^2 = 1, \tag{4.51}$$

$$\mathbb{E}\left\{|s(k)|^2\right\} = 1, \tag{4.52}$$

$$\mathbb{E}\left\{|z(k)|^2\right\} = 1. \tag{4.53}$$

When this is done, Eq. 4.50 must be rewritten as follows:

$$r(k) = \sqrt{\rho} \sum_{l=0}^{L-1} h^{(l)}(k) s(k-l) + z(k). \tag{4.54}$$

Table 4.3 Definitions of slow and fast fading.

Channel type	Definition
Slow fading	$T_s < T_c$ and $B_d < W$
Fast fading	$T_s > T_c$ and $B_d > W$

For a MIMO system, the normalizations would be the same, except Eq. 4.52 is replaced by $\mathbb{E}\left\{|s(k)|^2\right\} = 1/N_t$. Under that assumption, the system equations in frequency-selective fading become

$$r_i(k) = \sqrt{\rho}\sum_{j=1}^{N_t}\sum_{l=0}^{L-1} h_{i,j}^{(l)}(k)s_j(k-l) + z_i(k), \quad i = 1, \ldots, N_r, \tag{4.55}$$

where $\{h_{i,j}^{(l)}(k)\}$ denotes the set of channel impulse response values on the path between transmit antenna j and receive antenna i at time k. We use this form of the system equation in Chapter 9 when we talk about the use of MIMO in frequency-selective channels.

4.5.3 Slow and fast fading

DEFINITION *When the channel changes slowly relative to the rate at which the signal varies, the channel is said to undergo slow fading. Conversely, if the channel changes rapidly compared to the signal, the channel is said to undergo fast fading.*

Since the rate that a channel changes is measured in terms of the reciprocal of its coherence time and the rate that a signal changes is equivalent to its baud rate (i.e., $1/T_s$), it follows from the definition that slow fading implies $1/T_c < 1/T_s$. Furthermore, since the bandwidth of a signal $W \approx 1/T_s$ and the Doppler bandwidth is proportional to $1/T_c$, slow fading is also equivalent to saying that $B_d < W$. Table 4.3 summarizes the definitions of slow and fast fading in terms of the intrinsic parameters of a fading channel.

4.6 Statistics of small-scale fading

In this section, we consider the statistics of small-scale fading in a flat fading environment. We show that small-scale fading often exhibits Rayleigh or Rician behavior. In this section, we define what this means and why the fading often exhibits these statistics.

4.6.1 Rayleigh fading

Rayleigh fading often occurs in multipath environments where there is no direct path between the transmitter and receiver. By Rayleigh fading, we mean that the magnitude of the channel response (i.e., $|h(k)|$) is a Rayleigh random variable. As we will see, this fact follows naturally from the mathematical definition of $h(k)$.

To see this, recall that for flat fading, $h(k) = C(t_k;f = 0)$ (see discussion leading up to Eq. 4.41). Furthermore, Eqs. 4.24 and 4.13 show that $C(t_k;f = 0) = \int_{-\infty}^{\infty} c(t_k;\tau)d\tau$ and that $c(t_k;\tau) = \sum_i \alpha(t_k)e^{-j2\pi f_c \tau_i(t_k)}\delta(\tau - \tau_i(t_k))$, respectively. It follows that

$$
\begin{aligned}
h(k) &= C(t_k;f = 0) \\
&= \int_{-\infty}^{\infty} \sum_i \alpha_i(t_k)e^{-j2\pi f_c \tau_i(t_k)}\delta(\tau - \tau_i(t_k))d\tau \\
&= \sum_i \int_{-\infty}^{\infty} \alpha_i(t_k)e^{-j2\pi f_c \tau_i(t_k)}\delta(\tau - \tau_i(t_k))d\tau \\
&= \sum_i \alpha_i(t_k)e^{-j2\pi f_c \tau_i(t_k)} \\
&= \left[\sum_i \alpha_i(t_k)\cos(2\pi f_c \tau_i(t_k))\right] - j\left[\sum_i \alpha_i(t_k)\sin(2\pi f_c \tau_i(t_k))\right].
\end{aligned}
\tag{4.56}
$$

Denoting the first bracketed term by X and the second by Y, we can write $|h(k)|$ as

$$
|h(k)| = \sqrt{X^2 + Y^2}. \tag{4.57}
$$

If we can show that X and Y are IID zero-mean Gaussian random variables with equal variances, σ^2, then, by definition, $|h(k)|$ is a Rayleigh random variable with a probability density function, $f_{|h(k)|}(x)$, given by

$$
f_{|h(k)|}(x) = \frac{x}{\sigma^2}e^{-x^2/2\sigma^2} \tag{4.58}
$$

To show that X and Y are IID zero-mean Gaussian random variables, we first note from the fourth equality in Eq. 4.56 that $h(k)$ is a sum of random complex phasors that have random amplitudes and random phase angles $\theta_i \triangleq 2\pi f_c \tau_i$. Since there is no reason to assume that these angles have preferential values, we can assume that the phase is uniformly distributed; hence, $h(k)$ is a sum of IID circularly symmetric complex random variables. It follows from the central limit theorem that if the number of scatterers is large (in practice > 10), then $h(k)$ is a complex Gaussian random variable. Furthermore, the circular symmetry implies that the real and imaginary components of $h(k)$ each have zero mean. Since $h(k)$ is a complex Gaussian random variable, its real and imaginary components are jointly Gaussian. The final step in the argument is to appeal to the fact that uncorrelated jointly Gaussian random variables are independent.[1] Therefore, if we can show that the two terms in the rectangular brackets above are uncorrelated, it

[1] It is sometimes incorrectly stated without qualification that uncorrelated Gaussian random variables are independent. In general, however, that is not true. It is only true if the Gaussian random variables are jointly Gaussian, meaning that they are individually Gaussian *and* they have a joint pdf that is Gaussian. It is possible to find examples where random variables are individually Gaussian but their joint pdf is not

follows that $|h(k)|$ is the square root of the sum of the squares of two independent zero-mean Gaussian random variables, which is the definition of Rayleigh random variables. Since it is possible to show that the two terms in brackets are, in fact, uncorrelated (see Problem 4.7 at end of chapter), we conclude that $h(k)$ is Rayleigh.

It is possible to express $f_{|h(k)|}(x)$ in terms of the transmission gain (or attenuation) of the channel. To do this, we start by noting that $|h(k)|^2 = X^2 + Y^2$. Since we have argued that the means of X and Y are zero, it follows that the variance of $h(k)$, which we denote by σ_h^2, can be expressed as

$$\begin{aligned}\sigma_h^2 &= \mathbb{E}\left\{|h(k)|^2\right\} \\ &= \mathbb{E}\left\{X^2\right\} + \mathbb{E}\left\{Y^2\right\} \\ &= 2\sigma^2. \end{aligned} \tag{4.59}$$

Since we are considering flat fading, the non-normalized version of the system equation is given by Eq. 4.41. In the absence of thermal noise, it follows that $|r(k)|^2 = |h(k)|^2|s(k)|^2$. We can, therefore, write the received signal power, $P_r \triangleq \mathbb{E}\left\{|r(k)|^2\right\}$, in terms of the transmit power, $P_t \triangleq \mathbb{E}\left\{|s(k)|^2\right\}$, as follows:

$$\begin{aligned}P_r &= \mathbb{E}\left\{|h(k)|^2\right\}\mathbb{E}\left\{|s(k)|^2\right\} \\ &= 2\sigma^2 P_t, \end{aligned} \tag{4.60}$$

where we have used the fact that $h(k)$ and $s(k)$ are independent of each other in the first line, and the relationship given in Eq. 4.59 in the second line. Combining Eqs. 4.60 and 4.58 allows us to rewrite the pdf of $|h(k)|$ in the following form:

$$f_{|h(k)|}(x) = \frac{2x}{(P_r/P_t)} \mathrm{e}^{-x^2/(P_r/P_t)} \tag{4.61}$$

$$= \frac{2x}{G} \mathrm{e}^{-x^2/G}, \tag{4.62}$$

where $G \triangleq P_r/P_t$ denotes the gain of the channel (or the reciprocal of the path loss).

4.6.2 Rician fading

Rician fading occurs when the received signal consists of two components: a random portion that fluctuates in time and a non-fluctuating part. One way in which this might occur is if some of the transmitted signal energy propagates directly to the receiver and some of it gets scattered off objects along the path between the transmitter and receiver. Under Rician fading conditions, it can be shown that the pdf of $|h(k)|$ has the following form:

$$f_{|h|}(x) = \frac{2x(K+1)}{G} \exp\left(-K - \frac{(K+1)\,x^2}{G}\right) I_0\left(2x\sqrt{\frac{K(K+1)}{G}}\right), \quad x > 0 \tag{4.63}$$

Gaussian. This is why we have gone to the trouble of arguing that X and Y have a 2-D Gaussian pdf (i.e., that they are jointly Gaussian).

where $I_0(\cdot)$ is the *zero*th order modified Bessel function of the first kind, and K is called the Rician K-factor, which is defined as

$$K \triangleq \frac{[\text{Power in direct component}]}{[\text{Power in scattered component}]}. \tag{4.64}$$

It follows from this definition that the Rician distribution should become identical to the Rayleigh distribution when $K \to 0$ since, in that limit, all the received energy would be associated with scattering, which is the assumption underlying Rayleigh fading. Because $I_0(0) = 1$, it is clear that in this limit Eq. 4.63 does, in fact, become equivalent to the Rayleigh distribution as defined in Eq. 4.62.

Problems

4.1 Prove Eq. 4.12.

4.2 Prove Eq. 4.27. [*Hint*: To do so, start with the right side of Eq. 4.26 and use the definition in Eq. 4.24.]

4.3 Prove that the integral of the Clark Doppler spectrum given in Eq. 4.34 is a constant equal to 1.5. (*Note*: The value of 1.5 is the theoretical gain of a quarter-wave vertical whip antenna over an ideal ground plane, which is the type of antenna assumed in the derivation of the Clark Doppler spectrum.)

4.4 Consider a radio in a vehicle traveling at 25 mph that receives a signal from a stationary transmitter with a carrier frequency equal to 2 GHz. Assume that this vehicle is surrounded by scatterers that can be approximated to be uniformly distributed in azimuth.

a) What is the maximum Doppler spread that will be observed by the receiver?
b) Compute and plot the Doppler spectrum assuming that the scatterers are uniformly surrounding the transmitter in a circle.
c) What is the median value of the coherence time of this channel?
d) What is the minimum data rate for which slow fading can be said to occur?

4.5 Prove Eq. 4.44.

4.6 Prove that when the normalizations defined in Eqs. 4.51, 4.52, and 4.53 are assumed, Eq. 4.50 must be rewritten in the form shown in Eq 4.54.

4.7 The correlation between two random variables, X and Y, is defined as follows:

$$\mathrm{Corr}(X, Y) = \frac{\mathbb{E}\left\{(X - \mu_X)(Y - \mu_Y)^*\right\}}{\sigma_X \sigma_Y},$$

where μ_X and μ_Y denote the mean values of X and Y, respectively, and $*$ denotes complex conjugation. Use this definition to prove that the two bracketed terms in Eq 4.56 are uncorrelated, and, hence, that they are independent Gaussian random variables.

5 MIMO channel models

A common assumption in MIMO analyses is that the channel matrix consists of independent identically distributed complex Gaussian gains (i.e., $\mathbf{H} = \mathbf{H}_w$). In the real world, however, this is not always the case. In particular, correlation between the received or emitted signals to or from antenna pairs, or the existence of a direct LOS component in the signal at each receive antenna causes $\mathbf{H} \neq \mathbf{H}_w$. In this chapter, we introduce several analytical models for $\mathbf{H}$ that incorporate the effects of antenna correlation and the impact of LOS propagation. These models often appear in the MIMO literature and provide a convenient means to compute the impact of antenna correlation and Rician fading (as opposed to pure Rayleigh fading) on the capacity of a MIMO system.

5.1 MIMO channels in LOS geometry[1]

We begin this chapter by considering the case where the transmit and receive antenna arrays are within line-of-sight of each other and there is no scattering in the channel, as illustrated in Figure 5.1. Under this assumption, we seek a mathematical expression for $\mathbf{H}$ and a criterion that assures a high MIMO capacity, despite the fact that scattering is normally assumed to be required to support spatial multiplexing.

Denote the channel response between the kth transmit antenna and the receive array by $\mathbf{h}_k$, which we call the *signature vector* associated with the kth transmit antenna. We assume that the distance between the transmit and receive arrays is much larger than the dimension of either of the antenna arrays. Under this assumption, it is appropriate to model the received signal as a plane wave that arrives with an incidence angle equal to θ_k. The channel has three effects on the signal: first, it attenuates it by an amount equal to the free space loss; secondly, it introduces a common phase shift due to the distance D; and thirdly, it introduces a relative phase shift between the receive antennas due to the angle of arrival. We denote the free space attenuation by $\alpha = (4\pi fD/c)^2$ and the common phase shift by $\phi = 2\pi D/\lambda$. The relative phase shift at the rth receive antenna from the kth transmit antenna is $\phi_r = -2\pi(r-1)\sin(\theta_k)d_r/\lambda$. It follows that

$$\mathbf{h}_k = \frac{e^{-j\phi}}{\alpha}\left[1\ e^{-j2\pi\sin(\theta_k)d_r/\lambda}\ \ldots e^{-j2\pi(N_r-1)\sin(\theta_k)d_r/\lambda}\right]^T. \tag{5.1}$$

[1] This section is based on material in [34].

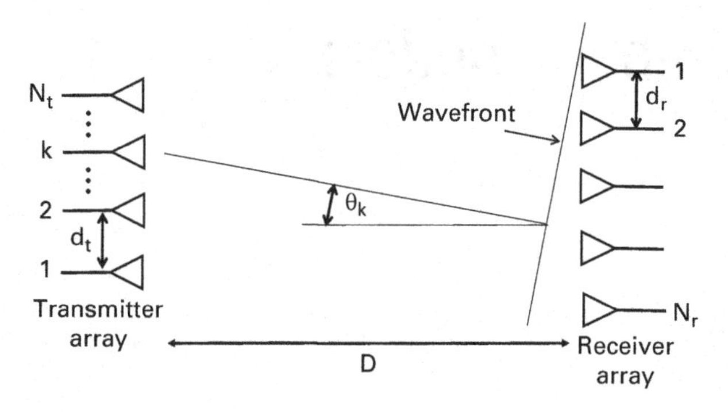

Figure 5.1 LOS MIMO geometry.

Therefore, the full channel matrix, $\mathbf{H}$, can be expressed as

$$\mathbf{H} = \left[\mathbf{h}_1\mathbf{h}_2 \ldots \mathbf{h}_{N_t}\right]. \tag{5.2}$$

It follows from Eqs. 5.1 and 5.2 that when θ_k approaches zero, the elements in $\mathbf{H}$ approach the same value, which results in $\mathbf{H}$ having rank 1 in that limit. This condition occurs when D is large. When this happens the capacity of the MIMO system approaches that of a SISO system, as we saw in Chapter 3. But are there conditions in a LOS configuration where the rank of $\mathbf{H}$ is large and maximum spatial multiplexing is still possible? The answer is yes; although the conditions under which this is possible are rather restricted, as we see below.

Intuitively, one expects that when the signature vectors for adjacent pairs of transmit antennas are orthogonal, the receiver should be able to decouple the signals from the adjacent antennas, resulting in the ability to transmit separate streams from each of the antennas in the pair over the same bandwidth. This suggests that spatial multiplexing may be possible as long as the correlation between all adjacent signature vectors is zero. That is, the high-capacity criterion is given by

$$\mathbb{E}\left\{\mathbf{h}_{k+1}^H\mathbf{h}_k\right\} = \sum_{m=0}^{N_r-1} e^{j2\pi[\sin(\theta_{k+1})-\sin(\theta_k)]m(d_r/\lambda)} = 0, \ \forall k = 1, \ldots, N_t - 1. \tag{5.3}$$

For practical values of D, d_t, and d_r, orthogonality occurs for small θ_k. Under that assumption, $\sin\theta_k \approx (k-1)d_t/D$ $(k = 1, 2, \ldots, N_t - 1)$ and Eq. 5.3 becomes

$$\sum_{m=0}^{N_r-1} e^{j2\pi m(d_t d_r/(\lambda D))} = 0. \tag{5.4}$$

It follows that the maximum value of D for which this equality holds is given by

$$D_{\max} = \frac{N_r d_t d_r}{\lambda}. \tag{5.5}$$

It should be kept in mind that the high-capacity criterion is based on ensuring that the receiver is able to separate the received signals from adjacent transmit antennas, which may not be sufficient to ensure that the entire channel matrix has full rank. Therefore, this criterion is a necessary but not sufficient condition for full rank. It is, however, a useful rule-of-thumb for determining if a LOS channel has full rank.

Example 5.1 Let $f = 3$ GHz, $d_t = d_r = 1$ m, and $N_r = 4$. What is the maximum value of D for which full rank is possible?

Answer $D_{\max} = N_r d_t d_r/\lambda = N_r d_t d_r f/c = 4 \times 1 \times 13 \times 10^9/3 \times 10^8 = 40$ meters.

This section shows that in principle it is possible to achieve a high rank channel, and potentially high spatial multiplexing gain, even in a non-scattering LOS environment; however, the maximum communication ranges over which this is possible are restricted to short distances, seldom exceeding a few tens of meters.

5.2 General channel model with correlation

In this section, we consider the most general form for an analytical model of a random MIMO channel matrix that explicitly incorporates correlation between pairs of antennas. This, of course, is in contrast to the form of $\mathbf{H}$ described in the previous section, which does not include either random effects or correlation, as well as to $\mathbf{H}_w$, which, while random, does not include correlation.

In general, the correlation properties between pairs of MIMO antennas are defined by the MIMO covariance matrix, $\mathbf{R}$, which is expressed mathematically as follows:

$$\mathbf{R} \triangleq \mathbb{E}\left\{\text{vec}(\mathbf{H})\text{vec}(\mathbf{H})^H\right\}, \tag{5.6}$$

where vec($\mathbf{H}$) denotes a column vector consisting of the columns of $\mathbf{H}$ concatenated on top of each other.

For readers not familiar with this technique, it is useful to consider a specific example. Assume that $\mathbf{H}$ is a generic 3×2 matrix as follows:

$$\mathbf{H} = \begin{pmatrix} h_{11} & h_{12} \\ h_{21} & h_{22} \\ h_{31} & h_{32} \end{pmatrix}. \tag{5.7}$$

It then follows that

$$\mathbf{R} = \mathbb{E}\left\{\text{vec}(\mathbf{H})\text{vec}(\mathbf{H})^H\right\}$$

$$= \mathbb{E}\left\{\begin{pmatrix} h_{11} \\ h_{21} \\ h_{31} \\ h_{12} \\ h_{22} \\ h_{32} \end{pmatrix} \begin{pmatrix} h_{11}^* & h_{21}^* & h_{31}^* & h_{12}^* & h_{22}^* & h_{32}^* \end{pmatrix}\right\}$$

$$= \mathbb{E}\left\{\begin{pmatrix} |h_{11}|^2 & h_{11}h_{21}^* & h_{11}h_{31}^* & h_{11}h_{12}^* & h_{11}h_{22}^* & h_{11}h_{32}^* \\ h_{21}h_{11}^* & |h_{21}|^2 & h_{21}h_{31}^* & h_{21}h_{12}^* & h_{21}h_{22}^* & h_{21}h_{32}^* \\ h_{31}h_{11}^* & h_{31}h_{21}^* & |h_{31}|^2 & h_{31}h_{12}^* & h_{31}h_{22}^* & h_{31}h_{32}^* \\ h_{12}h_{11}^* & h_{12}h_{21}^* & h_{12}h_{31}^* & |h_{12}|^2 & h_{12}h_{22}^* & h_{12}h_{32}^* \\ h_{22}h_{11}^* & h_{22}h_{21}^* & h_{22}h_{31}^* & h_{22}h_{12}^* & |h_{22}|^2 & h_{22}h_{32}^* \\ h_{32}h_{11}^* & h_{32}h_{21}^* & h_{32}h_{31}^* & h_{32}h_{12}^* & h_{32}h_{22}^* & |h_{32}|^2 \end{pmatrix}\right\}. \tag{5.8}$$

This demonstrates two properties of $\mathbf{R}$: a) it has dimensions of $(N_tN_r) \times (N_tN_r)$; and b) its elements consist of the correlations between all pairs of antennas. Clearly, $\mathbf{R}$ contains all the correlation information inherent in a MIMO system.

Now that we have a term that captures the correlation properties, we seek to express $\mathbf{H}$ in terms of $\mathbf{R}$. We do this by first defining the following matrix:

$$\mathbf{g} \triangleq \mathbf{R}^{-1/2}\text{vec}(\mathbf{H}), \tag{5.9}$$

where $\mathbf{R}^{-1/2}$ denotes the square root of the inverse of $\mathbf{R}$. The square root of a matrix $\mathbf{A}$ is defined implicitly as follows: $\mathbf{A} = \mathbf{A}^{1/2}\mathbf{A}^{H/2}$. It follows from Eq. 5.9 that

$$\text{vec}(\mathbf{H}) = \mathbf{R}^{1/2}\mathbf{g}. \tag{5.10}$$

Since $\mathbf{R}$ is $N_tNr \times N_tN_r$, Eq. 5.10 implies that $\mathbf{g}$ is dimensioned $N_tN_r \times 1$. We now define a new matrix $\mathbf{G}$ dimensioned $N_r \times N_t$, where

$$\mathbf{g} \triangleq \text{vec}(\mathbf{G}). \tag{5.11}$$

Thus, we can rewrite Eq. 5.10 in terms of $\mathbf{G}$ as follows:

$$\text{vec}(\mathbf{H}) = \mathbf{R}^{1/2}\text{vec}(\mathbf{G}). \tag{5.12}$$

Next, we examine the covariance matrix of $\mathbf{g}$. Denoting the covariance matrix of $\mathbf{g}$ by $\mathbf{R}_{gg}$, it follows that

$$\begin{aligned}\mathbf{R}_{gg} &\triangleq \mathbb{E}\left\{\mathbf{g}\mathbf{g}^H\right\} \\ &= \mathbb{E}\left\{\mathbf{R}^{-1/2}\text{vec}(\mathbf{H})\text{vec}(\mathbf{H})^H\mathbf{R}^{-H/2}\right\} \\ &= \mathbf{R}^{-1/2}\mathbb{E}\left\{\text{vec}(\mathbf{H})\text{vec}(\mathbf{H})^H\right\}\mathbf{R}^{-H/2} \\ &= \mathbf{R}^{-1/2}\mathbf{R}\mathbf{R}^{-H/2} \\ &= \mathbf{R}^{-1/2}\mathbf{R}^{1/2}\mathbf{R}^{H/2}\mathbf{R}^{-H/2} \\ &= \mathbf{I},\end{aligned} \tag{5.13}$$

which implies that the elements of $\mathbf{G}$ are uncorrelated. Since each of the elements of $\mathbf{G}$ is associated with different transmit and receive antenna combinations, the fact that the elements of $\mathbf{G}$ are uncorrelated means that $\mathbf{G}$ is spatially white.

The following theorem summarizes what we have just shown:

THEOREM 5.1 *Let* $\mathbf{H}$ *denote an* $N_r \times N_t$ *channel matrix where* $\mathbf{R} \triangleq \mathbb{E}\{vec(\mathbf{H})\, vec(\mathbf{H})^H\}$ *defines the overall covariance matrix of the MIMO channel. It follows that the channel matrix,* $\mathbf{H}$, *is related to* $\mathbf{R}$ *in the following way:*

$$\mathrm{vec}(\mathbf{H}) = \mathbf{R}^{1/2}\mathrm{vec}(\mathbf{G}), \tag{5.14}$$

where $\mathbf{G}$ *is spatially white (i.e.,* $\mathbb{E}\left\{[\mathbf{G}]_{ij}[\mathbf{G}]_{mn}\right\} = \delta_{i,m}\delta_{j,n}$*) and is dimensioned* $N_r \times N_t$.

Note that if the MIMO channel is uncorrelated (i.e., if $\mathbf{R} = \mathbf{I}_{N_r N_t}$) then Eq. 5.14 simplifies to

$$\mathbf{H} = \mathbf{G}. \tag{5.15}$$

It is important to keep in mind that Eq. 5.14 is completely general in that it applies to any random MIMO channel. It does not, however, imply anything about the probability distribution of the elements of $\mathbf{G}$, only that $\mathbf{G}$ is a spatially white matrix. If the MIMO channel undergoes Rayleigh fading, however, the elements of $\mathbf{H}$ are complex Gaussian and, therefore, so too are the elements of $\mathbf{G}$ since $\mathbf{G}$ is related to $\mathbf{H}$ through a linear transformation. We conclude that if $\mathbf{H}$ is Rayleigh, then $\mathbf{H}$ can be expressed as follows:

$$\mathrm{vec}(\mathbf{H}) = \mathbf{R}^{1/2}\mathrm{vec}(\mathbf{H}_w). \tag{5.16}$$

Similarly, if the channel is uncorrelated, then $\mathbf{H} = \mathbf{H}_w$.

In this section we have considered the most general form for $\mathbf{H}$ that includes the effects of correlation. In the next section, we consider a slightly more restricted set of assumptions that results in a commonly used channel model called the Kronecker model.

5.3 Kronecker channel model

The channel model in Eq. 5.14 has the advantage of being universally applicable and of explicitly incorporating the effects of correlation; however, it has the disadvantage of being relatively complex since it requires $(N_t N_r)^2/2$ unique correlation values to characterize $\mathbf{R}$. An alternative model that is often used in MIMO literature because of its comparative simplicity is the Kronecker channel model. Rather than using a single covariance matrix, the Kronecker model employs separate covariance matrices for the transmitter and receiver, denoted by $\mathbf{R}_t$ and $\mathbf{R}_r$, respectively, which have the following forms:

$$\mathbf{R}_t = \begin{pmatrix} \rho_{t_{1,1}} & \cdots & \rho_{t_{N_t,1}} \\ \vdots & \ddots & \vdots \\ \rho_{t_{N_t,1}} & \cdots & \rho_{t_{N_t,N_t}} \end{pmatrix} \tag{5.17}$$

and

$$\mathbf{R}_r = \begin{pmatrix} \rho_{r_{1,1}} & \cdots & \rho_{r_{N_r,1}} \\ \vdots & \ddots & \vdots \\ \rho_{r_{N_r,1}} & \cdots & \rho_{r_{N_r,N_r}} \end{pmatrix}. \tag{5.18}$$

The elements of these matrices are defined as follows:

$$\rho_{t_{m,j}} \triangleq \mathbb{E}\left\{h_{k,j}h^*_{k,m}\right\} \; k = 1, \ldots, N_t \tag{5.19}$$

and

$$\rho_{r_{i,n}} \triangleq \mathbb{E}\left\{h_{i,k}h^*_{n,k}\right\} \; k = 1, \ldots, N_r. \tag{5.20}$$

Thus, the elements of $\mathbf{R}_t$ are the correlations between all pairs of transmit antennas at receiver k, and the elements of $\mathbf{R}_r$ are the correlations between all pairs of receive antennas when the signal is emitted by transmitter k. It follows from the definitions in Eq. 5.19 and 5.20 that the top and bottom halves of the $\mathbf{R}_r$ and $\mathbf{R}_t$ matrices are simply the complex conjugates of each other. That is,

$$\rho_{t_{m,j}} = \rho^*_{t_{j,m}} \text{ and } \rho_{r_{i,n}} = \rho^*_{r_{n,i}}. \tag{5.21}$$

A key assumption in the Kronecker model is that the correlations in Eqs. 5.19 and 5.20 are independent of k. Under this assumption, the following theorem holds:[2]

THEOREM 5.2 *Let covariance matrices $\mathbf{R}_t$ and $\mathbf{R}_r$ be defined according to Eqs. 5.17–5.20, and let $\mathbf{G}$ denote an $N_r \times N_t$ spatially white matrix (i.e., $\mathbb{E}\left\{[\mathbf{G}]_{ij}[\mathbf{G}]_{mn}\right\} = \delta_{i,m}\delta_{j,n}$). It follows that*

$$\mathbf{H} = \mathbf{R}_r^{1/2}\mathbf{G}\mathbf{R}_t^{H/2} \iff \mathbf{R} = \mathbf{R}_r \otimes \mathbf{R}_t^T, \tag{5.22}$$

where $\iff$ denotes "if and only if" (i.e., iff) and $\otimes$ denotes the Kronecker product (see Section 1.9.1 for definition).[3]

This theorem states that if the channel matrix is assumed to have the form $\mathbf{H} = \mathbf{R}_r^{1/2}\mathbf{G}\mathbf{R}_t^{H/2}$ with $\mathbf{G}$ spatially white, it follows that the overall channel covariance matrix defined in Eq. 5.6 is equal to the Kronecker product of the receive and transmit covariance matrices. It also states that if the channel covariance matrix can be expressed as $\mathbf{R} = \mathbf{R}_r \otimes \mathbf{R}_t^T$, it follows that the channel matrix must have the form $\mathbf{H} = \mathbf{R}_r^{1/2}\mathbf{G}\mathbf{R}_t^{H/2}$, where $\mathbf{G}$ is spatially white. It should be emphasized that it is always possible to write the channel matrix in the form $\mathbf{H} = \mathbf{R}_r^{1/2}\mathbf{G}\mathbf{R}_t^{H/2}$ if $\mathbf{G}$ is not constrained to be white simply by defining $\mathbf{G} = \mathbf{R}_r^{-1/2}\mathbf{H}\mathbf{R}_t^{-H/2}$. The significance of this theorem is that it states that $\mathbf{G}$ is white as long as the full covariance matrix can be written as the Kronecker product shown above.

This theorem shows why writing the channel in the form $\mathbf{H} = \mathbf{R}_r^{1/2}\mathbf{G}\mathbf{R}_t^{H/2}$ is referred to as the Kronecker model of the channel. The MIMO literature is full of references

[2] See Appendix B for a proof of Theorem 5.2.

[3] This theorem is also true if the equation on the left side of the *iff* symbol in Eq. 5.22 is replaced by $\mathbf{H} = \mathbf{R}_r^{1/2}\mathbf{G}\mathbf{H}_t^{1/2}$. Some authors use this alternative form of the theorem.

Table 5.1 Comparison between complexity in general MIMO covariance channel model and the Kronecker model.

N_t	N_r	$(N_tN_r)^2/2$	$(N_t^2+N_r^2)/2$
2	2	8	4
2	3	18	6.5
2	4	32	10
3	3	40.5	9
3	4	72	12.5
4	4	128	16
8	8	2048	64
16	16	32,768	256

to this model and there are empirical studies that show that practical channels have the Kronecker property. The first paper to propose the Kronecker model was by Kermoal *et al.* in 2002 [46]. This paper also presents measurements that validate the Kronecker model for a variety of picocell and microcell indoor environments. A second paper by Stridh *et al.* [69] also presents indoor propagation measurements that show the MIMO channel covariance matrix can be well approximated by the Kronecker product.

A special form of the Kronecker model occurs when, in addition to requiring **G** to be spatially white, the elements of **G** are also assumed to undergo Rayleigh fading (i.e., the elements of **G** are complex Gaussian). In that case, the Kronecker model simplifies to

$$\mathbf{H} = \mathbf{R}_r^{1/2}\mathbf{H}_w\mathbf{R}_t^{H/2}. \tag{5.23}$$

Note that when the transmitted and received signals are uncorrelated, $\mathbf{H} = \mathbf{H}_w$, which is the same as we saw for the general channel model in the previous section.

Because the Kronecker channel model implies that $\mathbf{R} = \mathbf{R}_r \otimes \mathbf{R}_t^T$, the complexity of the overall covariance matrix is reduced from $(N_tN_r)^2/2$ correlation values to only $(N_t^2 + N_r^2)/2$. Table 5.1 compares the complexity for various combinations of antenna numbers. These results show that for practical numbers of antennas, the complexity of the Kronecker model is significantly less than the general channel model given in Eq. 5.14. The Kronecker model is used throughout the MIMO literature and has been adopted as the official link-level channel model for characterizing and modeling performance in 3rd Generation Partnership Project (3GPP) MIMO systems [1], [76].

5.4 Impact of antenna correlation on MIMO capacity

As stated earlier, channel models are useful because they provide a convenient means of analytically estimating the impact that correlation and LOS propagation have on MIMO performance. In this section, we use the Kronecker model to examine the impact that antenna correlation has on MIMO capacity.

We begin by assuming that

$$\mathbf{H} = \mathbf{R}_r^{1/2}\mathbf{H}_w\mathbf{R}_t^{H/2}. \tag{5.24}$$

If we now assume CSIR only, substituting Eq. 5.24 into the expression for the MIMO capacity given in Eq. 3.9 yields the following:

$$\begin{aligned} C &= \log_2 \det\left(\mathbf{I} + \frac{\rho}{N_t}\mathbf{H}\mathbf{H}^H\right) \\ &= \log_2 \det\left(\mathbf{I} + \frac{\rho}{N_t}\mathbf{R}_r^{1/2}\mathbf{H}_w\mathbf{R}_t^{H/2}\mathbf{R}_t^{1/2}\mathbf{H}_w^H\mathbf{R}_r^{H/2}\right) \\ &= \log_2 \det\left(\mathbf{I} + \frac{\rho}{N_t}\mathbf{R}_r^{1/2}\mathbf{H}_w\mathbf{R}_t\mathbf{H}_w^H\mathbf{R}_r^{H/2}\right). \end{aligned} \tag{5.25}$$

It follows that in the limit as the signal-to-noise ratio gets large, the capacity becomes

$$\lim_{\rho\to\infty} C = \log_2 \det \frac{\rho}{N_t}\mathbf{H}_w\mathbf{H}_w^H + \log_2 \det \mathbf{R}_t + \log_2 \det \mathbf{R}_r, \tag{5.26}$$

where the step in going from Eq. 5.25 to Eq. 5.26 employs the property that the determinant of a product of matrices is equal to the product of the determinants of those matrices.

It is straightforward to show that when $\mathbb{E}\left\{|[\mathbf{H}]_{ij}|^2\right\} = 1$, which is the normalization that was introduced in Chapter 3 (see Eq. 3.2), it follows that

$$\det(\mathbf{R}_r) \text{ and } \det(\mathbf{R}_t) \leq 1, \tag{5.27}$$

with equality occurring only when the correlation matrices are diagonal (i.e., only when the antennas are uncorrelated). Equation 5.26 clearly demonstrates that antenna correlation reduces MIMO capacity. The amount that the capacity is reduced, ΔC, is equal to[4]

$$\Delta C = -\log_2 \det \mathbf{R}_t - \log_2 \det \mathbf{R}_r. \tag{5.28}$$

Example 5.2 Consider a 2 × 2 MIMO system where there is correlation between antennas at the receiver but no corrrelation at the transmitter. Thus, $\mathbf{R}_t = \mathbf{I}_{N_t}$ and

$$\mathbf{R}_r = \begin{pmatrix} 1 & \rho_r \\ \rho_r^* & 1 \end{pmatrix}. \tag{5.29}$$

What is the degradation in capacity caused by this correlation when $|\rho_r| = 0.8$?

Answer $\Delta C = \log_2 \det \mathbf{R}_r = \log_2\left(1 - |\rho_r|^2\right) = \log_2(1 - 0.64) = \log_2(0.36) = -1.47$ bps/Hz.

[4] It should be noted that the expression for the degradation in capacity is the same regardless of the form of the Kronecker model (i.e., when $\mathbf{H} = \mathbf{R}_r^{1/2}\mathbf{H}_w\mathbf{G}\mathbf{H}_t^{H/2}$ as well as when $\mathbf{H} = \mathbf{R}_r^{1/2}\mathbf{H}_w\mathbf{H}_t^{1/2}$).

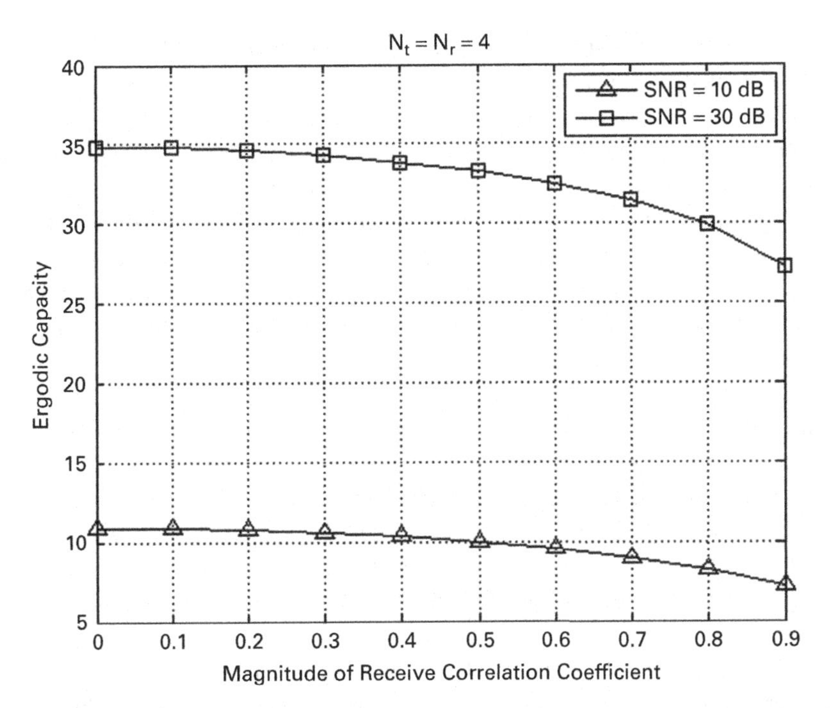

Figure 5.2 Ergodic capacity of a 4 × 4 MIMO system in Rayleigh fading as a function of $|\rho_r|$. The transmit antennas in this plot are assumed to be perfectly uncorrelated.

Figure 5.2 illustrates the sensitivity of MIMO capacity to antenna correlation. This figure shows the ergodic capacity of a 4 × 4 MIMO system operating in Rayleigh fading plotted versus $|\rho_r|$; where ρ_r denotes the off-diagonal elements of $\mathbf{R}_r$, which are assumed to have equal magnitudes. Curves are shown assuming SNR = 10 and 30 dB. These results show that relatively little capacity reduction occurs as long as the correlation is less than 0.5. Equation 5.28 predicts that the capacity reduction should be 8.1 bps/Hz when $|\rho_r| = 0.9$ and 1.7 dB when $|\rho_r| = 0.5$. The curve at SNR = 30 dB appears to match these predictions quite closely; however, the curve at SNR = 10 dB shows capacity reductions that are smaller than these predicted values. The reason for this is that Eq. 5.28 is valid only when ρ is large (i.e., $\rho \gtrsim 30$ dB). For smaller values of ρ, the predicted capacity reduction will be larger than what actually occurs. For this reason, Eq. 5.28 should be regarded as an upper limit on the capacity reduction.

5.5 Dependence of $\mathbf{R}_t$ and $\mathbf{R}_r$ on antenna spacing and scattering angle

Up to this point in our discussion we have treated the transmit and receive covariance matrices in the Kronecker model in largely an abstract manner in the sense that we have not attempted to explain how their properties depend on the physical parameters of the MIMO system. In this section, we consider the dependence of $\mathbf{R}_t$ and $\mathbf{R}_r$ on two key factors: the spacing of the antennas and the scattering angles at the transmitter and receiver.

As a general rule, $\mathbf{R}_t$ and $\mathbf{R}_r$ become more diagonal (i.e., the antennas become more uncorrelated) as either the antenna spacing or the scattering angle increases. This makes sense from an intuitive standpoint. As the antennas become more closely spaced, electromagnetic coupling increases and the differences in the channel paths associated with separated antennas decrease. This effect tends to increase the correlation between pairs of antennas, particularly adjacent pairs. In addition, as the angle over which radio waves arrive at the receiver (referred to as angle-of-arrival, abbreviated AoA) increases, multipath causes the signal to appear increasingly dissimilar at different antennas. At the opposite extreme, when AoA = 0°, the signal arrives at the receive array as a plane wave moving perpendicular to the array. When this happens the signal at each receive antenna is the same, so the antennas are perfectly correlated as long as they are spaced far enough apart that the electromagnetic coupling is negligible. Similar logic would lead one to conclude that when the angle-of-departure (AoD) approaches zero degrees, the transmitted signals also become perfectly correlated. The AoD = AoA = 0° condition corresponds to the LOS geometry analyzed in Section 5.1. As we saw in that discussion, when the arrival direction is perpendicular to the axis of the receive array, the receive signature vectors become identical, which results in perfect correlation between receive antenna pairs.

A relatively simple expression for $\mathbf{R}_r$ occurs when the scattered energy at the receiver is assumed to arrive over a uniformly distributed range of angles. In this model, which is depicted in Figure 5.3, S randomly located scatterers are assumed to be located in the vicinity of the receiver causing the energy from the transmitter to arrive over a total range of angles denoted by θ_r. Under this assumption, it can be shown (see [8] and [22]) that

$$[\mathbf{R}_r(\theta_r, d_r)]_{m,k} = \frac{1}{S} \sum_{i=(S-1)/2}^{(S-1)/2} \mathrm{e}^{-\mathrm{j}2\pi(k-m)d_r \cos(\pi/2+\theta_{r,i})}, \tag{5.30}$$

where $\theta_{r,i}$ denotes the angle from the ith scatterer to the receive array and d_r denotes the spacing between receive antennas.

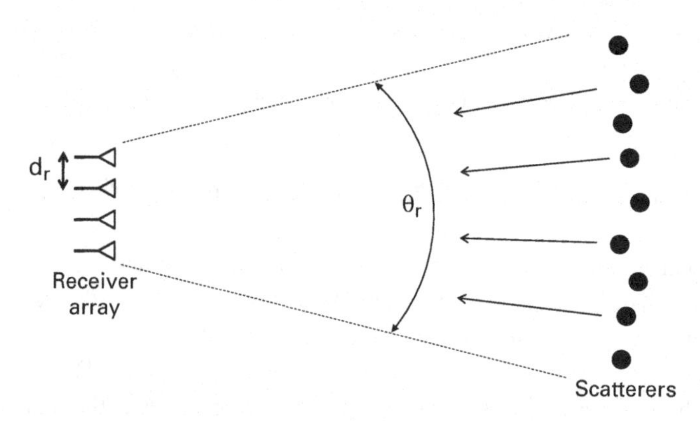

Figure 5.3 Uniform scattering geometry assumed in Eq. 5.30.

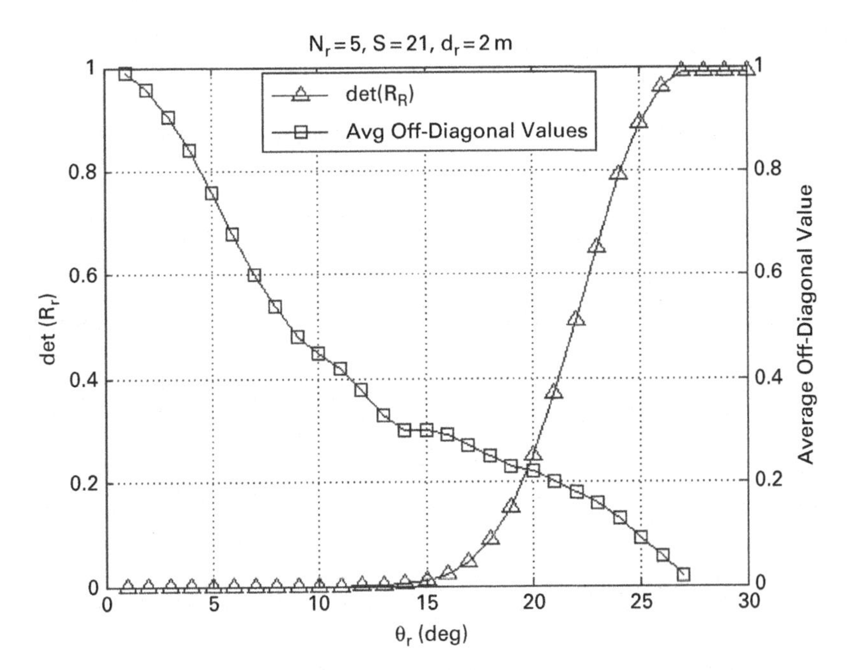

Figure 5.4 Variation of $\det \mathbf{R}_r$ and the average off-diagonal values of $\mathbf{R}_r$ with the scattering angle θ_r based on Eq. 5.30. This plot assumes $N_r = 5,\ S = 21,$ and $d_r = 2$ m.

Although it may not be obvious from Eq. 5.30, it can easily be demonstrated that $\mathbf{R}_r(\theta_r, d_r)$ becomes diagonal when d_r or θ_r are large, and becomes increasingly non-diagonal as these parameters become small. Figures 5.4 and 5.5 show the results of a computer simulation that demonstrates these properties. In the simulation, $\mathbf{R}_r$ was computed for different values of θ_r and d_r using Eq. 5.30. In the simulation the numbers of receive antennas and scatterers were arbitrarily chosen to be 5 and 21, respectively. For each $\mathbf{R}_r$ matrix, its determinant was computed (shown by the curves with triangle symbols and specified by the left axis) as well as the average value of the magnitude of its off-diagonal elements (shown by the curves with square symbols and right axis). The average off-diagonal values show that $\mathbf{R}_r$ becomes increasingly diagonal (i.e., the average values of the off-diagonal elements decrease) as the scattering angle increases (for a given antenna spacing), or as the antenna spacing increases (for a given scattering angle). The simulation also shows the effect that varying θ_r and d_r has on the value of $\det(\mathbf{R}_r)$. We see, as expected, that $\det(\mathbf{R}_r)$ is small when θ_r or d_r is small, which results in a large capacity reduction according to Eq. 5.28.

5.6 Pinhole scattering

Another type of MIMO channel that is often referred to in MIMO literature and that models certain types of conditions found in practice is the so-called *pinhole channel* (sometimes also referred to as the keyhole channel). The pinhole condition occurs when

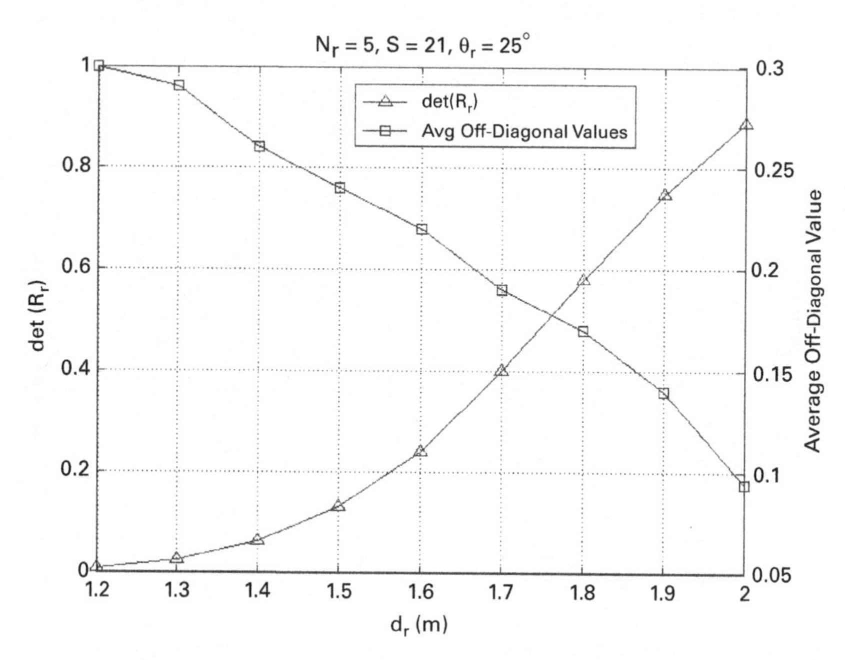

Figure 5.5 Variation of the det $\mathbf{R}_r$ and the average off-diagonal values of $\mathbf{R}_r$ with the antenna spacing d_r based on Eq. 5.30. This plot assumes $N_r = 5,\ S = 21,$ and $\theta_r = 25°$.

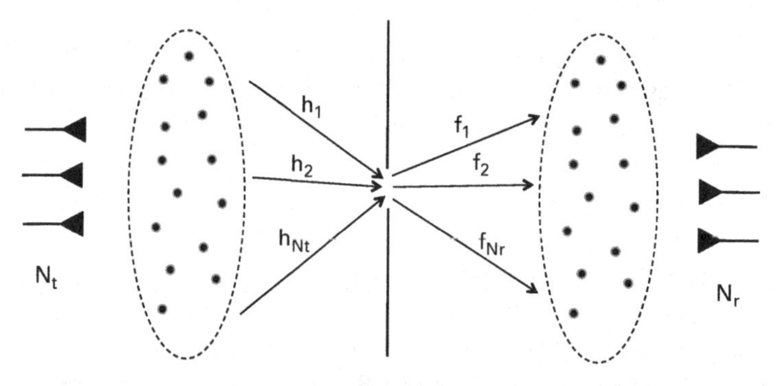

Figure 5.6 A pinhole channel.

the MIMO transmitter and receiver are each surrounded by local scatterers but the path between the transmitter and receiver arrays is blocked except for a relatively small opening – the pinhole. A depiction of a pinhole channel is shown in Figure 5.6.

The pinhole condition can occur when cell phones or user equipment (UE) devices, which are located indoors, communicate with a cell tower. In that situation, if the walls of the building are relatively opaque to the signal, then most of the RF energy flowing between the transmitter and receiver will pass through small openings such as windows or doors, thus mimicking the geometry shown in Figure 5.6.

The analysis of the pinhole channel is straightforward. We begin by assuming that the scattering near the transmitter prevents a LOS path between the transmit array and the pinhole; thus, the channel response from each transmit antenna to the pinhole aperture is Rayleigh. Similarly, the channel response from the pinhole to each of the receive antennas is also Rayleigh. If we denote the transmitted signal vector by $\mathbf{s} = \left[s_1, s_2, \ldots, s_{N_t}\right]^T$ and the channel response between the jth transmit antenna and the pinhole by h_j, then the received signal at the pinhole, r_p, is

$$\begin{aligned} r_p &= [h_1, h_2, \ldots, h_{N_t}]\mathbf{s} \\ &= \mathbf{h}_t\mathbf{s}. \end{aligned} \tag{5.31}$$

Similarly, if we denote the channel response between the pinhole and the ith receive antenna by f_i, then the received signal vector at the receive array, $\mathbf{r} = [r_1, r_2, \ldots, r_{N_t}]^T$, is

$$\mathbf{r} = \begin{bmatrix} f_1 \\ f_2 \\ \vdots \\ f_{N_r} \end{bmatrix} r_p = \mathbf{h}_r\mathbf{h}_t\mathbf{s}, \tag{5.32}$$

where $\mathbf{h}_r \triangleq [f_1, f_2, \ldots, f_{N_r}]^T$. It follows that the overall pinhole channel response, $\mathbf{H}_p$, is given by

$$\mathbf{H}_p = \mathbf{h}_r\mathbf{h}_t. \tag{5.33}$$

Since $\mathbf{h}_r$ and $\mathbf{h}_t$ are independent Rayleigh vectors, the statistics of the overall pinhole channel is said to be *double Rayleigh.*

How does the MIMO performance of a pinhole channel compare with the performance of a pure Rayleigh channel? The answer to this question can be at least qualitatively ascertained by appealing to a simple intuitive argument. Recall that the capacity of a MIMO system arises from scattering, and that the greater the amount of scattering – the so-called scattering richness of the channel – the greater the MIMO capacity. As we discussed in Chapter 3, the channel rank is a useful way to quantify the scattering richness, so the capacity increases with the rank of the channel, all else being equal (see Eq. 3.11). In a Rayleigh channel, the maximum rank is equal to $\min\{N_t, N_r\}$, but what about a pinhole channel? In a pinhole channel, the channel is essentially divided into a MISO channel followed by a SIMO channel, each of which have ranks only equal to 1. The pinhole in effect reduces the degrees of freedom of the channel, and we would expect that this reduction would result in a corresponding diminution in the capacity relative to a Rayleigh channel. This, in fact, is easily borne out by computer simulations.

Figure 5.7 shows the results of a computer simulation that compares the predicted ergodic capacities of a 4×4 MIMO system in Rayleigh and pinhole environments for a range of signal-to-noise ratios. These results show that the pinhole channel capacity is significantly less than the capacity in a Rayleigh fading environment. This suggests that MIMO performance between a cell tower and a mobile indoor device is expected to be worse than if the device were outdoors, even if the signal-to-noise ratio were the same since the energy arrives at the receiver primarily through windows, which mimic a pinhole geometry.

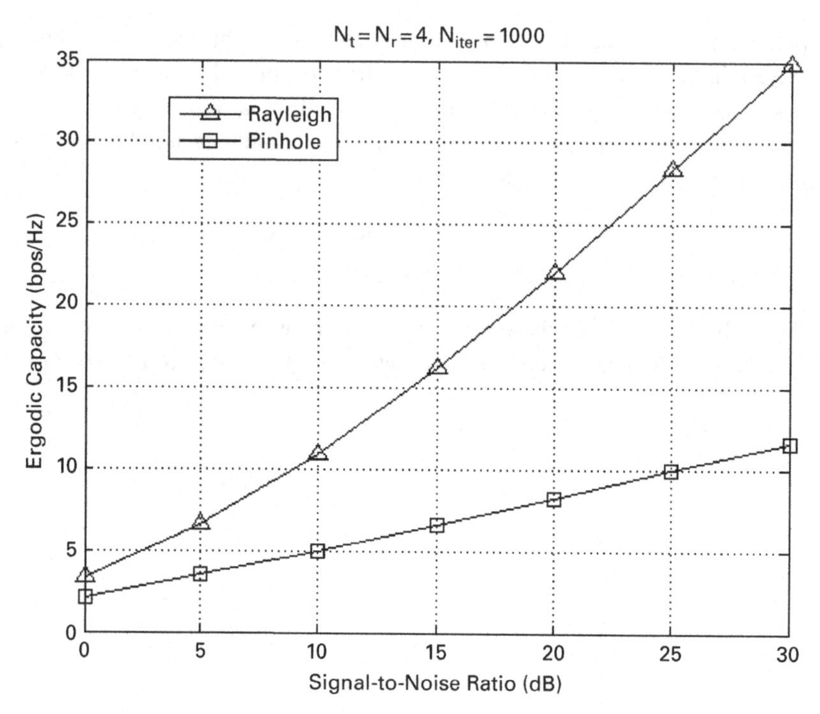

Figure 5.7 Comparison of the ergodic capacities of Rayleigh and pinhole channels for a 4×4 MIMO system.

5.7 Line-of-sight channel model

The third channel model that we consider in this chapter attempts to capture the effects of having both Rayleigh fading and LOS propagation present at the same time, which, as discussed in Chapter 4, is called Rician fading. Under this condition, the channel matrix is modeled as follows [29]:

$$\mathbf{H} = \sqrt{\frac{K}{1+K}}\mathbf{H}_{\mathrm{LOS}} + \sqrt{\frac{1}{1+K}}\mathbf{H}_w, \tag{5.34}$$

where $\mathbf{H}_w$ is the component of the channel matrix due to scattering and $\mathbf{H}_{\mathrm{LOS}}$ is the component of the channel matrix due to line-of-sight propagation. The elements of $\mathbf{H}_w$ are IID complex Gaussian, each with zero mean and variance equal to 1. $\mathbf{H}_{\mathrm{LOS}}$ has the properties defined by Eqs. 5.1 and 5.2.

Examination of Eq. 5.34 shows that it has properties we expect. For example, as the Rician K-factor approaches 0, implying that the direct component of the received signal is vanishing, $\mathbf{H} \longrightarrow \mathbf{H}_w$. Similarly, as K becomes large, implying that the scattered component is vanishing, $\mathbf{H} \longrightarrow \mathbf{H}_{\mathrm{LOS}}$.

In general, as the Rician K-factor increases, the MIMO capacity decreases when the communications path length is much larger than the size of the transmit antenna array,

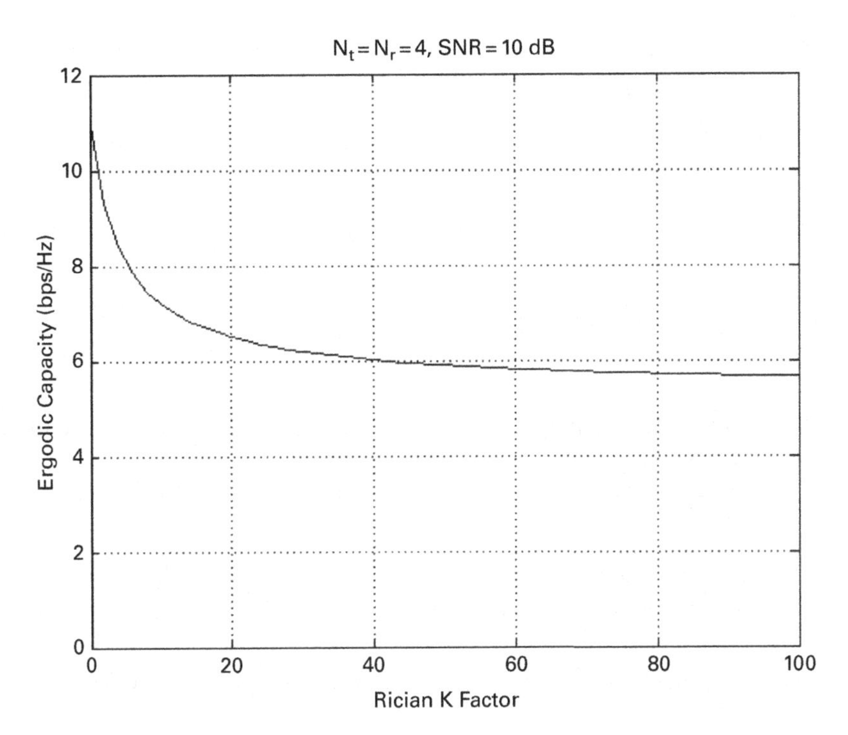

Figure 5.8 Ergodic capacity plotted versus the Rician K-factor for a 4×4 MIMO system. This plot assumes $\rho = 10$ dB and that $\mathbf{H}_{\text{LOS}}$ is the all ones matrix.

which is almost always the case. Under this assumption, $\theta_1 \simeq \theta_2 \simeq \cdots \simeq \theta_{N_t}$ in Figure 5.1, which implies according to Eqs. 5.1 and 5.2 that the columns of $\mathbf{H}$ become nearly identical. When this happens, the rank of the channel approaches 1, and the capacity decreases (see Eq. 3.11).

Figure 5.8 shows the ergodic capacity of a 4×4 MIMO system plotted as a function of the Rician K-factor. This plot shows that the ergodic capacity decreases monotonically with K and that it appears to approach an asymptote. When $K = 0$, the fading is Rayleigh, so the capacity should match the Rayleigh curve in Figure 5.7 for $\rho = 10$ dB, which is readily confirmed. The asymptote occurs because when $K \longrightarrow \infty$, $\mathbf{H} \longrightarrow \mathbf{H}_{\text{LOS}}$, which in this example is assumed to be the all-ones matrix. Since we are assuming CSIR only for these results, we can appeal to the capacity formula in Eq. 3.11. If $\mathbf{H}$ is an all-ones matrix, then the channel rank is 1, so the instantaneous capacity approaches $\log_2(1 + (\rho/N_t)\lambda)$; where λ is the sole eigenvalue. It follows from matrix theorem 1.9.2-(p) that $\text{Tr}(\mathbf{H}\mathbf{H}^H) = \lambda$. Since $\mathbf{H}$ is all ones and $\mathbf{H}\mathbf{H}^H$ is dimensioned $N_r \times N_r$, it follows that $\lambda = N_r N_t$. We conclude that the asymptotic capacity in Figure 5.8 should equal $\log_2(1 + \rho N_r)$. When $\rho = 10$ and $N_t = N_r = 4$, this expression equals 5.4 bps/Hz compared to 5.6 bps/Hz in the plot, the small disparity being due to the fact that the ergodic capacity is an average value whereas the capacity formula in Eq. 3.11 is an instantaneous value.

Problems

5.1 Consider a 2×3 MIMO communication system with the following covariance matrices:

$$\mathbf{R}_t = \begin{pmatrix} 1 & 0.4 \\ 0.4 & 1 \end{pmatrix}$$

and

$$\mathbf{R}_r = \begin{pmatrix} 1 & 0.5 & 0.9 \\ 0.5 & 1 & 0.8 \\ 0.9 & 0.8 & 1 \end{pmatrix}.$$

What is the reduction in the capacity of this system caused by correlation in the limit when the signal-to-noise ratio becomes large?

5.2 Prove Eq. 5.5.

5.3 Prove the inequality expressions in Eq. 5.27, and prove that equality occurs only when the correlation matrices are diagonal.

Hints:

a) Show that if $\mathbb{E}\left\{|[\mathbf{H}]_{i,j}|^2\right\} = 1$, then the diagonal elements of $\mathbf{R}_t$ and $\mathbf{R}_r$ are equal to 1.

b) Use matrix theorem 1.9.2-(p) and the results in (a) to find expressions for the sums of the eigenvalues of $\mathbf{R}_t$ and $\mathbf{R}_r$.

c) Complete the proof using the fact that the determinant of a square matrix is equal to the product of its eigenvalues and the following *arithmetic mean–geometric mean inequality theorem*:

For any set of N real numbers $\{a_i\}$,

$$\frac{a_1 + a_2 + \cdots + a_N}{N} \geq \left(\prod_{i=1}^{N} a_i\right)^{1/N}.$$

5.4 Equation 5.34 presents a channel model that incorporates Rayleigh scattering and LOS propagation. Use Eqs. 5.1 and 5.2 to define $\mathbf{H}_{\mathrm{LOS}}$ and show that each element of $\mathbf{H}$ in Eq. 5.34 has a variance equal to 1.

5.5 Write a Matlab program called ErgodicCapacity_Corr.m that computes the ergodic capacity of a MIMO system with CSIR only and antenna correlation. The program should have the following format:

C = ErgodicCapacity_Corr(SNRdBvec,Nt,Nr,Rt,Rr,Niter),

where C is the ergodic capacity, SNRdBvec is a vector of SNR values in dB, Nt and Nr are the number of transmit and receive antennas, respectively, Rt and Rr are the transmit and receive covariance matrices, respectively, and Niter is the number of Monte Carlo iterations. Use this program to duplicate Figure 5.2.

5.6 Write a Matlab function that computes the MIMO ergodic capacity as a function of the Rician K-factor under the assumption of CSIR only. Call this function ErgodicCapacity_Rician.m and use the following format:

C = ErgodicCapacity_Rician(SNRdB,Nt,Nr,Kvec,HLOS,Niter),

where SNRdB is the SNR value in dB, Kvec is a vector of K values at which the capacity is to be computed, and HLOS is the LOS component of the channel matrix. Use this function to duplicate the curve in Figure 5.8.

6 Alamouti coding

This chapter describes the Alamouti space-time coding scheme [6] for achieving transmit diversity. As discussed in Chapter 1, Alamouti coding was one of the first space-time codes to be developed, and it is now included in the definition of all modern wireless standards that employ MIMO techniques. Although other transmit diversity techniques were proposed in the 1990s prior to Alamouti's seminal paper (e.g. see [81], [82], and [79]), Alamouti's technique has the following advantages over alternative schemes: a) it requires CSIR only (as opposed to requiring both CSIT and CSIR); b) it does not involve any bandwidth expansion, which some of the competing techniques do; and c) Alamouti coding has relatively low computational complexity due to the fact that its decoding rules are quite simple.

Prior to the development of transmit diversity techniques in the 1990s, of which Alamouti coding is the most famous example, diversity benefits in fading environments were achieved using receive diversity only. In cellular applications, this meant that it was only possible to perform diversity combining at the base station because of the impracticality (due to lack of physical space and battery power limitations) of having multiple antennas on small hand-held devices. Therefore, prior to the development of space-time coding, which made transmit diversity possible, diversity gains were only available on the reverse links of cellular systems. The motivation for developing transmit diversity methods was that it would extend the benefits of diversity combining to the forward link as well without having to add multiple antennas and receive RF chains to mobile devices.

Although the purpose of this chapter is to describe Alamouti coding, it is important that we first review maximal ratio receive combining (MRRC), since MRRC is the optimum linear receive combining technique; thus, it is the standard against which Alamouti coding is to be compared. We will see that Alamouti coding matches the performance of MRRC combining, and is, therefore, optimal.

This chapter is organized as follows. We begin by analyzing the performance of maximum likelihood detection applied to a MRRC receiver. Next, we describe the Alamouti combining scheme for the case of a 2×1 MISO system and analyze its performance using maximum likelihood detection. By comparing these results with those for MRRC we demonstrate that Alamouti coding achieves the same diversity order as MRRC. The results for a 2×1 system are generalized to a $2 \times N_r$ configuration. We conclude the chapter by presenting an analysis of the bit error rate of the Alamouti scheme, followed by performance results comparing the simulated and theoretical performance predictions for Alamouti coding in Rayleigh fading.

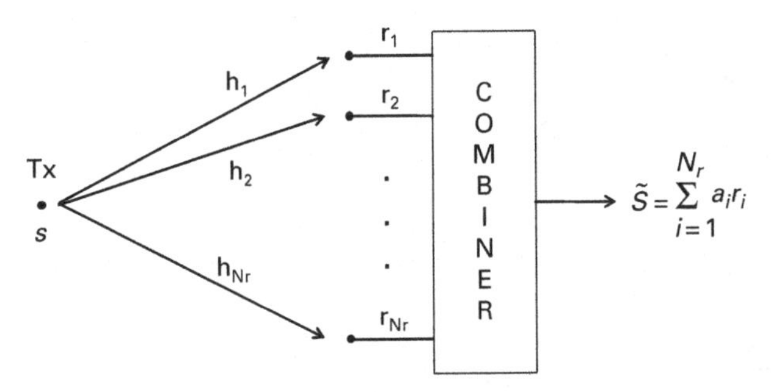

Figure 6.1 Architecture of a communication system with receive diversity combining.

6.1 Maximal ratio receive combining (MRRC)

We begin by reviewing the architecture of a MRRC receiver. In general, a receiver that employs diversity combining has the architecture shown in Figure 6.1, where the output of the combiner, $\tilde{s}$, consists of a linear weighted combination of the inputs. That is,

$$\tilde{s} = \sum_{i=1}^{N_r} a_i r_i, \tag{6.1}$$

where the $\{a_i\}$ denote complex weights and $\{r_i\}$ denote the received signals on each of the diversity channels.

In MRRC, each combining weight is equal to the complex conjugate of the respective channel gain; thus, in MRRC Eq. 6.1 becomes

$$\tilde{s} = \sum_{i=1}^{N_r} h_i^* r_i. \tag{6.2}$$

It is a well-known result (see [14]) that the combining defined in Eq. 6.2 results in the signal-to-noise ratio at the output of the combiner being maximized. Furthermore, the resulting signal-to-noise ratio, ρ, is given by

$$\rho = \sum_{i=1}^{N_r} \rho_i, \tag{6.3}$$

where ρ_i denotes the signal-to-noise ratio on the ith diversity channel.

In addition to performing the combining operation, a diversity receiver must also be capable of operating on the output of the combiner to estimate the information symbols that are being transmitted. For this purpose, we assume that the receiver performs maximum-likelihood (ML) detection on $\tilde{s}_1$. Figure 6.2 shows a block diagram of a 1×2 MRRC receiver with ML detection. As illustrated, we assume that during some symbol period, a symbol s is transmitted, which is chosen from a symbol alphabet $\{s_1, s_2, \ldots, s_M\}$, and that n_1 and n_2 denote the noise samples at the two receivers at this instant. Furthermore, we assume in this discussion that the channels between

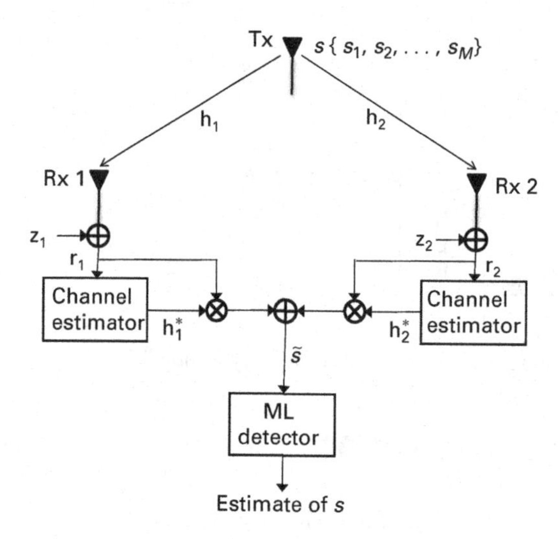

Figure 6.2 A 1 × 2 MRRC receiver with ML detection.

the transmitter and the two receivers undergo independent, uncorrelated flat fading. It follows that the two received signals are given by

$$r_i = h_i s + z_i,\ i = 1, 2 \tag{6.4}$$

and that the output of the combiner is given by

$$\begin{aligned}\tilde{s} &= h_1^* r_1 + h_2^* r_2 \\ &= h_1^* \left(h_1 s + z_1\right) + h_2^* \left(h_2 s + z_2\right) \\ &= \left(|h_1|^2 + |h_2|^2\right) s + h_1^* z_1 + h_2^* z_2.\end{aligned} \tag{6.5}$$

For a $1 \times N_r$ MRRC system, Eq. 6.5 generalizes to

$$\begin{aligned}\tilde{s} &= \sum_{i=1}^{N_r} h_i^* r_i \\ &= \left(\sum_{i=1}^{N_r} |h_i|^2\right) s + \sum_{i=1}^{N_r} h_i^* z_i.\end{aligned} \tag{6.6}$$

There are several important implications of Eq. 6.6, as follows:

1. $\tilde{s}$ has the form: [**Constant**] × [**Signal**] + [**Noise**].
2. Because there are no *signal* × *noise* cross terms, the maximum-likelihood detector logic is simple.
3. Notice that even if $(N_r - 1)$ channel gains are small because of fading, it may still be possible to successfully detect the transmitted symbol s based on $\tilde{s}$. This is why this system achieves a diversity order equal to N_r.

4. Lastly, this equation implies that the SNR associated with $\tilde{s}$ is given by:

$$\rho = \frac{\left(\sum_{i=1}^{N_r} |h_i|^2\right) \mathbb{E}\left\{|s|^2\right\}}{\sigma_z^2}; \tag{6.7}$$

where σ_z^2 denotes the variance of the noise terms, and we have assumed that a) the noise is uncorrelated on different receivers, and b) the noise and channel elements are statistically independent.

Now that we have briefly reviewed MRRC, we are ready to use the expressions in this section to compare against similar expressions when Alamouti coding is used. By doing so, we will show that Alamouti systems achieve the same performance as systems that employ maximal ratio receive combining. Before doing that, however, we briefly consider the challenges inherent in achieving transmit diversity to help appreciate the creativity that goes into developing space-time coding methods, of which Alamouti is only one example.

6.2 Challenges with achieving transmit diversity

Developing transmit diversity techniques that achieve the same amount of diversity as MRRC is not a trivial undertaking. Two examples help illustrate this point. First, consider a 2×1 MISO system in which we attempt to achieve transmit diversity by pre-multiplying the symbol to be transmitted, s, by complex numbers a_1 and a_2 at the two transmitters. The received signal at the single receiver is, therefore, given by

$$r = h_{11}a_1 s + h_{12}a_2 s + z, \tag{6.8}$$

where h_{1j} denotes the channel response between transmitter j and the one receiver, and n denotes noise. In order to make the signal-to-noise ratio in the above equation equal to the SNR achieved with MRRC given in Eq. 6.7, we set $a_j = h_{1j}^*/|h_{1j}|$, $j = 1, 2$. Clearly, this is a way to achieve transmit diversity that achieves the same performance as MRRC for $N_2 = 2$. The obvious problem with this solution, however, is that is requires CSIT! In MRRC, of course, only the receiver needs to know the channel in order to perform the required combining. As a result, although this proposed transmit diversity technique has the same performance as MRRC, its complexity is greater because it requires CSIT.

A second method that can be used to achieve transmit diversity is to use more than one symbol period to transmit s. For example, if we were to transmit s on transmitter 1 during the first symbol period and then transmit s from transmitter 2 during the second period, then the received signal is given by the following pair of equations:

$$\begin{aligned} \text{Period 1:}\quad r_1 &= h_{11}s + z_1, \\ \text{Period 2:}\quad r_2 &= h_{12}s + z_2. \end{aligned} \tag{6.9}$$

Since these equations are mathematically equivalent to Eq. 6.4 for MRRC (except that the subscript in these equations refers to time and in Eq. 6.4 the subscript refers to receiver number), it follows that this transmit diversity scheme achieves the same performance as MRRC. The advantage of this method over the first is that it doesn't require CSIT; however, the problem with this technique is that it takes two symbol periods to transmit one symbol – thus, the data rate is halved compared to MRRC.

These examples show that it is relatively easy to achieve transmit diversity if a sacrifice in data rate or the use of CSIT is allowed. The unique achievement of Alamouti coding is that it achieves transmit diversity without requiring sacrifices in either of these areas.

6.3 2 × 1 Alamouti coding

The most basic form of the Alamouti scheme is designed to transmit symbols using two transmit antennas to a single receive antenna, as illustrated in Figure 6.3.

The Alamouti block diagram illustrates that an Alamouti system involves four fundamental functions: a) space-time coding; b) diversity combining; c) channel state estimation at the receiver (i.e., CSIR); and d) maximum likelihood decoding. It should be noted that MRRC involves all of these functions except space-time coding, although as we will see, the details of Alamouti combining are different than the combining used in MRRC.

Alamouti space-time coding is defined in Table 6.1. This table shows how a sequence of two symbols, s_1 and s_2, is encoded using Alamouti coding. At some instant of time, t, symbol s_1 is transmitted from transmit antenna 1 and symbol s_2 is transmitted from the second transmit antenna. During the next symbol interval period at time $t+T_s$, the symbols $-s_2^*$ and s_1^* are transmitted from antennas 1 and 2, respectively. This demonstrates how Alamouti coding involves coding in both the spatial and time dimensions; hence, it is an example of a space-time code.

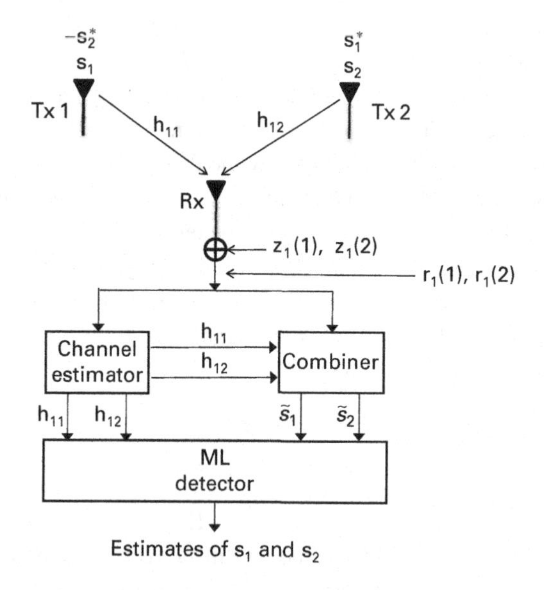

Figure 6.3 A 2 × 1 Alamouti communication system.

Table 6.1 Alamouti space-time coding.

Transmit time	Antenna 1	Antenna 2
t	s_1	s_2
$t+T_s$	$-s_2^*$	s_1^*

The conventions used in Figure 6.3 are as follows: a) subscripts on the received and noise signals denote the receiver number;[1] and b) the index in parentheses denotes the time interval. For example, $r_i(1)$ denotes the received signal at receiver i at time t and $r_i(2)$ denotes the received signal at receiver i at time $t+T_s$. Similarly, $z_i(1)$ denotes the noise signal at receiver i at time t and $z_i(2)$ denotes the noise signal at receiver i at time $t+T_s$.

The combining rules for 2 × 1 Alamouti space-time coding are defined as follows:

$$\begin{aligned}\tilde{s}_1 &= h_{11}^* r_1(1) + h_{12} r_1^*(2),\\ \tilde{s}_2 &= h_{12}^* r_1(1) - h_{11} r_1^*(2).\end{aligned} \tag{6.10}$$

Note that there are three differences between Alamouti and MRRC combining:

1. There are two Alamouti combining rules instead of one MRRC rule for a 2 × 1 antenna configuration.
2. Alamouti combining results in two combined signals compared to only one with MRRC.
3. In Alamouti coding, signals from two different times are combined instead of from two different receivers in MRRC.

Despite these differences, Alamouti coding has the same behavior as MRRC processing. To show this, we start by computing the received signals at t and $t+T_s$. If we assume that the channel gains are constant across two consecutive symbols, it follows that

$$\begin{aligned}r_1(1) &\triangleq r_1(t) = h_{11}s_1 + h_{12}s_2 + z_1(1),\\ r_1(2) &\triangleq r_1(t+T_s) = -h_{11}s_2^* + h_{12}s_1^* + z_1(2).\end{aligned} \tag{6.11}$$

Substituting Eq. 6.11 into Eq. 6.10 yields

$$\begin{aligned}\tilde{s}_1 &= h_{11}^*\left(h_{11}s_1 + h_{12}s_2 + z_1(1)\right) + h_{12}\left(-h_{11}s_2^* + h_{12}s_1^* + z_1(2)\right)^*\\ &= |h_{11}|^2 s_1 + h_{11}^* h_{12} s_2 + h_{11}^* z_1(1) - h_{12}h_{11}^* s_2 + |h_{12}|^2 s_1 + h_{12} z_1^*(2)\\ &= \left(|h_{11}|^2 + |h_{12}|^2\right) s_1 + h_{11}^* z_1(1) + h_{12} z_1^*(2)\\ &= \left(\sum_{i=1}^{2} |h_{1,i}|^2\right) s_1 + h_{11}^* z_1(1) + h_{12} z_1^*(2).\end{aligned} \tag{6.12}$$

[1] This figure only depicts one receiver, but in general there can be an arbitrary number of receivers with Alamouti coding. We consider more general cases subsequently.

Similarly,

$$\tilde{s}_2 = \left(|h_{11}|^2 + |h_{12}|^2\right) s_2 + h_{12}^* z_1(1) - h_{11} z_1(2)^*$$
$$= \left(\sum_{i=1}^{2} |h_{1,i}|^2\right) s_2 + h_{12}^* z_1(1) - h_{11} z_1(2)^*. \tag{6.13}$$

Examination of Eqs. 6.12 and 6.13 reveals the following:

1. The combined signals $\tilde{s}_1$ and $\tilde{s}_2$ are each a function of only one transmit symbol.
2. Each combined symbol has the same form as Eq. 6.5 for MRRC[2] and the same signal-to-noise expression as in Eq. 6.7 for $N_r = 2$.
3. Since two symbols are transmitted every two symbol periods, the symbol rate is the same as it is with MRRC.
4. CSIT is not required.

These observations imply that 2×1 Alamouti coding is mathematically equivalent to 1×2 MRRC. This shows that the Alamouti scheme achieves what the simple examples in the previous section did not, which is to achieve transmit diversity with optimum performance (i.e., performance equivalent to MRRC) without a reduction in the data rate or the need for CSIT.

6.4 $2 \times N_r$ Alamouti coding

The Alamouti technique can easily be generalized to an arbitrary number of receive antennas without changing the space-time coding performed by the transmitter. In going from 1 to an arbitrary number of receive antennas, the receiver architecture changes, but the transmit ST coding remains the same regardless of N_r. We start by considering a 2×2 configuration, which, when compared to the 2×1 case, will show us how to extrapolate to an arbitrary number of receive antennas.

6.4.1 The 2×2 case

Figure 6.4 shows a block diagram of a 2×2 Alamouti system. The main differences between this figure and the previous one for a 2×1 configuration are that there are two receive antennas and four channel gains. Recall that the subscripts on n and r in the figure denote the receive antenna number and the indices in parentheses denote the symbol period. The combining rules for a 2×2 Alamouti system are defined as follows:

[2] The term *same form* in this context simply means that the expressions for the Alamouti combiner outputs consist of the sum of the squares of the magnitudes of channel responses multiplied by a transmitted symbol plus two noise terms that consist of noise samples multiplied by channel responses. The fact that the complex conjugate operation is applied differently in these expressions than in Eq. 6.5 is a superficial difference that does not affect the SNR associated with $\tilde{s}_1$ and $\tilde{s}_2$.

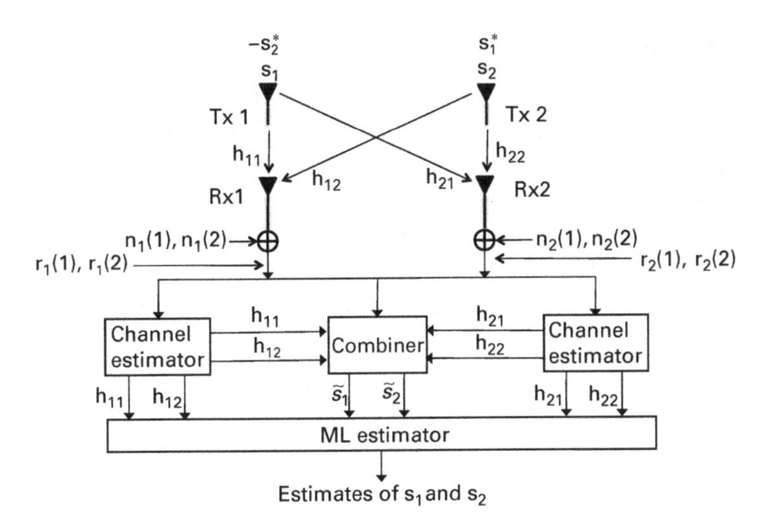

Figure 6.4 A 2 × 2 Alamouti system.

$$\tilde{s}_1 = h_{11}^* r_1(1) + h_{12} r_1^*(2) + h_{21}^* r_2(1) + h_{22} r_2^*(2),$$
$$\tilde{s}_2 = h_{12}^* r_1(1) - h_{11} r_1^*(2) + h_{22}^* r_2(1) - h_{21} r_2^*(2). \tag{6.14}$$

Examination of Figure 6.4 shows that the received signals are equal to the following:

$$r_1(1) = h_{11}s_1 + h_{12}s_2 + z_1(1),$$
$$r_1(2) = -h_{11}s_2^* + h_{12}s_1^* + z_1(2),$$
$$r_2(1) = h_{21}s_1 + h_{22}s_2 + z_2(1),$$
$$r_2(2) = -h_{21}s_2^* + h_{22}s_1^* + z_2(2). \tag{6.15}$$

Substituting Eq. 6.15 into Eq. 6.14 yields

$$\tilde{s}_1 = \left(\sum_{i=1}^{2} \sum_{j=1}^{2} |h_{ij}|^2 \right) s_1 + h_{11}^* z_1(1) + h_{12} z_1^*(2) + h_{21}^* z_2(1) + h_{22} z_2^*(2) \tag{6.16}$$

and

$$\tilde{s}_2 = \left(\sum_{i=1}^{2} \sum_{j=1}^{2} |h_{ij}|^2 \right) s_2 - h_{11}^* z_1^*(2) + h_{12}^* z_1^*(1) - h_{21}^* z_2^*(2) + h_{22}^* z_2^*(2). \tag{6.17}$$

Note that each of these equations has the same form as Eq. 6.6 for $N_r = 4$, and the signal-to-noise ratios associated with $\tilde{s}_1$ and $\tilde{s}_2$ are the same as Eq. 6.7 for the case where $N_r = 4$. We conclude from this that a 2 × 2 Alamouti system performs the same as a 1 × 4 MRRC receiver.

6.4.2 The $2 \times N_r$ case

Examination of the Alamouti combining rules in Eqs. 6.10 and 6.14 for the cases where N_r equals 1 and 2, respectively, makes it clear how to extrapolate to an arbitrary number of receive antennas. Note that the first two terms on the right side of the equal signs in Eqs. 6.14 specify the combining rule that would occur if receiver 1 were the only receiver. Similarly, the third and fourth terms correspond to the combining rule that would be used if receiver 2 were the only receiver. Using this pattern, we conclude that the Alamouti combining rules for a general $2 \times N_r$ system are as follows:

$$\tilde{s}_1 = \sum_{i=1}^{N_r} h_{i1}^* r_i(1) + h_{i2} r_i^*(2),$$
$$\tilde{s}_2 = \sum_{i=1}^{N_r} h_{i2}^* r_i(1) - h_{i1} r_i^*(2), \tag{6.18}$$

where

$$r_i(1) = h_{i1} s_1 + h_{i2} s_2 + z_i(1),$$
$$r_i(2) = -h_{i1} s_2^* + h_{i2} s_1^* + z_i(2), \; i = 1, \ldots, N_r. \tag{6.19}$$

Substituting Eqs. 6.19 into Eqs. 6.18 results in the following expressions for the Alamouti combiner output in the general case:

$$\tilde{s}_1 = \left(\sum_{i=1}^{N_r} \sum_{j=1}^{2} |h_{ij}|^2 \right) s_1 + \sum_{i=1}^{N_r} h_{i1}^* z_i(1) + h_{i2} z_i^*(2) \tag{6.20}$$

and

$$\tilde{s}_2 = \left(\sum_{i=1}^{N_r} \sum_{j=1}^{2} |h_{ij}|^2 \right) s_2 + \sum_{i=1}^{N_r} h_{i2}^* z_i(1) - h_{i1} z_i^*(2). \tag{6.21}$$

Note that these equations have the same form as Eq. 6.6, and the signal-to-noise ratios associated with $\tilde{s}_1$ and $\tilde{s}_2$ are the same as Eq. 6.7 for an arbitrary value of N_r. We conclude from this that a $2 \times N_r$ Alamouti system performs the same as a $1 \times 2N_r$ MRRC receiver.

We have now defined the space-time coding used in an Alamouti transmitter, described how combining is performed at the receiver, and shown that Alamouti processing results in the same performance as MRRC. In so doing, we have demonstrated that the combiner outputs of an Alamouti receiver, $\tilde{s}_1$ and $\tilde{s}_2$, each have the form [Constant]×[Signal] + [Noise] (see Eqs. 6.20 and 6.21). This, of course, is the same general form that an MRRC combiner output has (see Eq. 6.6). When signals have this form, the ML detection algorithm for demodulating the symbols is very simple. The next section reviews the ML algorithm and presents the decoding logic for both the MRRC and Alamouti schemes.

6.5 Maximum likelihood demodulation in MRRC and Alamouti receivers

As we have seen, the outputs of both the MRRC and Alamouti combiners have the following form:

$$\tilde{s} = Ks + z, \tag{6.22}$$

where $\tilde{s}$ represents the output(s) of the combiner, K is a constant that depends on the channel gain values between the transmit and receive antennas, s is the symbol we wish to detect, and z is a noise term, which we assume is complex Gaussian with zero mean and variance equal to σ_z^2 (i.e., $n \sim \mathcal{CN}(0, \sigma_z^2)$). We further assume that the symbol being transmitted is chosen from an M-ary alphabet $\mathcal{A} \triangleq \{a_1, a_2, \ldots, a_M\}$. The *maximum a-posteriori probability* (MAP) estimate of s, which we denote by $\hat{s}$, is expressed as follows:

$$\hat{s} = \arg\max_{\{s\}} P(s|\tilde{s}), \tag{6.23}$$

where $P(s|\tilde{s})$ denotes the probability that symbol s was sent, given that the output of the combiner is $\tilde{s}$. The symbol that maximizes this probability is the MAP estimate of the transmitted symbol. Using Bayes theorem, it follows that if the *a-priori probabilities* (APP) of the symbols being transmitted are all the same (i.e., $P(s = a_i) = 1/M$, $i = 1, \ldots, M$), then Eq. 6.23 can be rewritten as

$$\hat{s} = \arg\max_{\{s\}} P(\tilde{s}|s), \tag{6.24}$$

where $P(\tilde{s}|s)$ is called the *likelihood probability* and Eq. 6.24 defines the maximum likelihood (ML) criterion.

Since $n \sim \mathcal{CN}(0, \sigma_z^2)$), it follows that

$$P(\tilde{s}|s) = \frac{1}{\sqrt{2\pi\sigma_z^2}} \mathrm{e}^{-|\tilde{s}-Ks|^2/2\sigma_z^2}; \tag{6.25}$$

therefore, the value of s that maximizes $P(\tilde{s}|s)$ is the symbol that minimizes the exponent in this equation. That is,

$$\hat{s} = \arg\min_{\{s\}} |\tilde{s} - Ks|^2. \tag{6.26}$$

The preceding ML detection rule can be rewritten in a convenient form as follows:

$$\begin{aligned}
\hat{s} &= \arg\min_{\{s\}} \left[|\tilde{s} - Ks|^2 \right] \\
&= \arg\min_{\{s\}} \left[(\tilde{s} - Ks)(\tilde{s} - Ks)^* \right] \\
&= \arg\min_{\{s\}} \left[|\tilde{s}|^2 - \tilde{s}Ks^* - Ks\tilde{s}^* + K^2|s|^2 \right] \\
&= \arg\min_{\{s\}} \left[K^2|s|^2 - \tilde{s}Ks^* - Ks\tilde{s}^* \right] \\
&= \arg\min_{\{s\}} \left[K|s|^2 - \tilde{s}s^* - s\tilde{s}^* \right].
\end{aligned} \tag{6.27}$$

Next, we use the fact that $|\tilde{s}-s|^2 = |\tilde{s}|^2 - \tilde{s}s^* - s\tilde{s}^* + |s|^2$, which implies that $-\tilde{s}s^* - s\tilde{s}^* = |\tilde{s}-s|^2 - |\tilde{s}|^2 - |s|^2$. Using this result, we can write Eq. 6.27 in the following form:

$$\begin{aligned}\hat{s} &= \arg\min_{\{s\}} \left[K|s|^2 + |\tilde{s}-s|^2 - |\tilde{s}|^2 - |s|^2\right] \\ &= \arg\min_{\{s\}} \left[(K-1)|s|^2 + |\tilde{s}-s|^2\right]. \end{aligned} \tag{6.28}$$

It is straightforward to apply the ML detection rule in Eq. 6.28 to MRRC and Alamouti-based receivers. Equation 6.6, which gives the expression for the combiner output of an $N_r \times 1$ MRRC receiver, indicates that $K = \sum_{i=1}^{N_r} |h_i|^2$. Therefore, the ML detection rule for a MRRC receiver with a combiner output $\tilde{s}$ is as follows:

$$\hat{s} = \arg\min_{\{s\}} \left[\left(\sum_{i=1}^{N_r} |h_i|^2 - 1\right)|s|^2 + |\tilde{s}-s|^2\right], \tag{6.29}$$

where the minimization is performed over the set of possible transmitted symbols $s \in \mathcal{A}$.

Similarly, with Alamouti coding it follows from Eqs. 6.20 and 6.21, which specify the two combiner outputs, that $K = \left(\sum_{i=1}^{N_r}\sum_{j=1}^{2} |h_{ij}|^2\right)$. Therefore, the Alamouti decoding rules are

$$\begin{aligned}\hat{s}_1 &= \arg\min_{\{s_1\}} \left[\left(\sum_{i=1}^{N_r}\sum_{j=1}^{2} |h_{ij}|^2 - 1\right)|s_1|^2 + |\tilde{s}_1-s_1|^2\right], \\ \hat{s}_2 &= \arg\min_{\{s_2\}} \left[\left(\sum_{i=1}^{N_r}\sum_{j=1}^{2} |h_{ij}|^2 - 1\right)|s_2|^2 + |\tilde{s}_2-s_2|^2\right]. \end{aligned} \tag{6.30}$$

Equations 6.29 and 6.30 show that the detection rules for both MRRC and Alamouti systems have extremely low computational complexity. In each case, the estimate of the transmitted symbol is obtained by substituting in the M possible symbols from $\mathcal{A}$ into one of the bracketed expressions in these equations and determining which symbol results in the smallest value for that expression. The symbol that results in the smallest value of the bracketed term is the ML estimate of the transmitted symbol. For example, if binary modulation is used, then each decoding rule simply involves substituting two symbol values into an expression and comparing their values. Similarly, quaternary modulation involves comparing 4 values, etc.

Example 6.1 Consider the case where the modulation involves equi-energy signal constellations (i.e., where $|a_i|^2 = |a_k|^2$, $i \neq k$). What do the Alamouti decoding rules simplify to under this assumption?

Answer Under the equi-energy assumption, the values of the first terms in the brackets in Eq. 6.30 for $\hat{s}_1$ and $\hat{s}_2$ have the same value, irrespective of which symbol is assumed.

Therefore, only the second terms need to be considered when minimizing the terms in brackets, which results in the following simplified decoding rules:

$$\hat{s}_1 = \arg\min_{\{s_1\}} \left[|\tilde{s}_1 - s_1|^2 \right],$$
$$\hat{s}_2 = \arg\min_{\{s_2\}} \left[|\tilde{s}_2 - s_2|^2 \right]. \tag{6.31}$$

6.6 Performance results

This section presents performance predictions for Alamouti coding. We begin by presenting theoretical closed-form expressions for the bit error rate associated with Alamouti and MRRC systems. Next, we discuss how to simulate the performance of these systems. And, finally, we present performance predictions for an Alamouti system based on computer simulations and the theoretical predictions. These results are compared with corresponding predictions for MRRC.

6.6.1 Theoretical performance analysis

We have shown that a $2 \times N_r$ Alamouti system performs the same as a $1 \times 2N_r$ MRRC receiver, which, we have reasoned, follows from the fact that the outputs of the Alamouti combiner, $\tilde{s}_1$ and $\tilde{s}_2$, in Eqs. 6.20 and 6.21 each have the same mathematical form as the MRRC combiner output, $\tilde{s}$. Therefore, applying ML detection logic to each of the Alamouti combiner outputs should result in the same bit error rate performance as occurs when applied to the MRRC combiner output. We can use this observation to obtain closed-form expressions for the bit error rate of the Alamouti receiver by appealing to well-established theoretical results for MRRC performance.

Before applying MRRC theory directly to Alamouti systems, however, there is one important difference between these two diversity schemes that needs to be considered. That difference is the simple fact that Alamouti systems have two transmitters, but MRRC systems only have one. In order to compare the theoretical predictions for MRRC and Alamouti systems, it is necessary to assume that the total transmitted power in both cases is the same. If we denote the total transmit power by P_t, and the resulting average signal-to-noise ratio at each MRRC receiver by ρ, then in order for the total transmit power to be P_t in the Alamouti case, it is necessary to reduce the transmit power on each of the two Alamouti transmitters to $P_t/2$, which results in the signal-to-noise ratio associated with each transmitter-to-receiver link (i.e., each diversity path) being $\rho/2$. This means that if the bit error rate of the MRRC system, BER_{MRRC}, is some function of the signal-to-noise ratio, say, $f_{\text{MRRC}}(\rho)$, then the bit error rate of the Alamouti scheme is related to the theoretical MRRC bit error rate expression as follows:

$$BER_{\text{Alamouti}}(\rho) = f_{\text{MRRC}}(\rho/2). \tag{6.32}$$

Table 6.2 Values of μ for selected binary modulation schemes (Note: $\bar{\rho}$ denotes the time averaged SNR).

Modulation	MRRC	Alamouti
BPSK	$\sqrt{\frac{\bar{\rho}}{1+\bar{\rho}}}$	$\sqrt{\frac{\bar{\rho}}{2+\bar{\rho}}}$
Coherent orthog. BFSK	$\sqrt{\frac{\bar{\rho}}{2+\bar{\rho}}}$	$\sqrt{\frac{\bar{\rho}}{1+\bar{\rho}}}$
Noncoh. DPSK	$\frac{\bar{\rho}}{1+\bar{\rho}}$	$\frac{\bar{\rho}}{2+\bar{\rho}}$
Noncoh. orthog. BFSK	$\frac{\bar{\rho}}{2+\bar{\rho}}$	$\frac{\bar{\rho}}{4+\bar{\rho}}$

Table 6.3 Values of μ for selected QPSK and 4-DPSK assuming Gray coding.

Modulation	MRRC	Alamouti
QPSK	$\sqrt{\frac{\bar{\rho}}{1+\bar{\rho}}}$	$\sqrt{\frac{\bar{\rho}}{2+\bar{\rho}}}$
4-DPSK	$\frac{\bar{\rho}}{1+\bar{\rho}}$	$\frac{\bar{\rho}}{2+\bar{\rho}}$

This tells us that the bit error rate expression for Alamouti coding is obtained by taking the corresponding expression for MRRC and replacing each occurrence of ρ with $\rho/2$.

Proakis and Salehi [59] present theoretical expressions for the bit error rate of MRRC in flat Rayleigh fading for a variety of different modulation types. They show that the bit error rate, P_b, for coherently demodulated BPSK and orthogonal BFSK, and for non-coherently demodulated DPSK and orthogonal BFSK all have the following form:

$$P_b = \left[\frac{1}{2}(1-\mu)\right]^L \sum_{k=0}^{L-1} \binom{L-1+k}{k} \left[\frac{1}{2}(1+\mu)\right]^k, \tag{6.33}$$

where μ is given by the expressions in Table 6.2 and L is given by

$$L = \begin{cases} N_r, & \text{MRRC}, \\ 2N_r & \text{Alamouti}. \end{cases} \tag{6.34}$$

Proakis and Salehi [59] also derive the following expression for the probability of bit error rate for QPSK and 4-ary DPSK assuming Gray coding:

$$P_b = \frac{1}{2}\left[1 - \frac{\mu}{\sqrt{2-\mu}} \sum_{k=0}^{L-1} \binom{2k}{k} \left(\frac{1-\mu^2}{4-2\mu^2}\right)^k\right], \tag{6.35}$$

where L is given by Eq. 6.34 and μ is given by the expressions in Table 6.3

We close this section by noting that in the limit as $\bar{\rho}$ becomes large, the expressions in Eqs. 6.33 and 6.35 become linear when P_b is plotted on a log scale and $\bar{\rho}$ is expressed in dB. Furthermore, the slope is equal to $-L$ indicating that full diversity is achieved. This fact can be demonstrated empirically by evaluating these expressions numerically and plotting the results. It is also straightforward to show this analytically for Eq. 6.33 (see Problem 6.2 at end of chapter). Although Eq. 6.35 also has this property, as can be easily demonstrated numerically, a straightforward analytical proof does not exist to this author's knowledge.

6.6.2 Simulating Alamouti and MRRC systems

The theory presented above can be used to predict the bit error rate for MRRC and Alamouti systems for certain modulation schemes in Rayleigh fading environments; however, if the channel fading is not Rayleigh or if the modulation being used is not one of the methods listed, then it is necessary to develop computer simulations to assess performance. This section describes the steps that would need to be taken in order to simulate the bit error rate performance of MRRC and Alamouti communication systems.

Step 1 **Compute the transmit symbols**

Compute a sequence of symbols to be transmitted. When simulating Alamouti coding, each iteration of the simulation should begin by taking a pair of transmit symbols, assigning each of the symbols to s_1 and s_2, then generating a 2×2 Alamouti space-time code block using the definition of Alamouti space-time coding given in Table 6.1. When simulating MRRC, each iteration of the simulation involves assigning a single transmit symbol to s, and, of course, no space-time coding is needed.

Step 2 **Compute the received signals**

Alamouti: For each pair of transmitted symbols, s_1 and s_2, compute the received signals, $r_i(1)$ and $r_i(2)$, $i = 1, \ldots N_r$ using a modified version of Eq. 6.19 described below.

Although Eq. 6.19 could be used as is, it is important to keep in mind that this equation does not assume any normalization constraints. To examine the symbol or bit error rate as a function of SNR, it is useful to use the standard normalizations: $\mathbb{E}\{|h_{ij}|^2\} = \mathbb{E}\{|z_i|^2\} = 1$ and $\mathbb{E}\{|s|^2\} = 1/N_t = 1/2$. When doing this, Eq. 6.19 needs to be modified to the following:

$$\begin{aligned} r_i(1) &= \sqrt{\rho} h_{i1} s_1 + \sqrt{\rho} h_{i2} s_2 + z_i(1), \\ r_i(2) &= -\sqrt{\rho} h_{i1} s_2^* + \sqrt{\rho} h_{i2} s_1^* + z_i(2), \; i = 1, \ldots, N_r \end{aligned} \tag{6.36}$$

where ρ denotes the SNR at each receiver.

MRRC: Similarly, when simulating the performance of an MRRC system, Eq. 6.4 should be modified as follows:

$$r_i = \sqrt{\rho} h_i s + z_i, \; i = 1, \ldots, N_r, \tag{6.37}$$

where ρ denotes the SNR at each receiver and the following normalizations are assumed to hold: $E\{|h_i|^2\} = E\{|z_i|^2\} = \mathbb{E}\{|s|^2\} = 1$.

Step 3 **Compute the combiner output(s)**

Alamouti: For each pair of transmitted symbols, compute $\tilde{s}_1$ and $\tilde{s}_2$ using Eqs. 6.18 by substituting the expressions for $r_i(1)$ and $r_i(2)$ defined in Eq. 6.36.

MRRC: When simulating MRRC, compute $\tilde{s}$ by substituting Eq. 6.37 into the top righthand expression in Eq. 6.6.

Step 4 **Demodulate the received symbols using ML detection**

Alamouti: When simulating Alamouti coding, demodulate each pair of symbols using Eqs. 6.30.

MRRC: When simulating MRRC, demodulate each pair of symbols using Eq. 6.29.

6.6.3 Results

Figure 6.5 shows the bit error rate performance associated with BPSK modulation for both MRRC and Alamouti coding in Rayleigh fading. The curves with symbols are based on simulation results in which each bit error probability value was estimated from 200 000 bits. Associated with each simulated curve is a solid curve that passes near the simulated values based on the theoretical predictions described above. The simulated and theoretical predictions agree quite well with each other, particularly at higher bit error rates. At lower bit error rate levels, the disparity is due to inaccuracies in the simulated predictions caused by there not being a sufficiently large number of iterations in the simulation. In addition to showing the performance of MRRC and Alamouti processing, a reference curve (labeled "No diversity") is included that shows the theoretical bit error probability that occurs when no diversity exists. This curve is included to show the performance improvement that diversity combining provides.

The middle set of curves compares the performance of a 2×1 Alamouti system with a 1×2 MRRC-based scheme. Since both of these systems have diversity order equal to 2 (i.e., $N_d = 2$), we would expect the slopes of these curves to be equal to –2 at large SNR, which is easily confirmed by examination of the figure. The left-most set of curves shows results for a 2×2 Alamouti system and a 1×4 MRRC system. Examination of these curves shows that their slopes approach -4 at large SNR in accordance with what we would predict given that the diversity order is 4.

It has been argued in this chapter that Alamouti coding achieves the same performance as maximal ratio receive combining, yet the Alamouti curves in Figure 6.5 are shifted 3 dB to the right of the MRRC curves, suggesting that Alamouti performance is actually 3 dB worse than MRRC. This difference in performance is due to the fact that we have assumed for the purpose of this analysis that the total transmitted power is the same for both the Alamouti and MRRC schemes. This is one reasonable way to compare Alamouti and MRRC, and it is the convention that Alamouti uses in his paper

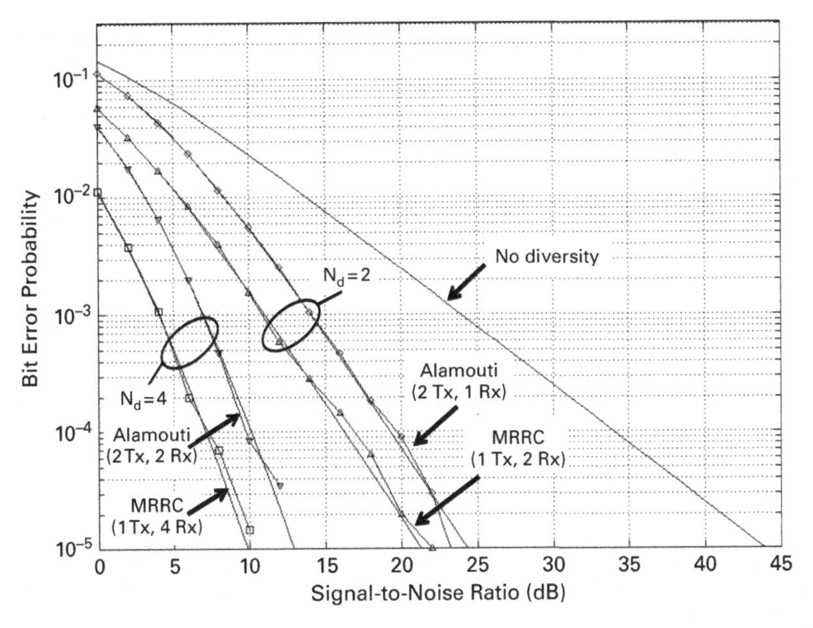

Figure 6.5 Bit error rate performance of MRRC and Alamouti coding in Rayleigh fading assuming BPSK modulation. Solid curves are based on theory and the curves with symbols are based on computer simulation results.

[6]; however, it is not the only way. Another "fair" way to compare these two schemes is to assume that the average transmit power per antenna is the same in both cases. If we were to do this, the two curves would fall on top of each other. Because the Alamouti scheme involves two transmit antennas and MRRC only uses one, there is an inherent and unavoidable ambiguity in how to compare these two schemes. When comparing a system that has multiple transmit antennas with one that only has one, "identical" performance can mean one of two things depending on what we choose to hold constant: a) if the transmit power levels are assumed to be the same, then it means that the system with multiple transmit antennas will have BER-versus-SNR curves that are shifted $10\log_{10}(N_t)$ dB to the right of the system with only one antenna; or b) if the transmit power per antenna is assumed to be the same in both systems, the BER-versus-SNR curves will lie on top of each other.

Problems

6.1 Prove that the signal-to-noise ratio associated with each of the Alamouti combiner outputs for a $2 \times N_r$ MIMO system is equal to the signal-to-noise ratio of the combiner output for a $1 \times 2N_r$ MRRC combiner with $\tilde{s}_1$ and $\tilde{s}_2$ are equal to the SNR associated with the MRRC combiner output given in Eq. 6.7. What assumptions do you need to make in order for the MRRC and Alamouti SNRs to be the same?

6.2 Show that the bit error rate expression in Eq. 6.33 is proportional to $\bar{\rho}^{-L}$ (where L is specified in Eq. 6.34) in the limit as $\bar{\rho} \to \infty$ for all of the modulation schemes listed in Table 6.2, for both MRRC and Alamouti coding.

6.3 Using Matlab, write two functions that compute the bit error probability of MRRC and Alamouti systems in Rayleigh fading using BPSK modulation. Call these functions MRRC.m and Alamouti.m. Each of these functions should compute the BER using both Monte Carlo simulation techniques as well as the theory given in Eq. 6.33. Use these functions to duplicate Figure 6.5

7 Space-time coding

In the previous chapter, we considered Alamouti space-time coding, which enables transmit diversity with optimum performance. As we saw, although Alamouti coding works with any number of receive antennas, it is restricted to cases where there are only two transmit antennas. In this chapter, we broaden our discussion and consider space-time codes that support more than two transmitters.

Since Alamouti's code was introduced in 1998, much research has been conducted in the area of space-time coding for MIMO applications, and space-time coding is now a broad field encompassing many different types of coding schemes. Space-time codes fall into one of two primary classes: space-time *block* codes (STBCs) and space-time *trellis* codes (STTCs). STBCs, in turn, fall into two subclasses called orthogonal space-time block codes (OSTBCs) and non-orthogonal space-time block codes (NOSTBCs). In this chapter we focus on OSTBCs because they have simple decoding schemes and are used in practical wireless systems.

The chapter begins with a general discussion of space-time coding concepts and terminology, followed by a section that derives criteria for designing space-time codes. Next, we describe OSTBCs in detail and describe how to decode them. After that, we include a brief section on NOSTBCs, listing some specific ones that have been proposed and giving references for further reading. We conclude the chapter with a section on STTCs. Although we focus on OSTBCs in this chapter, we also include a section on STTCs, both because of their historical significance and because, under limited conditions, STTCs can achieve better performance than OSTBCs.

7.1 Space-time coding introduction

7.1.1 Definition of STBC code rate

Before describing specific space-time codes (STCs), it is helpful to consider some basic concepts and definitions that apply to STCs in general, and which we use in our discussion throughout the remainder of the chapter. Since it is assumed that the reader has at least a cursory familiarity with conventional error control coding concepts, it is perhaps useful to draw comparisons and contrasts between conventional error coding and STCs. We start by considering a simple block diagram of a conventional block encoded communication system, which is illustrated in Figure 7.1.

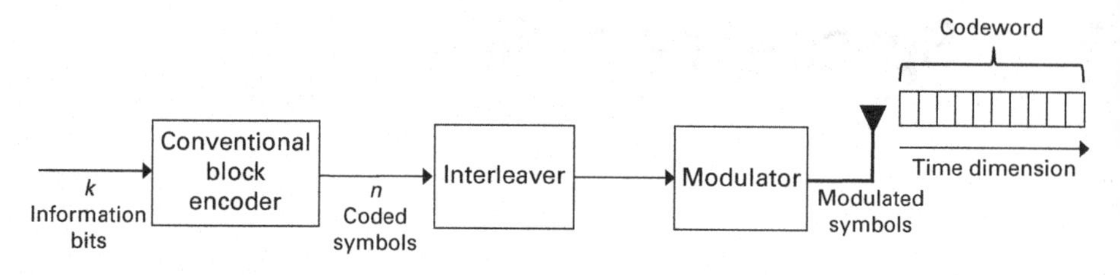

Figure 7.1 Block diagram of a conventional encoded communication system.

The primary point of this figure is that conventional block error control coding involves taking k information bits, and then using some method to encode those bits by introducing redundancy, which results in an encoded block of length $n > k$. The resulting block of encoded symbols, called a *codeword*, is spread out in the time dimension. The *code rate* of a conventional error control code, which we denote by r_t (the t subscript indicating that coding is performed in the time dimension), is normally defined as the ratio of the number of information bits divided by the length of the encoded block. This definition can be viewed in a slightly more general way as follows:

$$\begin{aligned} r_t &= \frac{k}{n} \\ &= \frac{\left[\text{No. of input information bits per codeword}\right]}{\left[\begin{array}{l}\text{No. of coded symbol periods in the time dimension} \\ \text{required to encode the } k \text{ information bits}\end{array}\right]}. \end{aligned} \tag{7.1}$$

Although the second equality, while clearly consistent with the usual definition, does not appear to provide additional insight, it will prove useful below when we define the code rate for space-time codes.

Next, consider the block diagram of a space-time coded system shown in Figure 7.2. This figure is identical to Figure 7.1 except that it includes a space-time encoder block that maps the modulated symbols onto different antennas, which are depicted by the output arrows from the space-time encoder block. The space-time encoder output, therefore, consists of a block of encoded symbols that span p symbol periods in the time dimension and N_t symbols in the spatial dimension. This $N_t \times p$ array of symbols is called a *space-time codeword*. It should be noted that in general $p \neq n$. We observe from this discussion that conventional codewords are represented by vectors and space-time codewords are represented by arrays.

Just as conventional error codes have code rates, so, too, do space-time codes, which we denote by r_s. If we assume that a single $N_t \times p$ ST codeword encodes k modulation symbols, then using the second equality in Eq. 7.1 to define r_s, it follows that

$$r_s = \frac{k}{p}. \tag{7.2}$$

We note that the smallest value of k is 1, which occurs when the ST codeword encodes only 1 modulation symbol, and the largest value of k occurs when all of the symbols in the codeword contain a unique modulated symbol. Thus, $1 \leq k \leq pN_t$, which implies

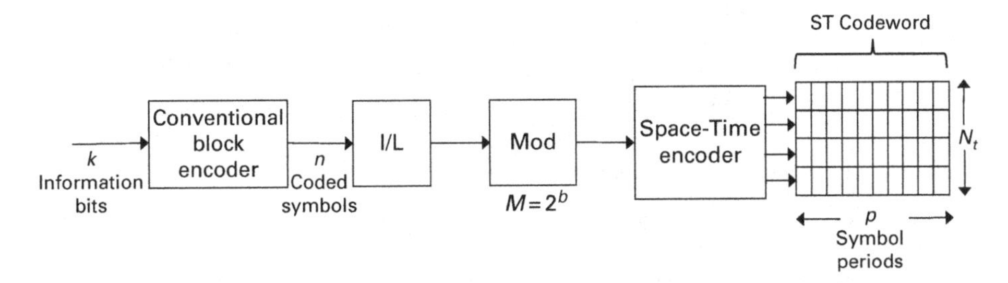

Figure 7.2 A space-time encoded communication system.

that $1/p \leq r_s \leq N_t$. This, of course, is different than with conventional time-domain only coding where $1/n \leq r_t \leq 1$. Despite the fact that ST codes can have code rates larger than 1, it is common in MIMO literature to refer to a ST code with $r_s = 1$ as being a *full-rate code*.

Another point that should be clarified has to do with the terminology that is often used to specify the dimension associated with r_s. Although r_s is a ratio of two numbers, and is, therefore, unitless, it has a physical interpretation. In information theory, it is common to refer to each of the p symbol periods in a ST codeword as a "channel useage" since the channel is being "used" to transmit information during each symbol period. Since k information symbols are transmitted during p uses of the channel, some authors denote the unit of r_s as being *symbols per channel use.*

7.1.2 Spectral efficiency of a STBC

In addition to the code rate, the spectral efficiency, η, is another important parameter used in characterizing communication systems that employ space-time block coding. Because systems that use space-time coding may also use conventional encoding, we seek an expression for the spectral efficiency of such a system in terms of the conventional code rate, r_t, as well as the STBC rate, r_s, and the modulation order, M.

Assume that k modulation symbols are transmitted in an $N_t \times p$ STBC codeword, where each modulation symbol is assumed to be of order M and, therefore, conveys $b = \log_2(M)$ bits of information. This corresponds to $kbr_t = k\log_2(M)r_t$ information bits per codeword; thus, the average information bit rate, R_b, is given by

$$R_b = \frac{kr_t \log_2(M)}{pT_s}, \tag{7.3}$$

where T_s denotes the modulation symbol period. Since the modulation symbol rate is $1/T_s$, it follows that the bandwidth is also approximately $1/T_s$. It follows that the spectral efficiency is given by the following expression:

$$\eta = \frac{R_b}{1/T_s} = \frac{kr_t \log_2(M)}{p} = r_s r_t \log_2(M) \text{ bps/Hz.} \tag{7.4}$$

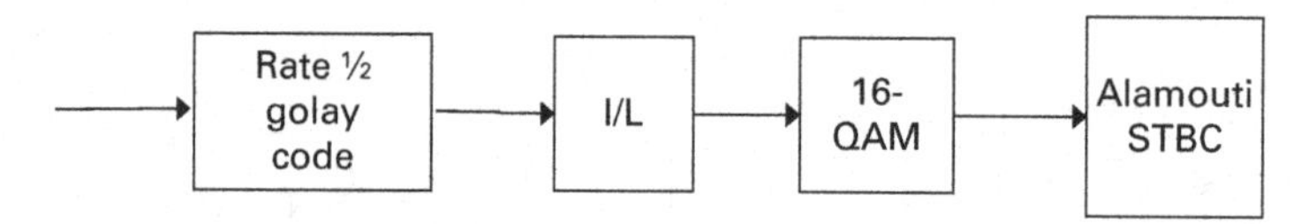

Figure 7.3 Transmitter block diagram for Example 7.1.

Example 7.1 Consider the communication transmitter depicted in Figure 7.3. What is the spectral efficiency of this system?

Answer From the figure, it follows that $r_t = 1/2$ and $M = 16$. Furthermore, since Alamouti space-time block coding is used, it follows that $r_s = 2/2 = 1$ since each Alamouti codeword encodes 2 modulation symbols (k) over 2 symbols in the time dimension (p). Therefore, the spectral efficiency is given by $\eta = r_s r_t \log_2(M) = 1 \times (1/2) \times \log_2(16) = 2$ bps/Hz.

7.1.3 A taxonomy of space-time codes

We conclude this introductory section with a brief description of the classification (i.e., taxonomy) of space-time codes. In this discussion, we examine the fundamental categories of space-time codes, compare their relative advantages and disadvantages, and discuss the fundamental tradeoff that exists between spatial diversity and spatial multiplexing.

Figure 7.4 depicts the taxonomy of space-time codes used in MIMO applications. The top of the figure shows a double-ended arrow that depicts the fundamental tradeoff that exists between diversity, which is focused on improving link reliability in the presence of fading, and capacity, which is achieved through the use of spatial multiplexing techniques. The left-most side of the diagram lists space-time codes that are designed to achieve maximum spatial diversity. Codes in this category include the two major classes of codes: STBCs and STTCs. These codes achieve maximum diversity gain, but have small spatial rates, and hence result in relatively low capacity.

STBCs are divided into two classes: OSTBCs and NOSTBCs. As stated earlier, the focus of this chapter is on OSTBCs because of their simple decoding schemes. In general, all of the space-time codes listed on the left side of the figure have code rates less than or equal to 1 (i.e., $r_s \leq 1$), which means that at most only 1 unique modulation symbol is transmitted per symbol period, resulting in relatively low throughput. However, OSTBCs and STTCs both achieve full diversity, meaning that their diversity gain $G_d = N_t N_r$, which is the maximum achievable value.

Although OSTBCs and STTCs both achieve full diversity, they differ in several important ways. One advantage that STTCs have over OSTBCs is that they are able to achieve coding gain in addition to diversity gain [73], which means that it is possible for STTCs to achieve smaller bit error rates than those associated with OSTBCs for a given average signal-to-noise ratio. The disadvantage of STTCs, however, is that their decoding complexity (measured by the number of trellis states in the decoder)

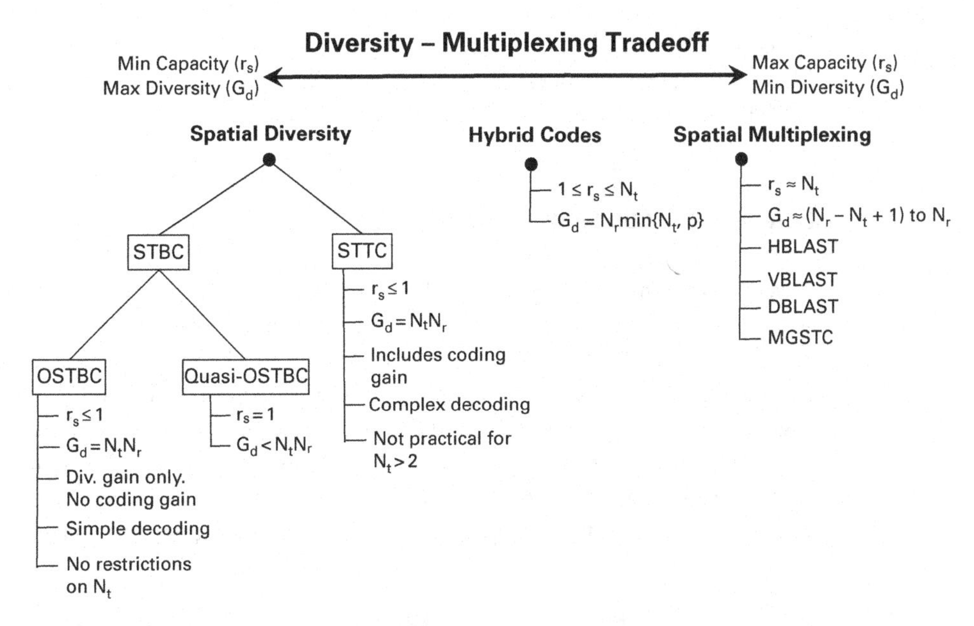

Figure 7.4 Space-time code taxonomy for MIMO applications.

increases exponentially as a function of both the diversity level and the transmission rate [72].

Another class of STBCs is non-orthogonal STBCs, which have been the subject of considerable research. One example of a NOSTBC is the so-called quasi-OSTBCs introduced by Jafarkhani [43]. These codes attempt to address one of the weaknesses of OSTBCs, which is that the only such code that is full rate (i.e., $r_s = 1$) when complex modulation symbols are used is the Alamouti code. The quasi-OSTBCs can achieve full-rate coding for configurations other than $N_t = 2$, which is the only condition with which Alamouti coding is compatible. Unfortunately, this advantage is achieved at the expense of diversity gain; thus, quasi-OSTBCs generally have $G_d < N_t N_r$.

The right-most side of the taxonomy lists techniques for achieving spatial multiplexing, which result in maximum capacity but low diversity gain. In contrast to the space-time coding methods on the left, which have $r_s \leq 1$ and $G_d = N_t N_r$ (at least for the OSTBCs and STTCs), the spatial multiplexing schemes typically have $r_s \approx N_t$ but G_d is at most N_r and is usually significantly smaller depending on the type of decoding scheme used at the receiver. Spatial multiplexing methods include the H-BLAST, V-BLAST, and D-BLAST spatial multiplexing techniques as well as the multi-group space-time coding scheme designated MGSTC in the figure. These techniques are the focus of Chapter 8 and will not be addressed further in this chapter.

In addition to codes and techniques that fall at one end or the other of the diversity–capacity spectrum, there has been research into codes that attempt to compromise between diversity and capacity. These codes have code rates that fall between the minimum, which is 1, and the maximum, which is N_t. For these hybrid codes, $1 < r_s < N_t$. As expected, diversity gain of such codes also falls between the two possible extremes.

In general, the diversity gain of these hybrid codes is given by $G_d = N_r \min\{N_t, p\}$ [57]. More information on codes that achieve a compromise between spatial diversity and spatial multiplexing can be found in [35], [40], and [9].

7.2 Space-time code design criteria

A variety of general design criteria has been developed to aid in the construction of space-time codes. This section describes two design criteria that were first developed by Tarokh, Seshadri, and Calderbank in 1998 [73]. Tarokh *et al.*'s criteria are based on minimizing the pairwise error probability (PEP) between two space-time codewords. We begin by developing a general expression for the PEP and then show how the expression for the PEP under the assumption of Rayleigh fading implies design criteria called the *rank criterion* and the *determinant criterion*. These criteria have been used to develop both STBCs and STTCs, examples of which are presented later in the chapter.

7.2.1 General pairwise error probability expression

The design criteria developed by Tarokh *et al.* in [73] are based on minimizing the PEP. What is the PEP and why is it often used in code design? The use of the PEP for code design is related to the union bound, which, in turn, is a commonly used upper bound on the probability of decoding error. If a ST codeword matrix $\mathbf{S}_i$ is transmitted and the received signal is mistaken to be more like another codeword $\mathbf{S}_k$, then a decoding error occurs and we denote this event by $\mathcal{E}_{ik}$. The probability of a decoding error, P_e, is, therefore, given by

$$P_e = \Pr\left\{\mathcal{E}_{i0} \text{ or } \mathcal{E}_{i1} \text{ or } \ldots \mathcal{E}_{i,i-1} \text{ or } \mathcal{E}_{i,i+1} \text{ or } \ldots \mathcal{E}_{i,K-1}\right\}. \tag{7.5}$$

In general, it can be shown that [83]

$$P_e \leq \sum_{k=0(k\neq i)}^{K-1} \Pr\{\mathcal{E}_{ik}\}, \tag{7.6}$$

which is called the *union bound* of the decoding error. Since $\Pr\{\mathcal{E}_{ik}\}$ depends only on the two codewords $\mathbf{S}_i$ and $\mathbf{S}_k$, an alternative way of writing $\Pr\{\mathcal{E}_{ik}\}$ is as $\Pr\{\mathbf{S}_i \to \mathbf{S}_k\}$, which is called the pairwise probability of error between the two codewords, $\mathbf{S}_i$ and $\mathbf{S}_k$. It follows that

$$P_e \leq \sum_{k=0(k\neq i)}^{K-1} \Pr\{\mathbf{S}_i \to \mathbf{S}_k\}. \tag{7.7}$$

Equation 7.7 shows that the total probability of decoding error is upper bounded by the sum of pairwise error probabilities between the transmitted codeword and all other codewords. It should be noted that when the signal-to-noise ratio is large, the sum in Eq. 7.7 is dominated by the pairwise probability of error between the transmitted codeword and the codeword that is its nearest neighbor in a Euclidean distance sense. It follows that good code design is based on determining those criteria that minimize the pairwise error probability between any two codewords, which we now derive. This derivation is based on the original work by Tarokh *et al.* [73] and the formulation presented in [21].

We begin by assuming that the receiver uses maximum-likelihood (ML) detection. We further assume that the matrix of received symbols, **R**, is given by

$$\mathbf{R} = \sqrt{\rho}\,\mathbf{HS} + \mathbf{Z}, \tag{7.8}$$

where **H** is the $N_r \times N_t$ channel matrix, **S** is the $N_t \times p$ transmitted space-time codeword, and **Z** denotes the $N_r \times p$ noise matrix at the receiver, which is assumed to consist of independent and identically distributed complex Gaussian random variables.

In Appendix C it is shown that the ML estimate of the transmitted codeword is given by

$$\hat{\mathbf{S}} = \arg\min_{\{\mathbf{S}\}} \|\mathbf{R} - \sqrt{\rho}\mathbf{HS}\|_F^2, \tag{7.9}$$

which is a generalization of the expression in Eq. 6.26 for the case where the signals are matrices. It follows from Eq. 7.9 that the PEP between the transmitted space-time codeword, **S**, and some other codeword, **E**, is expressed as

$$\Pr\{\mathbf{S} \to \mathbf{E}|\mathbf{H}\} = \Pr\left\{\|\mathbf{R} - \sqrt{\rho}\mathbf{HS}\|_F^2 > \|\mathbf{R} - \sqrt{\rho}\mathbf{HE}\|_F^2\right\}, \tag{7.10}$$

where the left-hand side of the equation denotes the PEP conditioned on a particular value of the channel matrix. Since $\mathbf{R} = \sqrt{\rho}\mathbf{HS} + \mathbf{Z}$, Eq. 7.10 can be rewritten as

$$\begin{aligned}\Pr\{\mathbf{S} \to \mathbf{E}|\mathbf{H}\} &= \Pr\left\{\|\mathbf{Z}\|_F^2 > \|\mathbf{Z} - \sqrt{\rho}\mathbf{H}(\mathbf{S} - \mathbf{E})\|_F^2\right\} \\ &= \Pr\left\{\|\mathbf{Z}\|_F^2 > \|\mathbf{Z} - \sqrt{\rho}\mathbf{D}\|_F^2\right\},\end{aligned} \tag{7.11}$$

where

$$\mathbf{D} \triangleq \mathbf{H}(\mathbf{S} - \mathbf{E}). \tag{7.12}$$

Since **Z** is dimensioned $N_r \times p$, it follows that

$$\|\mathbf{Z}\|_F^2 = \sum_{k=1}^{p}\sum_{i=1}^{N_r} |z_i(k)|^2, \tag{7.13}$$

where $z_i(k)$ denotes $[\mathbf{Z}]_{i,k}$. Similarly, since **D** has the same dimensions as **Z**, it follows that

$$\|\mathbf{Z} - \sqrt{\rho}\mathbf{D}\|_F^2 = \sum_{k=1}^{p}\sum_{i=1}^{N_r} |z_i(k) - \sqrt{\rho}d_i(k)|^2, \tag{7.14}$$

where $d_i(k)$ denotes the element $[\mathbf{D}]_{i,k}$. Combining Eqs. 7.11, 7.13, and 7.14 results in

$$\Pr\{\mathbf{S} \to \mathbf{E}|\mathbf{H}\} = \Pr\left\{\sum_{k=1}^{p}\sum_{i=1}^{N_r} |z_i(k)|^2 > \sum_{k=1}^{p}\sum_{i=1}^{N_r} |z_i(k) - \sqrt{\rho}d_i(k)|^2\right\}. \tag{7.15}$$

It is straightforward (see problems at end of chapter) to show that Eq. 7.15 can be rewritten as

$$\Pr\{\mathbf{S} \to \mathbf{E}|\mathbf{H}\} = \Pr\left\{\sum_{k=1}^{p}\sum_{i=1}^{N_r} 2\left(z_{Ri}(k)d_{Ri}(k) + z_{Ii}(k)d_{Ii}(k)\right) > \sqrt{\rho}\|\mathbf{D}\|_F^2\right\}, \tag{7.16}$$

where the subscripts "R" and "I" on the variables z and d denote the real and imaginary components of each, respectively.

Since we are assuming the noise is complex Gaussian and the variance of each noise element in $\mathbf{Z}$ is 1, it follows that $z_{Ri}(k)$ and $z_{Ii}(k) \sim \mathcal{N}(0, 1/2)$. Furthermore, we assume quasi-static fading in which $\mathbf{H}$ (and, accordingly, $\mathbf{D}$) are assumed to be constant over an entire codeword (i.e., over p symbol periods). It follows that the left side of the inequality in Eq. 7.16 is Gaussian with zero mean and variance given by

$$\begin{aligned}
&\mathrm{Var}\left\{\sum_{k=1}^{p}\sum_{i=1}^{N_r} 2\left(z_{Ri}(k)d_{Ri}(k) + z_{Ii}(k)d_{Ii}(k)\right)\right\} \\
&= 4\sum_{k=1}^{p}\sum_{i=1}^{N_r} \frac{1}{2}\left(d_{Ri}^2(k) + d_{Ii}^2(k)\right) \\
&= 2\sum_{k=1}^{p}\sum_{i=1}^{N_r} |d_i(k)|^2 \\
&= 2\|\mathbf{D}\|_F^2
\end{aligned} \tag{7.17}$$

We can now express the PEP in terms of the Q-function. Since $(2\pi\sigma^2)^{-1/2}$ $\int_x^\infty \exp(-\zeta^2/2\sigma^2)\,d\zeta = Q(x/\sqrt{\sigma})$, it follows from Eq. 7.16 that

$$\Pr\{\mathbf{S} \to \mathbf{E}|\mathbf{H}\} = Q\left(\frac{\sqrt{\rho}\|\mathbf{D}\|_F^2}{\sqrt{2\|\mathbf{D}\|_F^2}}\right) = Q\left(\sqrt{\rho/2}\|\mathbf{D}\|_F\right). \tag{7.18}$$

At this point in the derivation, Tarokh *et al.* appeal to a well-known upperbound for the Q-function, which states that $Q(x) \leq \frac{1}{2}e^{-x^2/2}$ [73]. It follows from this that an upper bound for the PEP can be expressed as

$$\Pr\{\mathbf{S} \to \mathbf{E}|\mathbf{H}\} \leq \frac{1}{2}\mathrm{e}^{-\rho\|\mathbf{D}\|_F^2/4}. \tag{7.19}$$

The next step in the derivation involves expressing $\|\mathbf{D}\|_F^2$ in terms of the Hermitian matrix $\mathbf{A} \triangleq (\mathbf{S}-\mathbf{E})(\mathbf{S}-\mathbf{E})^H$. We do this by recalling that $\mathbf{D} \triangleq \mathbf{H}(\mathbf{S}-\mathbf{E})$, and we denote each row of $\mathbf{H}$ by $\mathbf{h}_i,\ i = 1, \ldots, N_r$. We can, therefore, write $\mathbf{D}$ as

$$\mathbf{D} = \begin{pmatrix} \mathbf{h}_1(\mathbf{S}-\mathbf{E}) \\ \mathbf{h}_2(\mathbf{S}-\mathbf{E}) \\ \vdots \\ \mathbf{h}_{N_r}(\mathbf{S}-\mathbf{E}) \end{pmatrix}. \tag{7.20}$$

It follows that

$$\|\mathbf{D}\|_F^2 = \sum_{i=1}^{N_r} \|\mathbf{h}_i(\mathbf{S}-\mathbf{E})\|^2. \tag{7.21}$$

It is straightforward to show (see problems at end of chapter) that Eq. 7.21 can be rewritten as

$$\begin{aligned}\|\mathbf{D}\|_F^2 &= \sum_{i=1}^{N_r} \mathbf{h}_i(\mathbf{S}-\mathbf{E})(\mathbf{S}-\mathbf{E})^H\mathbf{h}_i^H \\ &= \sum_{i=1}^{N_r} \mathbf{h}_i\mathbf{A}\mathbf{h}_i^H,\end{aligned} \tag{7.22}$$

where we have defined a new matrix, $\mathbf{A}$, called the *codeword difference matrix*, as follows:

$$\mathbf{A} \triangleq (\mathbf{S}-\mathbf{E})(\mathbf{S}-\mathbf{E})^H. \tag{7.23}$$

Because $\mathbf{A}$ is Hermitian, it follows from matrix theorem Section 1.9.2-(u) that we can apply an eigenvalue decomposition to it and rewrite $\mathbf{A}$ as follows:

$$\mathbf{A} = \mathbf{U}\Lambda\mathbf{U}^H, \tag{7.24}$$

where $\mathbf{U}$ is a unitary matrix and Λ is a diagonal matrix with diagonal elements equal to the eigenvalues of the matrix $\mathbf{A}$. Substituting Eq. 7.24 into Eq. 7.22 results in

$$\|\mathbf{D}\|_F^2 = \sum_{i=1}^{N_r} \mathbf{h}_i\mathbf{U}\Lambda\mathbf{U}^H\mathbf{h}_i^H. \tag{7.25}$$

The last step in deriving a general expression for the upper bound of the PEP involves simplifying Eq. 7.25. To do this, we start by considering the term $\mathbf{U}^H\mathbf{h}_i^H$, which we can write out in long form as follows:

$$\begin{aligned}\mathbf{U}^H\mathbf{h}_i^H &= \begin{pmatrix} u_{11}^* & u_{21}^* & \cdots & u_{N_t,1}^* \\ u_{12}^* & u_{22}^* & \cdots & u_{N_t,2}^* \\ \vdots & \vdots & \ddots & \vdots \\ u_{1,N_t}^* & u_{2,N_t}^* & \cdots & u_{N_t,N_t}^* \end{pmatrix} \begin{pmatrix} h_{i,1}^* \\ h_{i,2}^* \\ \vdots \\ h_{i,N_t}^* \end{pmatrix} \\ &= \begin{pmatrix} \sum_{j=1}^{N_t} u_{j,1}^* h_{i,j}^* \\ \sum_{j=1}^{N_t} u_{j,2}^* h_{i,j}^* \\ \vdots \\ \sum_{j=1}^{N_t} u_{j,N_t}^* h_{i,j}^* \end{pmatrix} \triangleq \begin{pmatrix} \beta_{i,1}^* \\ \beta_{i,2}^* \\ \vdots \\ \beta_{i,N_t}^* \end{pmatrix},\end{aligned} \tag{7.26}$$

where $h_{i,k}$ denotes the kth element of $\mathbf{h}_i$ and $\beta_{i,k} \triangleq \sum_{j=1}^{N_t} u_{j,k} h_{i,j}$; thus, the beta terms are functions of the channel matrix. Using similar reasoning, it is straightforward to show that

$$\mathbf{h}_i \mathbf{U} = \left(\beta_{i,1}, \beta_{i,2}, \ldots, \beta_{i,N_t}\right). \tag{7.27}$$

Thirdly, we consider the term $\Lambda \mathbf{U}^H \mathbf{h}_i^H$. Since $\Lambda = \text{diag}\left\{\lambda_1, \ldots, \lambda_{N_t}\right\}$, it follows that

$$\Lambda \mathbf{U}^H \mathbf{h}_i^H = \begin{pmatrix} \lambda_1 \beta_{i,1}^* \\ \lambda_2 \beta_{i,2}^* \\ \vdots \\ \lambda_{N_t} \beta_{i,N_t} \end{pmatrix}. \tag{7.28}$$

Combining Eqs. 7.26, 7.27, and 7.28 yields

$$\begin{aligned} \|\mathbf{D}\|^2 &= \sum_{i=1}^{N_r} \left\{ \left(\beta_{i,1}, \beta_{i,2}, \ldots, \beta_{i,N_t}\right) \begin{pmatrix} \lambda_1 \beta_{i,1}^* \\ \lambda_2 \beta_{i,2}^* \\ \vdots \\ \lambda_{N_t} \beta_{i,N_t} \end{pmatrix} \right\} \\ &= \sum_{i=1}^{N_r} \left(\beta_{i,1} \lambda_1 \beta_{i,1}^* + \beta_{i,2} \lambda_1 \beta_{i,2}^* + \cdots + \beta_{i,N_t} \lambda_1 \beta_{i,N_t}^*\right) \\ &= \sum_{i=1}^{N_r} \sum_{j=1}^{N_t} \lambda_j |\beta_{i,j}|^2. \end{aligned} \tag{7.29}$$

Finally, combining Eqs. 7.19 and 7.29 results in the following general expression for the PEP upper bound:

$$\begin{aligned} \Pr\{\mathbf{S} \to \mathbf{E} | \mathbf{H}\} &\leq \frac{1}{2} \exp\left\{ \frac{\rho}{4} \sum_{i=1}^{N_r} \sum_{j=1}^{N_t} \lambda_j |\beta_{i,j}|^2 \right\} \\ &= \frac{1}{2} \prod_{i=1}^{N_r} \prod_{j=1}^{N_t} \exp\left(-\frac{\rho}{4} \lambda_j |\beta_{i,j}|^2\right). \end{aligned} \tag{7.30}$$

Next, we consider the form of the upper bound of the PEP under the assumption of Rayleigh fading. The resulting expression provides insight into how to develop criteria for designing space-time codes that achieve good performance in environments where Rayleigh fading exists.

7.2.2 Pairwise error probability in Rayleigh fading

In Rayleigh fading, the elements of $\mathbf{H}$ are independent, zero-mean, complex Gaussian random variables with variance equal to 1/2 per dimension. Thus,

$$\beta_{i,j} = \sum_{k=1}^{N_t} u_{k,j} h_{i,k} \sim \mathcal{CN}(0, 1), \tag{7.31}$$

which follows from the fact that a linear combination of Gaussian random variables is also Gaussian, as well as from the following properties:

$$\mathbb{E}\left\{\beta_{i,j}\right\} = \sum_{k=1}^{N_t} u_{k,j}\mathbb{E}\left\{h_{i,j}\right\} = 0 \tag{7.32}$$

and

$$\text{Var}\left\{\beta_{i,j}\right\} = \sum_{k=1}^{N_t} |u_{k,j}|^2 \text{Var}\left\{h_{i,j}\right\} = \sum_{k=1}^{N_t} |u_{k,j}|^2 = 1, \tag{7.33}$$

where the last equality follows from the property of unitary matrices.

Next, we consider the distribution of $|\beta_{i,j}|^2$, which follows from Eqs. 7.32 and 7.33, and from the fact that $|\beta_{i,j}|^2 = \mathfrak{Re}^2\left\{\beta_{i,j}\right\} + \mathfrak{Im}^2\left\{\beta_{i,j}\right\}$. In general, if $Y = \sum_{i=1}^{N} X_i^2$, where $X_i \sim \mathcal{N}(0, \sigma^2)$, then the distribution of $Y, f_Y(y)$, is given by

$$f_Y(y) = \frac{1}{\sigma^N 2^{N/2}\Gamma(N/2)} y^{N/2-1} \mathrm{e}^{-y/2\sigma^2}, \tag{7.34}$$

which is the chi-square (or gamma) distribution with N degrees of freedom. It follows that $|\beta_{i,j}|^2$ has the following distribution:

$$f_{|\beta_{i,j}|^2}(y) = \frac{1}{2\sigma^2}\mathrm{e}^{-y/2\sigma^2}. \tag{7.35}$$

Since σ^2 refers to the variance of a single dimension of $\beta_{i,j}$ (i.e., $\sigma^2 = 1/2$), Eq. 7.35 simplifies to

$$f_{|\beta_{i,j}|^2}(y) = \mathrm{e}^{-y}. \tag{7.36}$$

We can, therefore, use Eqs. 7.30 and 7.36 to obtain an expression for the *unconditional* upper bound for the PEP in Rayleigh fading under quasi-static conditions. We do this by integrating the right-hand side of the inequality in Eq. 7.30 over the term that contains channel effects (i.e., over $|\beta_{i,j}|^2$). It follows that the unconditional PEP upper bound is given by

$$\begin{aligned}
\Pr\{\mathbf{S} \rightarrow \mathbf{E}\} &\leq \frac{1}{2}\int_0^\infty \Pr\{\mathbf{S} \rightarrow \mathbf{E}\} f_{|\beta_{i,j}|^2}(y)\, dy \\
&= \frac{1}{2}\int_0^\infty \prod_{i=1}^{N_r}\prod_{j=1}^{N_t} \exp\left(-\frac{\rho}{4}\lambda_j|\beta_{i,j}|^2\right)\exp(-|\beta_{i,j}|^2)\, d|\beta_{i,j}|^2 \\
&= \frac{1}{2}\prod_{i=1}^{N_r}\prod_{j=1}^{N_t}\int_0^\infty \exp\left(-\frac{\rho}{4}\lambda_j y\right)\mathrm{e}^{-y}\, dy \\
&= \frac{1}{2}\prod_{i=1}^{N_r}\prod_{j=1}^{N_t}\left(\frac{1}{1+\frac{\rho\lambda_j}{4}}\right) \\
&= \frac{1}{2}\left[\frac{1}{\prod_{j=1}^{N_t}\left(1+\frac{\rho\lambda_j}{4}\right)}\right]^{N_r}.
\end{aligned} \tag{7.37}$$

We will use this bound for the PEP to develop code design principles for ST codes over quasi-static Rayleigh fading channels. Before we do that, however, we present the PEP bound for the general case of a Rician fading channel.

7.2.3 Pairwise error probability in Rician fading

A similar bound on the PEP is derived in Tarokh *et al.*'s paper [73] for the case where the fading is quasi-static Rician. The authors show that in Rician fading

$$\Pr\{\mathbf{S} \to \mathbf{E}\} \leq \frac{1}{2}\prod_{i=1}^{N_r}\left(\prod_{j=1}^{N_t}\frac{1}{1+\frac{\rho\lambda_j}{4}}\exp\left(-\frac{K_{i,j}\frac{\rho\lambda_j}{4}}{1+\frac{\rho\lambda_i}{4}}\right)\right), \tag{7.38}$$

where $K_{i,j} \triangleq |\mathbb{E}\left\{\beta_{i,j}\right\}|^2$, which, under the general Rician assumption, is not necessarily equal to zero. Clearly, Eq. 7.38 reduces to the Rayleigh expression in Eq. 7.37 when $K_{i,j} = 0$. We will use this expression, along with the corresponding result for Rayleigh fading, to derive the design criteria for space-time codes.

7.2.4 Summary of design criteria

The first design criterion is based on maximizing the diversity gain of a space-time code. Since diversity gain is defined as the slope of the decoding error versus signal-to-noise ratio curve at large signal-to-noise ratios, we consider the PEP bounds in the limit as $\rho \to \infty$. In this limit, the Rayleigh and Rician PEP bounds are given by the following:

Rayleigh Bound

$$\begin{aligned}\lim_{\rho\to\infty}\Pr\{\mathbf{S}\to\mathbf{E}\} &\leq \frac{1}{2}\left[\left(\frac{\rho}{4}\right)^r\prod_{j=1}^{r}\lambda_j\right]^{-N_r}\\ &= \frac{1}{2}\left[\frac{1}{4}\left(\prod_{j=1}^{r}\lambda_j\right)^{1/r}\rho\right]^{-rN_r}.\end{aligned} \tag{7.39}$$

Rician Bound

$$\begin{aligned}\lim_{\rho\to\infty}\Pr\{\mathbf{S}\to\mathbf{E}\} &\leq \frac{1}{2}\prod_{i=1}^{N_r}\prod_{j=1}^{r}\left(\frac{\rho}{4}\right)^{-1}\lambda_j e^{-K_{i,j}}\\ &= \frac{1}{2}\left[\frac{1}{4}\left(\prod_{j=1}^{r}\lambda_j\right)^{1/r}\left(\prod_{i=1}^{N_r}\prod_{j=1}^{r}e^{-K_{i,j}}\right)^{-1/rN_r}\rho\right]^{-rN_r}.\end{aligned} \tag{7.40}$$

Note that in taking the limits above, we have explicitly taken into account that in general there are only $r \leq N_t$ non-zero eigenvalues of $\mathbf{A}$.

Before proceeding, it is useful to recall the relationship between the limiting bounds listed above and the overall decoding error. Recall that in the large SNR regime Eq. 7.7 implies that the overall decoding error is bounded by the PEP between the transmitted codeword and its nearest neighbor. Therefore, if we interpret the Rayleigh and Rician bounds as PEPs between the transmitted codeword and its nearest neighbor, we can interpret the bounds in Eqs. 7.39 and 7.40 as expressions for the bounds of the overall decoding error (in Rayleigh and Rician fading) in the limit as ρ becomes large.

Next, we can use the expressions for the PEP to write down equations for the coding and diversity gains as defined by Eq. 1.3. Before doing so, however, we first need to establish the relationship between the PEP, which we have argued is equal to the probability of codeword error, P_w, and the bit error probability, since the coding and diversity gains are defined in terms of the bit error probability. We start by assuming that the codewords have N modulation symbols. It follows that the modulation symbol error, P_m, is related to P_w as follows:

$$\begin{aligned} P_w &= 1 - (1 - P_m)^N \\ &\approx NP_m, \end{aligned} \tag{7.41}$$

where the second approximation holds when $P_m << 1$, which occurs when the SNR is large. Furthermore, if the modulation order is M, then when Gray coding is used, the bit error probability is given by [59]

$$P_b \approx \frac{P_w}{N \log_2(M)}. \tag{7.42}$$

Thus, we see that the bit error probability is proportional to the codeword error probability, which means that we can use the expression for the PEP in the high SNR limit to infer the coding and diversity gains defined in Eq. 1.3, since the proportionality constant is not relevant for the purpose of estimating G_c and G_d.

Having established the significance of the limiting bounds, note that both expressions have the form given in Eq. 1.3, which allows us to immediately write expressions for the diversity and coding gains. It follows by inspection that the diversity gain is given by rN_r in both Rayleigh and Rician fading. Therefore, to achieve the maximum diversity, N_tN_r, r must equal N_t; that is, $\mathbf{A}$ must be full rank for all codeword pairs $\mathbf{S}$ and $\mathbf{E}$, $\mathbf{S} \neq \mathbf{E}$. This criterion, the first of two, is called the *rank criterion*, which addresses the design of codes that achieve maximum diversity order.

The second design criterion has to do with maximizing the coding gain, which is defined by the term G_c in Eq. 1.3. It follows from the form of Eq. 1.3 that the coding gain (at large SNR) for Rayleigh and Rician fading is given by:

Coding gain for Rayleigh fading

$$G_c = \frac{1}{4}\left(\prod_{j=1}^{r} \lambda_j\right)^{1/r}. \tag{7.43}$$

Coding gain for Rician fading

$$G_c = \frac{1}{4}\left(\prod_{j=1}^{r} \lambda_j\right)^{1/r} \left(\prod_{i=1}^{N_r}\prod_{j=1}^{r} \mathrm{e}^{-K_{i,j}}\right)^{-1/rN_r}. \tag{7.44}$$

From matrix theorem Section 1.9.2-(p), it follows that if $r = N_t$ (i.e., **A** is full rank), then the product of the eigenvalues in Eqs. 7.43 and 7.44 is equal to the determinant of **A**. We conclude that maximizing the coding gain involves maximizing the determinant of **A**. Combining these observations, we arrive at the following design criteria for space-time codes:

Design criteria for space-time codes

- ***The rank criterion*** This criterion applies when the fading is either Rayleigh or Rician. In order to achieve full diversity (i.e., $G_d = N_t N_r$), a space-time code must have the property that the matrix $\mathbf{A} = (\mathbf{S} - \mathbf{E})(\mathbf{S} - \mathbf{E})^H$ is full rank for all pairs of distinct codewords **S** and **E**. If the minimum rank of **A** over all pairs of codewords is only $r \leq N_t$, then the diversity gain will be rN_r.
- ***The determinant criterion (Rayleigh fading)*** This criterion applies when the fading is Rayleigh. Let the target diversity gain be rN_r. The design strategy is to first achieve the rank criterion, then, secondly, find a family of codewords such that the minimum value of $\prod_{j=1}^{r} \lambda_j$ is maximized. If full diversity is the goal, then this criterion reduces to finding a family of codewords that maximizes the minimum value of det(**A**) among all codeword pairs in the family.
- ***The coding advantage criterion (Rician fading)*** This criterion applies when the fading is Rician. Let the target diversity gain be rN_r. Then the design goal is to first achieve the rank criterion, then, secondly, find a family of codewords such that the minimum value of $\left(\prod_{j=1}^{r} \lambda_j\right)^{1/r} \left(\prod_{i=1}^{N_r} \prod_{j=1}^{r} \mathrm{e}^{-K_{i,j}}\right)^{1/rN_r}$ is maximized among all codeword pairs.

It should be noted that the rank criterion is the more important code design criterion since it has to do with the slope of the code error rate versus SNR curve. The code gain, in contrast, has to do with shifting the error curve to the left without changing its slope, which is not as beneficial as increasing the slope as the SNR gets large. These results tell us that when comparing multiple space-time codes, the first criterion is to check the minimum rank of **A** for each one. In general, the code with the largest minimum rank is preferable because that code will have the largest diversity gain. If two or more codes have full rank, then check the determinants of **A** for each one. The code with the largest minimum determinant among all codeword pairs is generally the better code in Rayleigh or Rician fading.

Example 7.2 Show that the Alamouti code meets the rank criterion and is, therefore, full rank when used in Rayleigh fading.

Answer Consider two Alamouti codewords **S** and **E** as follows:

$$\mathbf{S} = \begin{pmatrix} s_1 & -s_2^* \\ s_2 & s_1^* \end{pmatrix} \text{ and } \mathbf{E} = \begin{pmatrix} e_1 & -e_2^* \\ e_2 & e_1^* \end{pmatrix}. \tag{7.45}$$

Table 7.1 det(**A**) values for the Alamouti code using BPSK.

Bits in **S**	Bits in **E**	**S**	**E**	**A**	det(**A**)
$\{1,1\}$	$\{1,-1\}$	$\begin{pmatrix} 1 & -1 \\ 1 & 1 \end{pmatrix}$	$\begin{pmatrix} 1 & 1 \\ -1 & 1 \end{pmatrix}$	$\begin{pmatrix} 4 & 0 \\ 0 & 4 \end{pmatrix}$	16
$\{1,1\}$	$\{-1,1\}$	$\begin{pmatrix} 1 & -1 \\ 1 & 1 \end{pmatrix}$	$\begin{pmatrix} -1 & -1 \\ 1 & -1 \end{pmatrix}$	$\begin{pmatrix} 4 & 0 \\ 0 & 4 \end{pmatrix}$	16
$\{1,1\}$	$\{-1,-1\}$	$\begin{pmatrix} 1 & -1 \\ 1 & 1 \end{pmatrix}$	$\begin{pmatrix} -1 & 1 \\ -1 & -1 \end{pmatrix}$	$\begin{pmatrix} 8 & 0 \\ 0 & 8 \end{pmatrix}$	64

The **A** matrix is, therefore, given by

$$\mathbf{A} = (\mathbf{S} - \mathbf{E})(\mathbf{S} - \mathbf{E})^H$$

$$= \begin{pmatrix} |s_1 - e_1|^2 + |s_2 - e_2|^2 & 0 \\ 0 & |s_1 - e_1|^2 + |s_2 - e_2|^2 \end{pmatrix}. \tag{7.46}$$

Since the two codewords are assumed to be distinct, the diagonal elements in **A** are non-zero; thus, the rank of **A** is 2, which means that **A** is full rank and the diversity gain of the Alamouti code is $2N_r$, which is full diversity.

Example 7.3 Compute the coding gain of the Alamouti code at large SNR. Assume the modulation is BPSK with modulation symbols chosen from the alphabet $\{+1, -1\}$.

Answer To do this, evaluate Eq. 7.43 for the pair of codewords that yield the smallest product of eigenvalues. Since the previous example proved that the Alamouti code is full rank, under that condition, $r = N_t$ and, therefore, Eq. 7.43 can be expressed as follows:

$$G_c = \frac{1}{4} \min \left\{ [\det(\mathbf{A})]^{1/r} \right\}. \tag{7.47}$$

Without loss of generality, we assume that the two modulation symbols associated with the Alamouti codeword **S** are $\{1, 1\}$. We then compute the codewords associated with all the other bit combinations (i.e., $\{1, -1\}$, $\{-1, 1\}$, and $\{-1, -1\}$ and evaluate the determinants associated with **A**. The results are summarized in Table 7.1.

Since the rank of **A** is 2 for Alamouti coding, it follows that $r = 2$ and that $G_c = 1$, which is consistent with our claim that OSTBCs do not have coding gain.

As a side note, the results in Table 7.1 demonstrate that the rank of **A** for all distinct code pairs is 2, which confirms that Alamouti achieves full diversity.

As mentioned earlier, the design criteria presented in this section provide a framework for searching out good space-time codes; however, they do not explain how to find codes that meet those criteria. In the next section, we present a variety of orthogonal space-time block codes that achieve full diversity.

7.3 Orthogonal space-time block codes

An OSTBC is defined by a family of $N_t \times p$ codewords where:

a) The entries in each of the codeword matrices are linear combinations of the k encoded modulated symbols; and
b) The rows are orthogonal (i.e., $\text{row}_i \cdot \text{row}_j^H = 0,\ i \neq j$).

***An example*: Alamouti coding**
Alamouti coding is an example of an OSTBC. The transmission matrix for Alamouti coding is given by

$$\mathcal{G}_2 = \begin{pmatrix} s_1 & -s_2^2 \\ s_2 & s_1^* \end{pmatrix}, \tag{7.48}$$

where the variables s_1 and s_2 represent the two modulation symbols that are encoded in each codeword. It follows that

$$\begin{aligned} \text{row}_1 \cdot \text{row}_2^H &= \left(s_1, -s_2^*\right) \begin{pmatrix} s_2^* \\ s_1 \end{pmatrix} \\ &= s_1 s_2^* - s_2^* s_1 = 0. \end{aligned} \tag{7.49}$$

This shows that Alamouti coding is orthogonal.

A word of clarification regarding the definition of OSTBCs is in order. In the definition above, we have defined orthogonality to mean that the rows of the codeword are orthogonal. This is true as long as the number of columns is greater than or equal to the number of rows in the codeword, which occurs when the codeword is dimensioned $N_t \times p$, since, in general, $p \geq N_t$. Some authors use the opposite convention in which the codewords are dimensioned $p \times N_t$. When this is the case, the columns are orthogonal instead of the rows and item (b) above should be changed accordingly. If $p = N_t$, which is the case for Alamouti coding, then both the rows and columns are orthogonal.

Next, we present a variety of different examples of OSTBCs categorized according to shape (i.e., whether the codewords are square or non-square) and according to the type of symbols that can be used (i.e., whether the encoded symbols are real or complex). The examples presented in the following sections should be regarded as illustrative rather than comprehensive – indeed, there is an infinite number of possible OSTBC designs. The examples shown below have been developed by different authors using the design criteria presented in the previous section.

7.3.1 Real, square OSTBCs

This category of OSTBCs have codewords that are square (i.e., $p = N_t$) and only permit the use of real symbols (i.e., BPSK, PAM, and ASK). The following codes, which were first described by Tarokh, Jafarkhani, and Calderbank [71], are the only three known real, square OSTBC designs:

$$\mathcal{D}_{RS,2\times 2} = \begin{bmatrix} s_1 & -s_2 \\ s_2 & s_1 \end{bmatrix}, \tag{7.50}$$

$$\mathcal{D}_{RS,4\times 4} = \begin{bmatrix} s_1 & -s_2 & -s_3 & -s_4 \\ s_2 & s_1 & s_4 & -s_3 \\ s_3 & -s_4 & s_1 & s_2 \\ s_4 & s_3 & -s_2 & s_1 \end{bmatrix}, \tag{7.51}$$

$$\mathcal{D}_{RS,8\times 8} = \begin{bmatrix} s_1 & -s_2 & -s_3 & -s_4 & -s_5 & -s_6 & -s_7 & -s_8 \\ s_2 & s_1 & -s_4 & s_3 & -s_6 & s_5 & s_8 & -s_7 \\ s_3 & s_4 & s_1 & -s_2 & -s_7 & -s_8 & s_5 & s_6 \\ s_4 & -s_3 & s_2 & s_1 & -s_8 & s_7 & -s_6 & s_5 \\ s_5 & s_6 & s_7 & s_8 & -s_1 & -s_2 & -s_3 & -s_4 \\ s_6 & -s_5 & s_8 & -s_7 & s_2 & s_1 & s_4 & -s_3 \\ s_7 & -s_8 & -s_5 & s_6 & s_3 & -s_4 & s_1 & s_2 \\ s_8 & s_7 & -s_6 & -s_5 & s_4 & s_3 & -s_2 & s_1 \end{bmatrix}. \tag{7.52}$$

An examination of these matrices shows that in all cases, $k = p$, so $r_s = 1$ for this class of OSTBCs (i.e., they are full rate space-time codes). Despite the severe restrictions imposed by these codes, $D_{RS,4\times 4}$ and $D_{RS,8\times 8}$ permit the number of transmit antennas to be greater than two, which is the limitation associated with Alamouti codes.

7.3.2 Real, non-square OSTBCs

A second class of OSTBCs is similar to the first category except that the shape is non-square (i.e. $p \neq N_t$). It turns out that by lifting the restriction that the codeword be square, the number of codes increases from only three to an infinite number. In fact, it is proven by Tarokh *et al.* in [71] that real, non-square OSTBCs exist for any value of N_t. The following are some examples presented in [71];

$$\mathcal{D}_{RNS,3\times 4} = \begin{bmatrix} s_1 & -s_2 & -s_3 & -s_4 \\ s_2 & s_1 & s_4 & -s_3 \\ s_3 & -s_4 & s_1 & s_2 \end{bmatrix}, \tag{7.53}$$

$$\mathcal{D}_{RNS,5\times 8} = \begin{bmatrix} s_1 & -s_2 & -s_3 & -s_4 & -s_5 & -s_6 & -s_7 & -s_8 \\ s_2 & s_1 & -s_4 & s_3 & -s_6 & s_5 & s_8 & -s_7 \\ s_3 & s_4 & s_1 & -s_2 & -s_7 & -s_8 & s_5 & s_6 \\ s_4 & -s_3 & s_2 & s_1 & -s_8 & s_7 & -s_6 & s_5 \\ s_5 & s_6 & s_7 & s_8 & s_1 & -s_2 & -s_3 & -s_4 \end{bmatrix}, \tag{7.54}$$

$$\mathcal{D}_{RNS,6\times 8} = \begin{bmatrix} s_1 & -s_2 & -s_3 & -s_4 & -s_5 & -s_6 & -s_7 & -s_8 \\ s_2 & s_1 & -s_4 & s_3 & -s_6 & s_5 & s_8 & -s_7 \\ s_3 & s_4 & s_1 & -s_2 & -s_7 & -s_8 & s_5 & s_6 \\ s_4 & -s_3 & s_2 & s_1 & -s_8 & s_7 & -s_6 & s_5 \\ s_5 & s_6 & s_7 & s_8 & s_1 & -s_2 & -s_3 & -s_4 \\ s_6 & -s_5 & s_8 & -s_7 & s_2 & s_1 & s_4 & -s_3 \end{bmatrix}, \tag{7.55}$$

$$\mathcal{D}_{RNS,7\times 8} = \begin{bmatrix} s_1 & -s_2 & -s_3 & -s_4 & -s_5 & -s_6 & -s_7 & -s_8 \\ s_2 & s_1 & -s_4 & s_3 & -s_6 & s_5 & s_8 & -s_7 \\ s_3 & s_4 & s_1 & -s_2 & -s_7 & -s_8 & s_5 & s_6 \\ s_4 & -s_3 & s_2 & s_1 & -s_8 & s_7 & -s_6 & s_5 \\ s_5 & s_6 & s_7 & s_8 & s_1 & -s_2 & -s_3 & -s_4 \\ s_6 & -s_5 & s_8 & -s_7 & s_2 & s_1 & s_4 & -s_3 \\ s_7 & -s_8 & -s_5 & s_6 & s_3 & -s_4 & s_1 & s_2 \end{bmatrix}. \tag{7.56}$$

Tarokh *et al.* prove that the code rate for all real, non-square OSTBCs is 1. Examination of the examples above confirms that, indeed, $r_s = k/p = 1$.

For any given value of N_t, Tarokh *et al.* prove that a real, non-square OSTBC only exists for p greater than or equal to some minimum value, $p_{\min}$, given by the following expression:

$$p_{\min} = \min\left(2^{4c+d}\right), \tag{7.57}$$

where the minimization is taken over the set

$$\left\{c, d \mid c \geq 0, 0 \leq d < 4, 8c + 2^d \geq N_t\right\}. \tag{7.58}$$

The following example illustrates how to perform the minimization in Eq. 7.58.

Example 7.4 Assume $N_t = 3$. Compute the minimum value of p for which a real, non-square OSTBC exists.

Answer To answer this question, write down all the parameters involved in computing Eqs. 7.57 and 7.58 and determine the smallest value of the expression 2^{4c+d} for those values of c and d defined by Eq. 7.58. A systematic way to do this is to start with $c = 0$ and compute the values of 2^{4c+d} for $d = 0, 1, \ldots, 3$. Next, set $c = 1$ and repeat for all values of d. Repeat this process until the minimum value of 2^{4c+d} is obvious. For example, consider the following entries:

c	d	$8c + 2^d$	$4c + d$	2^{4c+d}
0	0	1	0	1
0	1	2	1	2
0	2	4	2	4
0	3	8	3	8
1	0	9	4	16
1	1	10	5	32

These entries show that the smallest value of 2^{4c+d} for which $8c+2^d \geq N_t$ is 4. Thus, $p_{\min} = 4$, which is consistent with the OSTBC in Eq. 7.53.

7.3.3 Complex OSTBCs

Complex OSTBCs are orthogonal codes that allow the codeword symbols to be complex, which includes M-PSK and QAM symbols. Tarokh *et al.* [71] show that complex OSTBCs exist for any value of N_t as long as the user is willing to tolerate code rates less than 1. In fact, they show that the Alamouti code is the only complex OSTBC that achieves a code rate equal to 1. OSTBCs that support $N_t > 2$ all have code rates less than 1.

The following show examples of complex OSTBCs:

$N_t = 2, r_s = 1$

$$\mathcal{G}_2 = \begin{bmatrix} s_1 & -s_2^* \\ s_2 & s_1^* \end{bmatrix}, \tag{7.59}$$

$N_t = 3, r_s = 3/4$

$$\mathcal{D}_{c,3\times 4} = \begin{bmatrix} s_1 & -s_2^* & s_3^* & 0 \\ s_2 & s_1^* & 0 & s_3^* \\ s_3 & 0 & -s_1^* & -s_2^* \end{bmatrix}, \tag{7.60}$$

$N_t = 4, r_s = 3/4$

$$\mathcal{D}_{c,4\times 4} = \begin{bmatrix} s_1 & -s_2* & s_3^* & 0 \\ s_2 & s_1^* & 0 & s_3^* \\ s_3 & 0 & -s_1^* & -s_2^* \\ 0 & s_3 & s_2 & -s_1 \end{bmatrix}, \tag{7.61}$$

$N_t = 3, r_s = 3/4$

$$\mathcal{H}_3 = \begin{bmatrix} s_1 & -s_2^* & \frac{s_3^*}{\sqrt{2}} & \frac{s_3^*}{\sqrt{2}} \\ s_2 & s_1^* & \frac{s_3^*}{\sqrt{2}} & \frac{-s_3^*}{\sqrt{2}} \\ \frac{s_3}{\sqrt{2}} & \frac{s_3}{\sqrt{2}} & \frac{(-s_1-s_1^*+s_2-s_2^*)}{2} & \frac{(s_2+s_2^*+s_1-s_1^*)}{2} \end{bmatrix}, \tag{7.62}$$

$N_t = 4, r_s = 3/4$

$$\mathcal{H}_4 = \begin{bmatrix} s_1 & -s_2 & \frac{s_3^*}{\sqrt{2}} & \frac{s_3^*}{\sqrt{2}} \\ s_2 & s_1 & \frac{s_3^*}{\sqrt{2}} & \frac{-s_3^*}{\sqrt{2}} \\ \frac{s_3}{\sqrt{2}} & \frac{s_3}{\sqrt{2}} & \frac{(-s_1-s_1^*+s_2-s_2^*)}{2} & \frac{(s_2+s_2^*+s_1-s_1^*)}{2} \\ \frac{s_3}{\sqrt{2}} & \frac{-s_3}{\sqrt{2}} & \frac{(-s_2-s_2^*+s_1-s_1^*)}{2} & \frac{-(s_1+s_1^*+s_2-s_2^*)}{2} \end{bmatrix}, \tag{7.63}$$

$N_t = 3, r_s = 1/2$

$$\mathcal{G}_3 = \begin{bmatrix} s_1 & -s_2 & -s_3 & -s_4 & s_1^* & -s_2^* & -s_3^* & -s_4^* \\ s_2 & s_1 & s_4 & -s_3 & s_2^* & s_1^* & s_4^* & -s_3^* \\ s_3 & -s_4 & s_1 & s_2 & s_3^* & -s_4^* & s_1^* & s_2^* \end{bmatrix}, \tag{7.64}$$

$N_t = 4, r_s = 1/2$

$$\mathcal{G}_4 = \begin{bmatrix} s_1 & -s_2 & -s_3 & -s_4 & s_1^* & -s_2^* & -s_3^* & -s_4^* \\ s_2 & s_1 & s_4 & -s_3 & s_2^* & s_1^* & s_4^* & -s_3^* \\ s_3 & -s_4 & s_1 & s_2 & s_3^* & -s_4^* & s_1^* & s_2^* \\ s_4 & s_3 & -s_2 & s_1 & s_4^* & s_3^* & -s_2^* & s_1^* \end{bmatrix}. \tag{7.65}$$

We see from this discussion that the Alamouti code is part of the larger class of complex OSTBCs. As we saw in Chapter 6, the maximum likelihood decoding logic of the Alamouti code is extremely simple. This simplicity is due to its orthogonal structure. As a result, all OSTBCs have extremely simple decoding algorithms, which is one of their primary attractions. In the next section, we discuss decoding OSTBCs in greater detail.

7.3.4 Decoding OSTBCs

In Chapter 6, we saw that maximum likelihood decoding of Alamouti codes results in extremely simple receiver algorithms. Since Alamouti codes are a subclass of a larger category of codes called OSTBCs, we might expect that all OSTBCs have simple decoding structures. Such is the case, as we discuss in this section. To show this, we consider the maximum likelihood decoding rule for a general $N_t \times p$ space-time codeword, $\mathbf{S}$, embedded in AWGN. For consistency with Alamouti's analysis in [6], and Tarokh *et al.*'s work in [72], which we use in Appendix D, we temporarily divert from our usual normalization convention and write the system equation without an explicit reference to the signal-to-noise ratio as follows:

$$\mathbf{R} = \mathbf{H}\mathbf{S} + \mathbf{Z}. \tag{7.66}$$

The maximum likelihood decoding rule for such a system is given by

$$\hat{\mathbf{S}} = \arg\min_{\{\mathbf{S}\}} \|\mathbf{R} - \mathbf{H}\mathbf{S}\|_F^2 \tag{7.67}$$

$$= \arg\min_{\{\mathbf{S}\}} \left[\sum_{k=1}^{p} \sum_{i=1}^{N_r} |r_i(k) - \sum_{j=1}^{N_t} h_{i,j} c_j(k)|^2 \right], \tag{7.68}$$

where we distinguish between the elements of $\mathbf{S}$, which we denote by $\{c_j(k)\}$ and the values they are assigned, which are chosen from the set $\{s_1, s_2, \ldots, s_K\}$ in a way that depends on the specific OSTBC.

We will now show that when we apply the Alamouti code to this general expression for the maximum likelihood estimate of the transmitted symbol block, we obtain the same decoding rules that we derived in Eq. 6.30.

With Alamouti coding, $p = N_t = 2$ and the following hold: $c_1(1) = s_1$, $c_1(2) = -s_2^*$, $c_2(1) = s_2$, and $c_2(2) = s_1^*$; where, s_1 and s_2 denote the information symbols that are being mapped to the elements of the Alamouti codeword. Maximum likelihood decoding of the Alamouti code involves minimizing the right side of Eq. 7.68, which is given by

$$\begin{aligned}
\mathcal{M} &= \sum_{i=1}^{N_r} \left[|r_i(1) - h_{i,1}s_1 - h_{i,2}s_2|^2 + |r_i(2) + h_{i,1}s_2^* - h_{i,2}s_1^*|^2 \right] \\
&= \sum_{i=1}^{N_r} \left[\left(r_i(1) - h_{i,1}s_1 - h_{i,2}s_2\right)\left(r_i(1)^* - h_{i,1}^* s_1^* - h_{i,2}^* s_2^*\right) \right. \\
&\qquad \left. + \left(r_i(2) + h_{i,1}s_2^* - h_{i,1}s_1^*\right)\left(r_i(2)^* + h_{i,1}^* s_2 - h_{i,1}^* s_1\right) \right] \\
&= \sum_{i=1}^{N_r} \left(|r_i(1)|^2 + |r_i(2)|^2\right) + \sum_{i=1}^{N_r} \left(|h_{i,1}|^2 + |h_{i,2}|^2\right) \sum_{k=1}^{2} |s_k|^2 \\
&\quad - \sum_{i=1}^{N_r} \left[r_i(1)h_{i,1}^* s_1^* + r_i(1)^* h_{i,1}s_1 + r_i(2)h_{i,2}^* s_1 + r_i(2)^* h_{i,2} s_1^* \right] \\
&\quad - \sum_{i=1}^{N_r} \left[r_i(1)h_{i,2}^* s_2^* + r_i(1)^* h_{i,2}s_2 - r_i(2)h_{i,1}^* s_2 - r_i(2)^* h_{i,1} s_2^* \right].
\end{aligned} \tag{7.69}$$

Since the first summation after the last equal sign only involves the received signals and not the coded symbols, we can ignore that part of the expression when performing the minimization over the coded symbols. Therefore, minimizing the above metric is equivalent to minimizing $\mathcal{M}' \triangleq \mathcal{M} - \sum_{i=1}^{N_r} \left(|r_i(1)|^2 + |r_i(2)|^2\right)$, where

$$\begin{aligned}
\mathcal{M}' &= -\sum_{i=1}^{N_r} \left[r_i(1)h_{i,1}^* s_1^* + r_i(1)^* h_{i,1}s_1 + r_i(2)h_{i,2}^* s_1 + r_i(2)^* h_{i,2} s_1^* \right] \\
&\quad + |s_1|^2 \sum_{i=1}^{N_r} \sum_{j=1}^{2} |h_{i,j}|^2
\end{aligned}$$

$$
\begin{aligned}
&-\sum_{i=1}^{N_r}\left[r_i(1)h_{i,2}^* s_2^* + r_i(1)^* h_{i,2} s_2 - r_i(2) h_{i,1}^* s_2 - r_i(2)^* h_{i,1} s_2^*\right] \\
&+|s_2|^2 \sum_{i=1}^{N_r}\sum_{j=1}^{2} |h_{i,j}|^2 \\
&= |\tilde{s}_1 - s_1|^2 + \left(\sum_{i=1}^{N_r}\sum_{j=1}^{2} |h_{i,j}|^2 - 1\right)|s_1|^2 \\
&+|\tilde{s}_2 - s_2|^2 + \left(\sum_{i=1}^{N_r}\sum_{j=1}^{2} |h_{i,j}|^2 - 1\right)|s_2|^2,
\end{aligned} \tag{7.70}
$$

where $\tilde{s}_1$ and $\tilde{s}_2$ are defined in Eq. 6.18, which we repeat below for convenience:

$$
\begin{aligned}
\tilde{s}_1 &= \sum_{i=1}^{N_r} h_{i1}^* r_i(1) + h_{i2} r_i^*(2), \\
\tilde{s}_2 &= \sum_{i=1}^{N_r} h_{i2}^* r_i(1) - h_{i1} r_i^*(2).
\end{aligned} \tag{7.71}
$$

This shows that the maximum likelihood estimates of the Alamouti coded symbols are obtained by separately minimizing the first and second expressions after the second equal sign in Eq. 7.70, which are equivalent to the decoding rules in Eq. 6.30. However, even without expressing $\mathcal{M}'$ in terms of $\tilde{s}_1$ and $\tilde{s}_2$, note that the first two lines in Eq. 7.70 involve the received signal, the channel matrix elements, and symbol s_1 only, which we can simply denote as a function $f(s_1)$. Similarly, we note that lines 3 and 4 only involve symbol s_2, which we can denote by a second function $g(s_2)$. This means that we have shown the maximum likelihood decoding rule for the Alamouti code consists of minimizing $\mathcal{M}' = f(s_1) + g(s_2)$, which is equivalent to separately minimizing $f(s_1)$ and $g(s_2)$.

The process that we have just described for the Alamouti code can be applied to any OSTBC. In general, simply use the structure of the OSTBC and apply that structure to Eq. 7.68. Doing so results in a decoding metric that is equal to the sum of k equations that are each a function of only one of the encoded symbols. For convenience, the decoding rules for codes $\mathcal{G}_2$, $\mathcal{G}_3$, $\mathcal{G}_4$, $\mathcal{H}_3$, and $\mathcal{H}_4$ are included in Appendix D.

The following example illustrates the computational savings associated with OSTBCs.

Example 7.5 Computational savings with OSTBCs Determine the number of computations needed to implement the maximum likelihood minimization in Eq. 7.68 for OSTBCs and compare these values with the number of computations needed for general space-time codes for $k = 2$ and 3 and $M = 2, 4, 8$, and 16 (i.e., for space-time block codes that encode $k = 2$ or 3 modulation symbols with modulation order 2, 4, 8, and 16).

Answer Due to the separability property of OSTBCs, the maximum likelihood minimization operation in Eq. 7.68 can be performed by plugging M symbols into each of k functions to find the symbol that minimizes each of those functions. Thus, the total number of computations with OSTBCs is kM. In contrast, with non-OSTBCs that require finding the minimum k-tuple of symbols that minimize $\|\mathbf{R}-\mathbf{HS}\|_F^2$, it is necessary to perform k^M substitutions into Eq. 7.68 in order to find the k symbols that minimize the decoding metric. The table compares the number of computations needed for both cases.

k	M	Non-OSTBCs	OSTBCs
2	2	4	4
2	4	16	8
2	8	256	16
2	16	65 536	32
3	2	9	6
3	4	81	12
3	8	6561	24
3	16	43 046 721	48

These results show that the savings are minimal for low modulation orders but become enormous when the modulation order is large. In general, the modulation order at which the difference becomes significant decreases as the number of symbols per space-time codeword, k, gets larger. This particular example applies when comparing Alamouti coding, which has $k=2$, with the codes $\mathcal{D}_{c,3\times4}$, $\mathcal{D}_{c,4\times4}$, $\mathcal{H}_3$, and $\mathcal{H}_4$ defined earlier, in which each has $k=3$.

7.3.5 Simulating OSTBC performance

It is straightforward to develop computer simulations that predict the error performance of OSTBCs in flat fading environments using the decoding expressions in Appendix D. When doing so, however, attention needs to be paid to the fact that these expressions assume the non-normalized system equation given in Eq. 7.66. As discussed, it is convenient to use the normalized version of the MIMO system equation when programming simulations because it allows the signal-to-noise ratio to be specified explicitly. If the normalized system equation is used, which has the form $\mathbf{R}=\sqrt{\rho}\mathbf{HS}+\mathbf{Z}$, then it is first necessary to convert it into the non-normalized form so that the decoding rules given in the appendix can be used directly. To do that, define $\mathbf{H}'\triangleq\sqrt{\rho}\mathbf{H}$ and use the elements of $\mathbf{H}'$ (given by $\left\{h'_{i,j}\right\}$) in place of the $\{h_{i,j}\}$ terms in the decoding equations in Appendix D.

7.3.6 OSTBC performance results

Figure 7.5 shows some example predictions of the modulation symbol error rate plotted versus signal-to-noise ratio (i.e. ρ) in flat Rayleigh fading for MIMO systems that

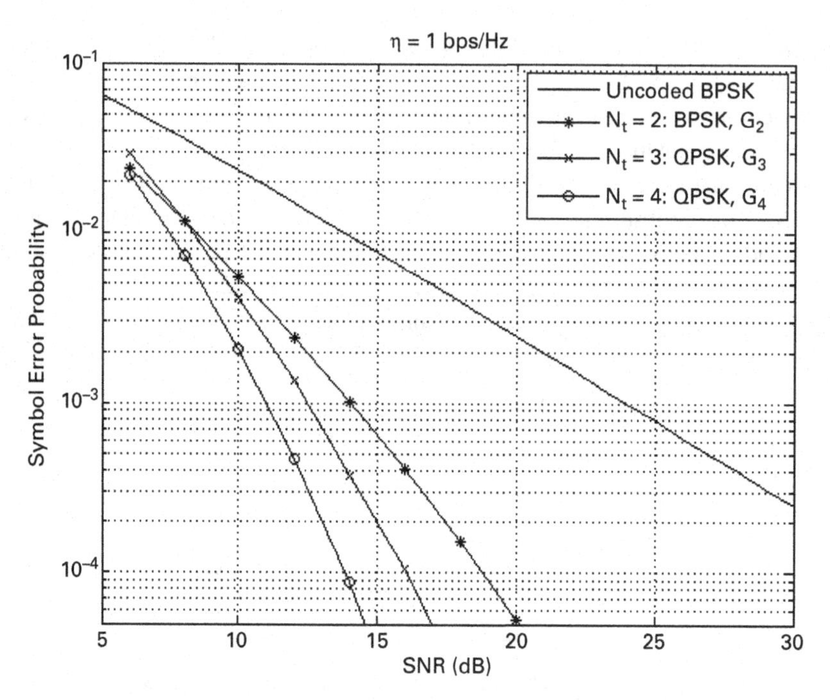

Figure 7.5 Symbol error probability versus SNR for orthogonal space-time codes. These results assume a spectral efficiency of 1 bps/Hz and $N_r = 1$.

have 1 receive antenna and 2, 3, and 4 transmit antennas. These results were obtained through computer simulation using the appropriate decoding equations in Appendix D. A separate curve is shown for each of the antenna configurations, along with a legend that shows the modulation and OSTBC type used. The combination of OSTBC type and modulation order are chosen such that the spectral efficiency is 1 bps/Hz. For comparison purposes, uncoded BPSK is included. The results show that significant improvement in performance occurs as the number of antennas is increased. A careful examination of the curves shows that at large ρ their slopes are approximately equal to $N_t N_r = N_t$, which is expected since OSTBCs are designed to achieve full diversity.

When designing a MIMO system that employs OSTBCs, it is necessary to ensure that the type of OSTBC and modulation scheme are compatible with the number of transmit antennas and the desired spectral efficiency. The number of rows in the OSTBC needs to be equal to N_t. Furthermore, only those combinations of OSTBCs, conventional codes, and modulation methods that result in the desired value of $\eta = r_s r_t \log(M)$ are permissible. For example, consider the curve in Figure 7.5 that corresponds to $N_t = 3$. The plot indicates that this curve assumes the OSTBC $\mathcal{G}_3$ defined in Eq. 7.64. The $\mathcal{G}_3$ code has three rows as required. In addition, its rate is 1/2, so to achieve a spectral efficiency of 1 bps/Hz, this code needs to be paired with a modulation scheme that has $M = 4$ since no conventional coding is used (i.e., $r_t = 1$). The curve associated with $N_t = 3$ assumes QPSK, which meets this requirement. Additional simulation results using other OSTBCs and modulation schemes are presented in Tarokh *et al.*'s paper [71].

Example 7.6 Assume that you need to design a MIMO system to achieve a spectral efficiency equal to 2 bps/Hz and that $N_t = 2$. If a rate 1/2 convolutional code is included in the design to enhance performance, what modulation order and OSTBC design combinations (chosen from the list of OSTBCs given in this chapter) can be used to implement this system?

Answer Equation 7.4 shows that the spectral efficiency $\eta = r_t r_s \log_2(M)$. Since $r_t = 1/2$ and we require that $\eta = 2$ bps/Hz, it follows that $r_s \log_2(M) = 4$. The following table lists the permissible combinations of OSTBCs designs and modulation orders based on examining the codes listed earlier.

r_s	M	OSTBC design
1	16	$\mathcal{D}_{\mathrm{RNS},3\times4}$
1	16	$\mathcal{G}_2$ (Alamouti)
1/2	256	$\mathcal{G}_3$
1/2	256	$\mathcal{G}_4$

7.4 Space-time trellis codes

So far we have focused on block-based space-time coding in general and on orthogonal space-time block codes in particular because of their simple decoding methods. As mentioned earlier in the chapter, however, a second class of space-time codes exists called space-time trellis codes (STTCs). In fact, the rank and determinant criteria were first developed for and applied to the design of STTCs in Tarokh *et al.*'s 1998 paper [73], and only later in 1999 to STBCs [71]. Although STTCs have a potential performance advantage over block codes because of their ability to achieve coding gain, the complexity of their decoding methods is greater than that of OSTBCs and is generally considered to be prohibitively high when $N_t > 2$. Despite this limitation, an introductory book on MIMO communications would not be complete without some coverage of these codes, due to their potential use for small numbers of transmit antennas and because of their historical importance in the development of space-time coding. Table 7.2 summarizes the key differences between STBCs and STTCs.

The next section describes the encoding method used with STTCs and presents several STTCs that have been developed using the rank and determinant design criteria described earlier. In the last section we present performance results and show that STTCs, like the STBCs we have considered in this chapter, also achieve full diversity in Rayleigh and Rician fading environments.

Table 7.2 Comparison summary of STBCs and STTCs.

Characteristic	STBC	STTC
Encoding	block-based	trellis-based
Decoding	linear processing	Viterbi-based
Decoding complexity	low	high
Coding gain	no	yes
N_t limitation	none	$N_t \lesssim 2$
Error performance	can be worse than STTC	can be better than STBCs

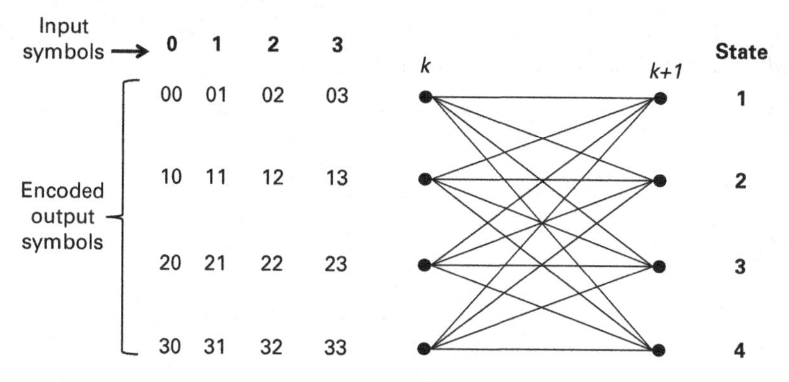

Figure 7.6 A four-state space-time trellis encoder for a MIMO system with $N_t = 2$ and a modulation order $M = 4$.

7.4.1 STTC encoding

Unlike STBCs, which are defined in terms of their design matrices (e.g. Eqs. 7.50–7.65), STTCs are normally described using encoding trellis diagrams. The concept of STTC encoding is best illustrated using a specific example. For this purpose, we assume that QPSK modulation is used and assign numbers to each of the four QPSK symbols, say, 0, 1, 2, and 3, which correspond to the bit pairs 00, 01, 10, and 11, respectively. Using that numbering scheme, we define a STTC using the trellis structure illustrated in Figure 7.6.

The trellis encoder in this figure consists of several components: the trellis structure, which consists of *states* denoted by the dots; permissible paths from the set of states on the left to those on the right, denoted by the connecting lines between the states; a list of input symbols; and a matrix listing of possible encoded output symbol combinations. As illustrated in the figure, the two sets of trellis states (i.e., the left and right states) are assumed to refer to two successive modulation symbol periods denoted by the integers k and $k+1$. The trellis encoder is normally assumed to begin in state 1 in the beginning (i.e., when $k = 0$); it then transitions from one state to another depending on which modulation symbol arrives at each symbol period. The paths are numbered from 0 to $M-1$ starting from the top; thus, there are M paths emanating from each left-side state. The encoded symbols at time index k are specified by the entry in the encoded output symbol matrix to the left of the trellis that corresponds to the modulation symbol and the trellis state at time k. Each entry has N_t digits that specify which symbol is assigned to each of the N_t antennas, reading left to right.

As an example, consider the following bit sequence:

$$\text{Input bit sequence: } 01\ 10\ 10\ 11\ 00\ 11\ 00\ 10\ \ldots$$

Based on our QPSK symbol numbering convention, this corresponds to the following modulation symbol sequence:

$$\text{Input modulation symbol sequence: } 1\ 2\ 2\ 3\ 0\ 3\ 0\ 2\ \ldots$$

At time $k = 0$ we start in state 1, and because the modulation symbol at this instant is symbol 1, the trellis diagram tells us that the encoded symbols are 01, which mean that we transmit QPSK symbol 0 on transmit antenna 1 and QPSK symbol 1 on antenna 2. The presence of QPSK symbol 1 at $k = 0$ also tells us that we should use the second path from the top to transition to the next state, which is state 2 according to the trellis diagram. After transitioning states, we increment the time index, resulting in $k = 1$, and repeat this process for the second input symbol. Repeating this process results in the following transmitted symbol sequences:

$$\text{Transmit antenna 1: } 0\ 1\ 2\ 2\ 3\ 0\ 3\ 0\ \ldots$$
$$\text{Transmit antenna 2: } 1\ 2\ 2\ 3\ 0\ 3\ 0\ 2\ \ldots$$

It should be noted that the spectral efficiency of STTCs, η_{STTC}, is determined by the modulation order since one modulation symbol is encoded during each symbol period. Thus, for STTCs

$$\eta_{\text{STTC}} = \log_2(M). \tag{7.72}$$

It follows from this that for the code depicted in Figure 7.6, $\eta = 2$ bps/Hz.

Figure 7.7 shows a second example of a space-time trellis code. This code is also designed to work with 4-ary modulation, two transmit antennas, and to achieve a spectral efficiency of 2 bps/Hz; however it has eight states instead of just four associated with the previous figure. In general, the larger the number of states, the larger the coding gain, but this comes at the expense of greater decoder complexity. Since STTCs are designed using the rank and determinant criteria, they achieve full diversity in Rayleigh and Rician fading. Therefore, the STTCs defined in Figures 7.6 and 7.7 both achieve a diversity gain of $2N_r$, which means that the slopes of their decoding error versus SNR curves are the same, but the curve for the 8-state code is shifted to the left of the 4-state curve.

7.4.2 STTC performance results

In this section, we show some performance results based on computer simulations of STTCs in quasi-static Rayleigh fading presented in Tarokh *et al.*'s paper on STTCs [73]. We start by examining the impact that the number of STTC encoder states has on performance. Figures 7.8 and 7.9 show plots of STTC frame error rate versus SNR

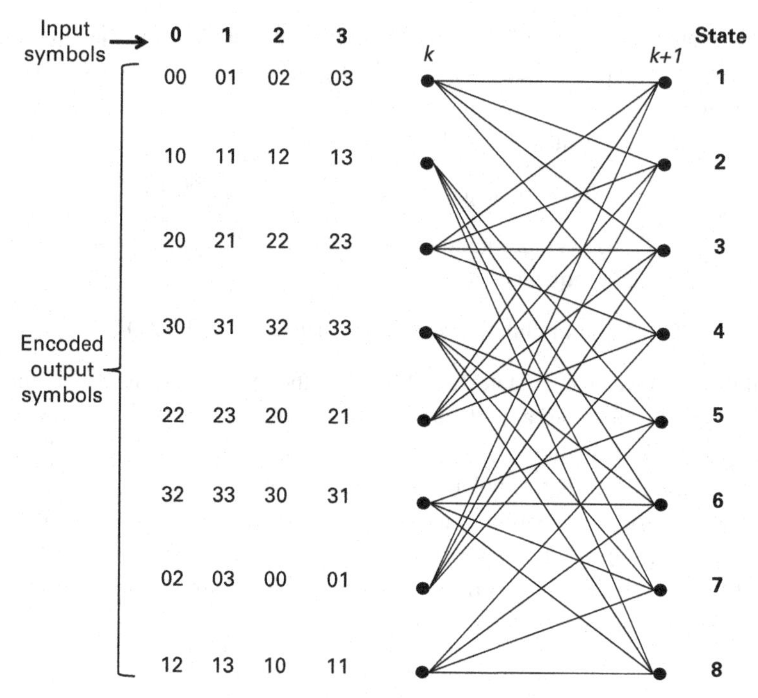

Figure 7.7 An eight-state space-time trellis encoder for a MIMO system with $N_t = 2$ and a modulation order $M = 4$.

for STTCs that have different numbers of encoder states. The various codes examined in these plots are designed for $N_t = 2$ and for 4-ary modulation (the specific results shown assume QPSK modulation). Furthermore, the frame lengths are fixed and equal to 130 modulation symbols. Figures 7.8 and 7.9 assume that the numbers of receive antennas are 1 and 2, respectively. The results in these figures demonstrate that the diversity gain is $N_t N_r$ (i.e., 2 and 4, respectively) and that the coding gain advantage obtained by increasing the number of states increases as the number of receive antennas is increased.

Next, we show results that compare STTC and STBC performance. For the purpose of this comparison, we use the frame error rate results shown earlier for 8-state and 32-state STTCs designed for QPSK modulation, $N_t = 2$, and $N_r = 1$. Since $N_t = 2$, the STBC that we use for comparison is the Alamouti code. The comparison is shown in Figure 7.10.

These results show that by increasing the number of states associated with the STTC, a coding gain advantage is realized over STBCs. We observe in this plot that when the number of states is small, the performance of the STTC is comparable to the STBC. When the number of states is large, however, the STTC performance advantage can be greater than 1 dB. As discussed above, both STTCs as well as the Alamouti code meet the rank criterion; therefore, they achieve full diversity. This fact is confirmed by the fact that the slopes of these curves are equal to $N_t N_r = 2$, so there is clearly no diversity advantage with the STTC codes in this plot. The advantage that the STTC code has over

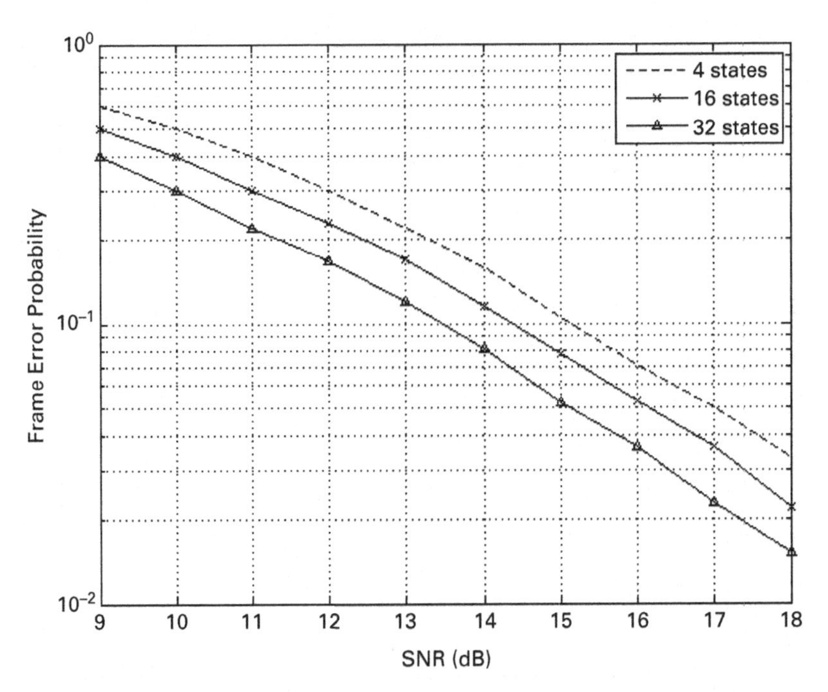

Figure 7.8 Frame error rate versus SNR for STTCs with different numbers of states. All codes in this plot assume QPSK modulation, $N_t = 2$, $N_r = 1$, and $\eta = 2$ bps/Hz.

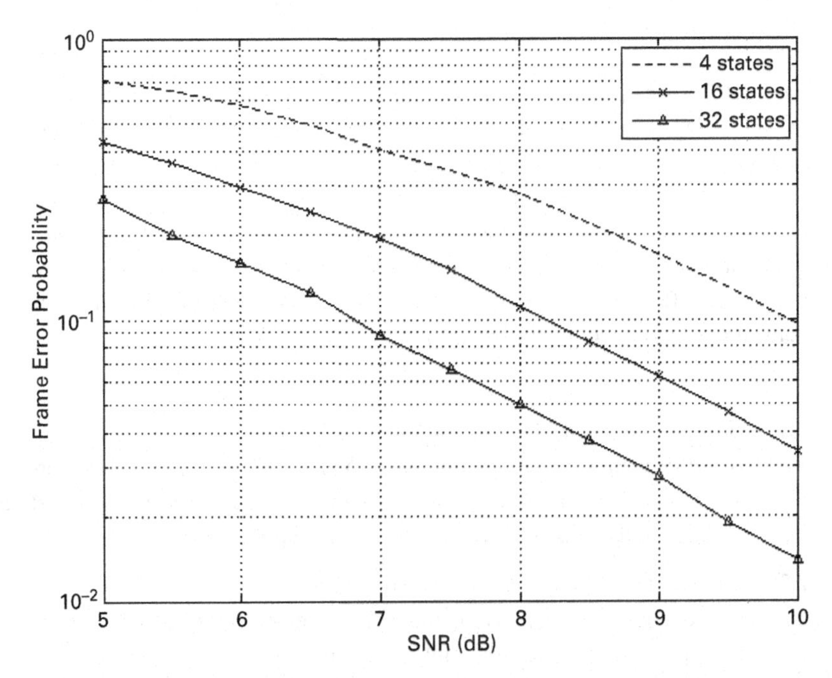

Figure 7.9 Frame error rate versus SNR for STTCs with different numbers of states. All codes in this plot assume QPSK modulation, $N_t = 2$, $N_r = 2$, and $\eta = 2$ bps/Hz.

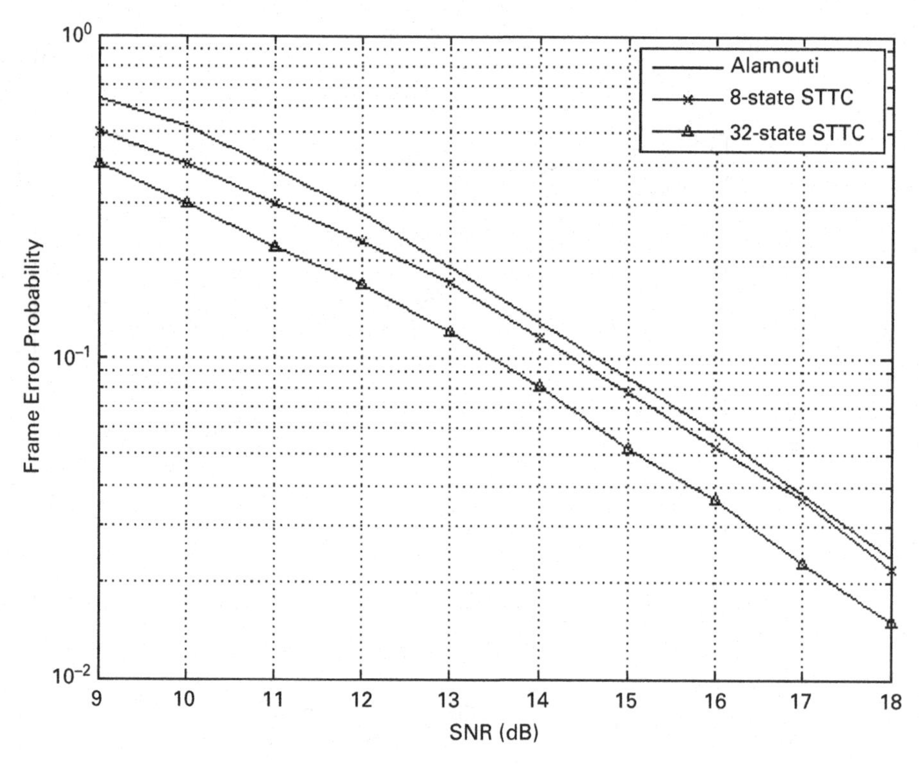

Figure 7.10 Frame error rate comparison of Alamouti coding and STTCs for $N_t = 2, N_r = 1$, assuming QPSK modulation.

the STBC, instead, is seen in the shift of its performance curve to the left, which is due to the coding gain. As mentioned earlier, this relatively modest coding gain advantage is achieved at the cost of increased decoding complexity.

It should be mentioned that many other STTCs have been developed besides the two examples given in Figures 7.6 and 7.7. In Tarokh *et al.*'s seminal paper on STTCs [73], numerous other STTCs are presented for both 4-ary and 8-ary modulation and for a variety of different numbers of encoder states. The interested reader is encouraged to consult that reference for further information on STTCs.

Problems

7.1 Assume you are designing a MIMO system that has seven transmit antennas for the purpose of achieving spatial diversity. What is the minimum codeword block size for which a real, non-square OSTBC exists that could be used for this system? Which OSTBC listed in this chapter can be used for this application?

7.2 Consider the code $\mathcal{H}_3$ defined in Eq. 7.62:

a) Prove that $\mathcal{H}_3$ is an orthogonal space-time block code;
b) Show that $\mathcal{H}_3$ meets the rank criterion.

7.3 Prove that the expression after the first equal sign in Eq. 7.70 is equivalent to the expression after the second equal sign.

7.4 A key step in deriving the rank and determinant criteria was going from Eq. 7.21 to Eq. 7.22. Prove that Eq. 7.21 can be written in the form given by Eq. 7.22.

7.5 Assume that you desire to design a MIMO system to achieve spatial diversity while maintaining a spectral efficiency of 2 bps/Hz. If $N_t = 2$ and no conventional coding is used, list possible modulation types and orthogonal space-time code designs (chosen from those listed in this chapter) that could be used.

7.6 Using Matlab, simulate the three OSTBCs in Figure 7.5. Duplicate the curves in that plot for $N_t = 2$, 3, and 4 antennas.

8 Spatial multiplexing

In this chapter, we turn our attention to the second major class of MIMO processing techniques: spatial multiplexing. As we discussed in Chapter 1, spatial multiplexing refers to transmitting multiple independent data streams over multipath channels, without the need to increase the bandwidth. Unlike space-time coding, which is used to achieve spatial diversity and which transmits at most one modulation symbol per modulation symbol period (i.e., $r_s \leq 1$), spatial multiplexing techniques are capable of achieving spatial rates equal to $\min\{N_t, N_r\}$; that is, rather than only transmitting one or fewer modulation symbols per symbol period, spatial multiplexing involves transmitting up to $\min\{N_t, N_r\}$ modulation symbols per symbol period, resulting in a concomitant increase in throughput relative to spatial diversity schemes. This improvement in throughput, however, is achieved at the expense of diversity gain, so the diversity gains associated with spatial multiplexing methods are normally significantly less than N_tN_r. This chapter describes several fundamental, practical techniques that are used to achieve spatial multiplexing.

The chapter is divided into four sections. The first section presents an overview of spatial multiplexing concepts and reviews the major types of spatial multiplexing methods that have been proposed. The second section describes the transmit architectures associated with the class of SM schemes known as BLAST. Section 3 describes four spatial demultiplexing methods that can be used with H-BLAST and V-BLAST. The chapter concludes with a section that describes a hybrid technique, called *multi-group space-time coding*, that combines spatial multiplexing and space-time coding techniques.

8.1 Overview of spatial multiplexing

Spatial multiplexing is possible because of the underlying propagation physics in a multipath communications channel. In a multipath environment, energy from each transmit antenna arrives at each receive antenna after being scattered by various objects. If the antenna spacing at the transmitter and receiver is sufficiently large, then the characteristics of the scattering between each pair of transmit and receive antennas are sufficiently different that each one can be regarded as fading independently of the others. This property can be exploited by the receiver using techniques that we describe in this chapter to separate out the signals from each of the transmitters. A popular rule-of-thumb is that

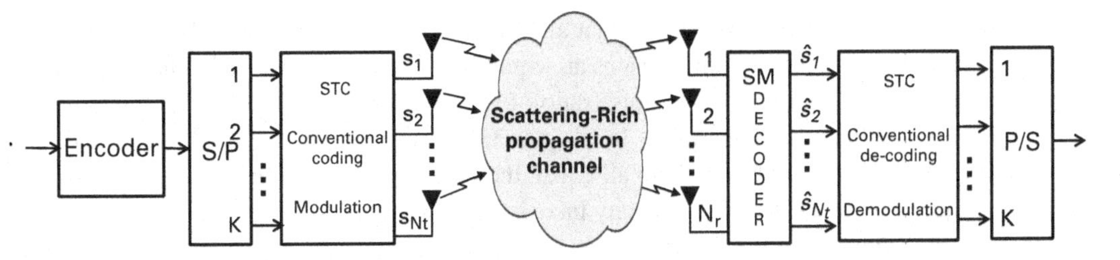

Figure 8.1 The basic components of a MIMO communication system that performs spatial multiplexing.

independent fading occurs as long as the antenna spacing at both the transmitter and receiver is at least half a wavelength.

Figure 8.1 shows a block diagram of a MIMO communication system that employs spatial multiplexing. This figure is similar to Figure 1.10 except that it includes slightly more detail. Information bits are assumed to arrive on the left and may first undergo optional conventional error control coding, which is performed by the encoder block. After encoding, the encoded bit stream is passed through a serial-to-parallel converter that splits the serial input stream into K parallel streams. The next block maps the K input streams into N_t streams that are fed to each of the transmit antennas. The block that performs this mapping may employ some combination of either space-time coding, additional conventional error coding, and modulation, depending on the particular type of spatial multiplexing scheme. Specific transmitter architectures used for achieving spatial multiplexing are presented later in the next section.

At the receiver, the signal at each antenna consists of the sum of the signals from all the transmitters. Since spatial multiplexing involves transmitting multiple distinct data streams, the receiver must be capable of demultiplexing (or decoupling) the individual data streams from each other. That is the purpose of the "SM decoder" block shown in the diagram.[1] The SM decoder performs signal processing on the signals from the receive antennas and generates estimates of the individual transmitted streams. A variety of different decoding techniques has been developed for this purpose, and the bulk of this chapter is devoted to describing and analyzing some of the most common schemes.

As we saw in Eqs. 3.11 and 3.27, we can express the capacity of a MIMO system as a sum of $r = \text{rank}(\mathbf{H})$ SISO channels. We conclude from this that the maximum number of independent data streams that can be transmitted over a spatial multiplexing system, N_{stream}, is

$$N_{\text{stream}} = \min\{N_t, N_r\}. \tag{8.1}$$

It should be noted that in order to decode the separate data streams it is necessary that $N_r \geq N_t$. All of the decoding techniques that we consider in this chapter have this requirement. It follows that the maximum number of data streams that can be spatially multiplexed is equal to

$$N_{\text{stream}} = N_t. \tag{8.2}$$

[1] In this chapter, we use the terms decoder and demultiplexer interchangeably.

Another point to be made is that although it is theoretically possible to support N_t data streams, not all of the streams are equivalent in terms of their individual capacities since the effective signal-to-noise ratio of each spatial channel is different, depending on the value of its eigenvalue. In Chapter 3, we saw that when $N_t = N_r = N$ and the eigenvalues of the channel are all equal, then the capacities of the spatial channels are equivalent and the overall capacity increases linearly with N.

It follows from Eq. 8.2 that for an SM system, the maximum number of modulation symbols that can be transmitted per symbol period, $\max\{r_s\}$, is given by

$$\max\{r_s\} = N_t, \tag{8.3}$$

which implies that the maximum spectral efficiency of an SM system is given by

$$\eta_{\max} = N_t r_t \log_2(M)\,\text{bps/Hz}, \tag{8.4}$$

where r_t is the rate of any conventional coding used in the SM system and M is the modulation order. By comparison, for spatial diversity systems that use space-time coding, we saw in Chapter 7 that $r_s \leq 1$ and $\eta \leq r_t \log_2(M)$.

In general, spatial multiplexing is achieved using a concept called *layered space-time* (LST) coding, of which there is a variety of examples. In this context, a layer simply refers to a data stream from a single transmit antenna. Types of LST codes include the following specific schemes:

1. Bell Laboratory layered space-time (BLAST) family of techniques:
   a) Vertical BLAST (V-BLAST);
   b) Horizontal BLAST (H-BLAST);
   c) Diagonal BLAST (D-BLAST);
2. Multi-group space-time coding (MGSTC);
3. Threaded space-time coding (TSTC).

In addition to these LST methods, spatial multiplexing can also be achieved using eigenbeamforming, which was discussed in detail in Chapter 3. Eigenbeamforming is a practical SM technique that is used in most modern wireless communication systems; however, it is generally not regarded as being a LST coded SM method. In this chapter, we focus our attention on the BLAST and MGSTC schemes. Readers interested in learning more about TSTC are referred to [24].

The type of decoding algorithm that is used is an important consideration for LST coded SM systems. Four decoding schemes have been analyzed extensively in MIMO literature:

1. Zero forcing (ZF);
2. Zero forcing with interference cancellation (ZF-IC);
3. Linear minimum mean square error estimation (LMMSE);
4. LMMSE with interference cancellation (LMMSE-IC).

In this chapter, we analyze these four detection schemes in detail and compare the differences in their performance.

In addition to the four demultiplexing methods listed above, numerous other schemes have been proposed. The following is a list of other SM decoding methods with references for the interested reader: a) sphere decoding [87]; b) belief propagation detection [88]; c) greedy detection [89]; and d) turbo-BLAST detection [90].

In closing this section, it should be mentioned that wireless standards, such as those for WiFi (IEEE 802.11n), LTE, LTE-advanced, WiMAX (IEEE 802.16), etc., only define what the transmitter must do; they do not specify the details of the receiver. As a result, companies that build receivers for these applications are free to employ techniques of their choosing as long as they meet the overall system performance requirements specified by the standard. The choice of decoding method is a good example of this. There are many papers in the MIMO literature on SM decoding (some of which are listed above), and more will undoubtedly be written in the future. It will continue to be up to equipment vendors, however, to choose from among these published techniques or to use their own proprietary methods when designing receivers to comply with the various wireless standards.

8.2 BLAST encoding architectures

Before describing SM decoding in detail, we pause briefly in this section to consider the encoding architectures of the BLAST-based schemes. The concepts of LST coding in general, and of D-BLAST in particular were first described by Foschini in 1996 [30]. In 1998, Wolniansky *et al.* described a related SM architecture that they called V-BLAST [75], which was simpler to implement than D-BLAST. In that paper, they presented laboratory results of the first demonstration of a practical V-BLAST system. In 2000, Li *et al.* [48] described a third type of BLAST-type architecture, which is similar to V-BLAST except that it involves conventional coding on each layer. In that paper, they refered to this concept as simply "horizontal coding for BLAST." Finally, in 2003, Foschini *et al.* [28] published a paper that summarizes BLAST architectures and in that paper coined the term *horizontal BLAST* (H-BLAST) to describe the architecture first described in [48].

Although D-BLAST was the first BLAST technique to be described, it is the most complex of the BLAST architectures. In the following discussion, we go from least to most complex; thus, not in historical order.

8.2.1 Vertical-BLAST (V-BLAST)

Figure 8.2 is a block diagram of a V-BLAST encoder. In V-BLAST, the information bit stream is processed by an optional conventional error encoder and then split into N_t data streams, each of which is separately modulated before being passed to its respective antenna for transmission. The use of the adjective "vertical" in V-BLAST is reference to the fact that the input is split into parallel streams that are depicted vertically in most diagrams. Since each layer in a V-BLAST encoder employs its own modulator, the V-BLAST architecture is capable of accommodating applications where different data rates are applied to different layers. In such applications, layers with higher data

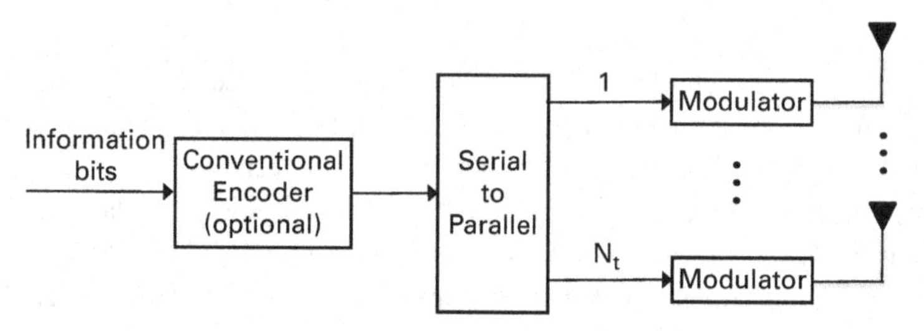

Figure 8.2 V-BLAST encoding architecture.

rates might use higher order modulation schemes so that each layer would have the same bandwidth.

Since distinct data streams are applied to each of the N_t layers, during each use of the channel (i.e., each modulation symbol period) there are N_t different modulation symbols transmitted. Therefore, the space-time code rate associated with the V-BLAST encoder is $r_s = N_t$ and the spectral efficiency is $N_t r_t \log_2(M)$ bps/Hz; where M is the modulation order (assuming that the same order is used on all layers). In general, the diversity gain in spatial multiplexing depends on the layer being detected. In the case of V-BLAST, Loyka and Gagnon [47] prove that the diversity order varies from $(N_r - N_t + 1)$ up to N_r, depending on which layer is being decoded. We prove later in this chapter that similar results apply to MGSTC. Despite the fact that the diversity gain associated with spatial multiplexing varies over a wide range, it is shown that the overall performance is dominated by the layer with the least spatial diversity, so the overall spatial diversity for V-BLAST is approximately equal to $(N_r - N_t + 1)$. We see from this that $N \times N$ V-BLAST only achieves a maximum diversity gain equal to 1, compared with $N_t N_r$ for systems with full spatial diversity.

8.2.2 Horizontal-BLAST (H-BLAST)

The H-BLAST encoding architecture, which is illustrated in Figure 8.3, is similar to V-BLAST except that it includes separate conventional error encoders on each of the transmit data streams instead of a single encoder prior to the serial-to-parallel converter. The term "horizontal" in the name refers to the fact that the encoders on each layer perform coding in the time domain (as opposed to the space domain), which can be pictured as being horizontal in the picture, compared with the space dimension that is depicted being vertical. The V-BLAST architecture has the same spatial rate, spectral efficiency, and spatial diversity properties as V-BLAST.

8.2.3 Diagonal-BLAST (D-BLAST)

Figure 8.4 illustrates the D-BLAST encoding architecture. As can be seen, D-BLAST is similar to H-BLAST except that it includes a block after the modulators that performs stream rotation. To understand what the stream rotator does, we consider an example where we assume that $N_t = 4$. The outputs from the four conventional encoders are

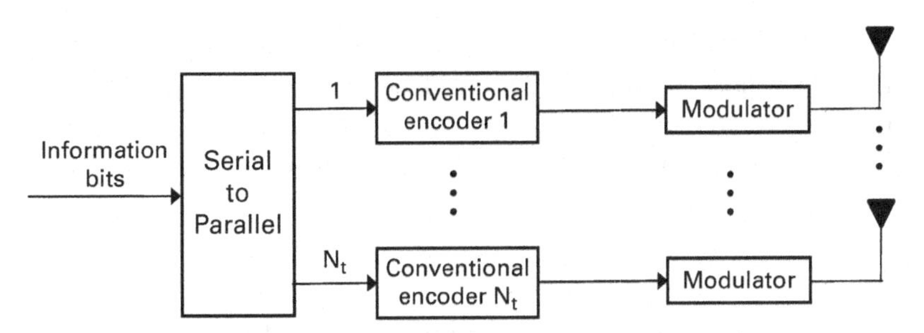

Figure 8.3 H-BLAST encoding architecture.

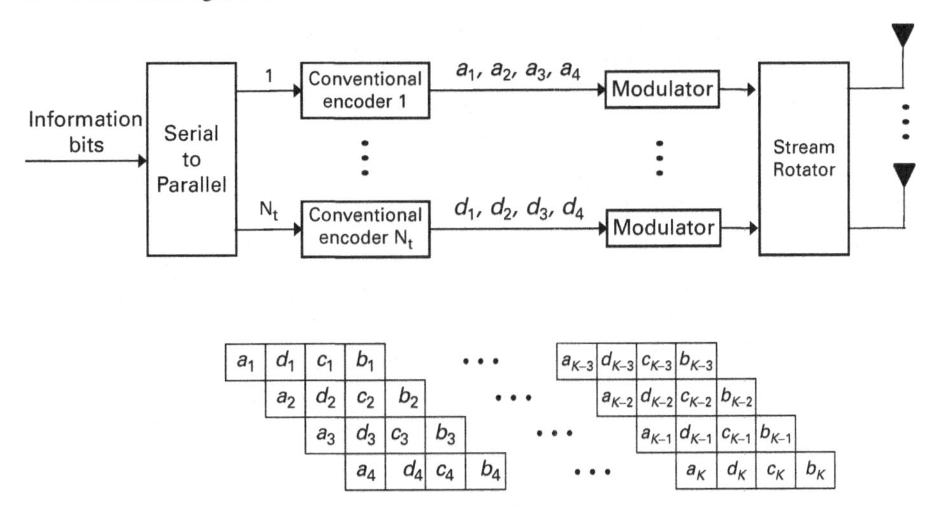

Figure 8.4 D-BLAST encoding architecture.

vectors denoted by **a**, **b**, **c**, and **d** and we assume that the outputs are divided into blocks consisting of N_t consecutive segments. We denote the first four encoded segments out of conventional encoder 1 by $\mathbf{a}_1$, $\mathbf{a}_2$, $\mathbf{a}_3$, and $\mathbf{a}_4$, the next set of four encoded segments by $\mathbf{a}_5$, $\mathbf{a}_6$, $\mathbf{a}_7$, and $\mathbf{a}_8$, and so forth. The outputs of the other encoders are denoted in a similar way.

Rather than simply passing the modulated outputs from each encoder onto its respective antenna, the stream rotator rotates the modulated segments in a round-robin fashion by performing two operations: a) it distributes consecutive sequences of N_t segments from each encoder onto each of the antennas; and b) the order of the encoders that it operates on is chosen in a circularly rotated manner rather than simply sequentially from encoder 1 to N_t. The result is depicted in the bottom of the figure.

The advantage of D-BLAST over the other BLAST schemes is that the outputs from each conventional encoder are distributed over space (i.e., transmitted over different antennas), which provides a greater spatial diversity. As a result, D-BLAST can achieve a spatial diversity gain greater than N_r, and with proper conventional coding, full diversity (i.e., diversity gain equal to $N_t N_r$) can be achieved [57].

This concludes our overview of the basic BLAST encoder architectures. In the next section, we describe four decoding methods that can be used with V-BLAST and H-BLAST encoders. Different, but similar, demultiplexing methods are used with D-BLAST; however, because D-BLAST decoding is more complex, without providing significantly greater insight, we elect not to discuss D-BLAST in further detail. The interested reader is referred to [21] for more information on how to decode D-BLAST.

8.3 Demultiplexing methods for H-BLAST and V-BLAST

This section describes the following four BLAST demultiplexing methods: ZF, ZF-IC, LMMSE, and LMMSE-IC. Each is discussed in a separate subsection. This section concludes by comparing the performance of these four techniques.

8.3.1 Zero-forcing (ZF)

8.3.1.1 Analysis

Consider an $N_t \times N_r$ MIMO system that uses either horizontal or vertical BLAST encoding. At any given point in time, each antenna will be transmitting a symbol. If we denote the symbol being transmitted from antenna j during symbol period k by $s_j(k)$, then we can express the received signal at receiver i, $r_i(k)$, by the following expression:

$$r_i(k) = \sqrt{\rho} \sum_{j=1}^{N_t} h_{i,j} s_j(k) + z_i(k), \tag{8.5}$$

where $z_i(k)$ denotes the noise amplitude at receiver i during symbol period k and ρ denotes the signal-to-noise ratio at the receiver. If we consider a group of symbol periods, say, $k = 1, \ldots, p$, then there are pN_r equations like Eq. 8.5, correponding to $i = 1, \ldots, N_r$ and $k = 1, \ldots, p$. For notational convenience, we write these equations using the following matrix format:

$$\mathbf{R} = \sqrt{\rho}\mathbf{HS} + \mathbf{Z}, \tag{8.6}$$

where $\mathbf{H}$ is the $N_r \times N_t$ channel matrix, and the other matrices are expressed as follows:

$$\mathbf{R} = \begin{pmatrix} r_1(1) & r_1(2) & \ldots & r_1(p) \\ r_2(1) & r_2(2) & \ldots & r_2(p) \\ \vdots & \vdots & \ddots & \vdots \\ r_{N_r}(1) & r_{N_r}(2) & \ldots & s_{N_r}(p) \end{pmatrix}, \tag{8.7}$$

$$\mathbf{S} = \begin{pmatrix} s_1(1) & s_1(2) & \ldots & s_1(p) \\ s_2(1) & s_2(2) & \ldots & s_2(p) \\ \vdots & \vdots & \ddots & \vdots \\ s_{N_t}(1) & s_{N_t}(2) & \ldots & s_{N_t}(p) \end{pmatrix}, \tag{8.8}$$

and

$$\mathbf{Z} = \begin{pmatrix} z_1(1) & z_1(2) & \dots & z_1(p) \\ z_2(1) & z_2(2) & \dots & z_2(p) \\ \vdots & \vdots & \ddots & \vdots \\ z_{N_r}(1) & z_{N_r}(2) & \dots & z_{N_r}(p) \end{pmatrix}. \tag{8.9}$$

Examination of Eq. 8.5 shows that each received signal consists of a linear superposition of signals from all of the transmit antennas. As a result, if receiver i attempts to detect the signal from a specific transmitter, say transmitter l, then the signals from the other transmitters constitute interference, I_l, given by

$$I_l \triangleq \sqrt{\rho} \sum_{j \neq l}^{N_t} h_{i,j} s_j(k). \; k = 1, \dots, p \tag{8.10}$$

It follows from Eqs. 8.5 and 8.10 that the signal-to-interference-plus-noise ratio at each receiver, SINR, is equal to

$$\text{SINR} = \frac{\rho}{\rho(N_t - 1) + N_t}. \tag{8.11}$$

This expression simply makes precise the intuitively obvious fact that when transmitting multiple data streams to a receiver, each stream is interfered with by $N_t - 1$ other streams, which poses a clear challenge for the receiver in a spatial multiplexing system. It should be noted that interference is much less of a problem with spatial diversity techniques, since the transmit antennas transmit redundant information. The purpose of the demultiplexing scheme in spatial multiplexing is to mitigate the effects of the interference.

In the case of zero-forcing, the effect of interference is reduced by premultiplying the received signal matrix by the *Moore–Penrose pseudo inverse* [56] of the channel matrix, which we denote by $\mathbf{H}^+$ and define as follows:

$$\mathbf{H}^+ \triangleq \left(\mathbf{H}^H \mathbf{H}\right)^{-1} \mathbf{H}^H. \tag{8.12}$$

To see how this operation reduces interference, we evaluate $\mathbf{H}^+\mathbf{R}$, which results in the following:

$$\begin{aligned} \tilde{\mathbf{R}} &\triangleq \mathbf{H}^+\mathbf{R} \qquad (8.13) \\ &= \left(\mathbf{H}^H \mathbf{H}\right)^{-1} \mathbf{H}^H \mathbf{R} \\ &= \sqrt{\rho} \left(\mathbf{H}^H \mathbf{H}\right)^{-1} \left(\mathbf{H}^H \mathbf{H}\right) \mathbf{S} + \left(\mathbf{H}^H \mathbf{H}\right)^{-1} \mathbf{H}^H \mathbf{Z} \\ &= \sqrt{\rho}\, \mathbf{S} + \tilde{\mathbf{Z}}, \qquad (8.14) \end{aligned}$$

where $\tilde{\mathbf{Z}} \triangleq \left(\mathbf{H}^H \mathbf{H}\right)^{-1} \mathbf{H}^H \mathbf{Z}$. If we denote the jth row of $\tilde{\mathbf{R}}$ by $\tilde{\mathbf{r}}_j$, it follows from Eq. 8.14 that

$$\tilde{\mathbf{r}}_j = \sqrt{\rho} \mathbf{s}_j + \tilde{\mathbf{z}}_j, \tag{8.15}$$

Table 8.1 ZF decoding algorithm for H-BLAST and V-BLAST.

1. Compute the pseudoinverse of the channel matrix, $\mathbf{H}^+$, using Eq. 8.12.
2. Premultiply the receive matrix by $\mathbf{H}^+$. Call the resulting matrix $\tilde{\mathbf{R}}$.
3. Estimate the symbols for each layer by performing maximum likelihood decoding on each row of $\tilde{\mathbf{R}}$ using Eq. 8.17.

where $\mathbf{s}_j$ and $\tilde{\mathbf{z}}_j$ denote the jth rows of $\mathbf{S}$ and $\tilde{\mathbf{N}}$, respectively. This equation shows that the jth row of $\tilde{\mathbf{R}}$ consists of a vector of transmitted symbols from the jth transmitter plus a noise vector *with no interference!*

If we denote the kth element of $\tilde{\mathbf{r}}_j$ by $\tilde{r}_j(k)$, it follows from Eq. 8.15 that

$$\tilde{r}_j(k) = \sqrt{\rho}\, s_j(k) + \tilde{z}_j(k), \tag{8.16}$$

where $\tilde{z}_j(k)$ denotes the kth element of $\tilde{\mathbf{z}}_j$. Comparison of this equation with Eq. 8.5 clearly shows the interference reduction that has occurred as a result of premultiplying by the pseudoinverse.

Estimates of the transmitted symbols from layer j can easily be obtained by performing maximum likelihood detection on the signals, $\{\tilde{r}_j(k)\}$, as follows:

$$\hat{s}_j(k) = \underset{\{s_j(k)\}}{\arg\min}\ |\tilde{r}_j(k) - \sqrt{\rho} s_j(k)|^2. \ k = 1, \ldots, p \tag{8.17}$$

The process of premultiplying the received matrix by the pseudoinverse of the channel matrix and then performing maximum likelihood detection on each of the rows of the resulting matrix constitutes zero-forcing detection. Table 8.1 summarizes the ZF detection algorithm for use with H-BLAST and V-BLAST.

We have shown that premultiplying $\mathbf{R}$ by $\mathbf{H}^+$ eliminates interference, or, equivalently, that it forces the interference to be zero; hence, the terminology *zero-forcing*. In the process, however, the noise term has also been premultiplied by $\mathbf{H}^+$, resulting in a modified noise matrix, $\tilde{\mathbf{Z}}$. Unfortunately, the elimination of interference in zero-forcing detection comes at the expense of a growth in noise power called *noise amplification*. The next subsection describes this phenomenon in detail.

8.3.1.2 Noise amplification in ZF detection

The phenomenon of noise amplification in ZF detection is a well-known effect that can be easily explained mathematically by computing the total power in the noise matrix, $\tilde{\mathbf{Z}}$. The total noise power, $P_{\tilde{Z}}$, is equal to the sum of the variances of the individual elements within the noise matrix. That is,

$$\begin{aligned} P_{\tilde{Z}} &= \sum_{k=1}^{p} \sum_{j=1}^{N_t} \mathrm{Var}\left\{\tilde{z}_j(k)\right\} \\ &= \sum_{k=1}^{p} \sum_{j=1}^{N_t} \mathbb{E}\left\{|\tilde{z}_j(k)|^2\right\} \end{aligned}$$

$$
\begin{aligned}
&= \mathrm{Tr}\left[\mathbb{E}\left\{\tilde{\mathbf{Z}}\tilde{\mathbf{Z}}^{H}\right\}\right] \\
&= \mathrm{Tr}\left[\mathbb{E}\left\{\left(\mathbf{H}^{H}\mathbf{H}\right)^{-1}\mathbf{H}^{H}\mathbf{Z}\mathbf{Z}^{H}\mathbf{H}\left(\mathbf{H}^{H}\mathbf{H}\right)^{-H}\right\}\right] \\
&= \mathrm{Tr}\left[\left(\mathbf{H}^{H}\mathbf{H}\right)^{-1}\mathbf{H}^{H}\mathbb{E}\left\{\mathbf{Z}\mathbf{Z}^{H}\right\}\mathbf{H}\left(\mathbf{H}^{H}\mathbf{H}\right)^{-H}\right] \\
&= N_r\sigma_z^2\mathrm{Tr}\left[\left(\mathbf{H}^{H}\mathbf{H}\right)^{-1}\left(\mathbf{H}^{H}\mathbf{H}\right)\left(\mathbf{H}^{H}\mathbf{H}\right)^{-H}\right] \\
&= N_r\sigma_z^2\mathrm{Tr}\left[\left(\mathbf{H}^{H}\mathbf{H}\right)^{-H}\right] \\
&= N_r\sigma_z^2\mathrm{Tr}\left[\left(\mathbf{H}^{H}\mathbf{H}\right)^{-1}\right].
\end{aligned}
\tag{8.18}
$$

To help see the implication of this equation, recall that the inverse of any square matrix, $\mathbf{A}$, is equal to $\mathrm{Adj}(\mathbf{A})/\det(\mathbf{A})$; where, $\mathrm{Adj}(\cdot)$ denotes the adjoint of $\mathbf{A}$. It follows that the term $\left(\mathbf{H}^{H}\mathbf{H}\right)^{-1}$ in Eq. 8.18 is equal to $\mathrm{Adj}\left(\mathbf{H}^{H}\mathbf{H}\right)/\det\left(\mathbf{H}^{H}\mathbf{H}\right)$, which implies that when $\det\left(\mathbf{H}^{H}\mathbf{H}\right)$ is small, $P_{\tilde{Z}}$ is large. This, in turn, is equivalent to any of the following statements:

a) $\det\left(\mathbf{H}^{H}\mathbf{H}\right)$ is small;
b) $\mathbf{H}^{H}\mathbf{H}$ is nearly singular;
c) At least one of the eigenvalues of $\mathbf{H}^{H}\mathbf{H}$ is small;
d) $\mathrm{Rank}(\mathbf{H}^{H}\mathbf{H}) \approx 1$ (i.e., low rank).

All four of these statements represent different ways of saying the same thing. The increase in the noise power that occurs when $\det\left(\mathbf{H}^{H}\mathbf{H}\right)$ is small is called ZF noise amplification. It can be shown that ZF noise amplification becomes infinite with high probability as N_t and N_r become large [91].

From a physical standpoint, the conditions that lead to ZF noise amplification include: a) antenna spacing that is too small; b) small angle spread associated with the received signal due to scattering; and c) LOS paths between the transmitter and the receiver. For example, as we discussed in Chapter 5 in connection with Eqs. 5.1 and 5.2, when the sizes of the transmit and receive antenna arrays are small compared to the path length,

$$
\mathbf{H} \to \frac{e^{-j\phi}}{\alpha}\begin{pmatrix} 1 & \cdots & 1 \\ \vdots & \ddots & \vdots \\ 1 & \cdots & 1 \end{pmatrix}, \tag{8.19}
$$

which results in $\det(\mathbf{H}^{H}\mathbf{H}) \to 0$.

In the next section, we discuss a second BLAST decoding technique, ZF-IC, that avoids the problem of noise amplification.

8.3.2 Zero-forcing with interference cancellation (ZF-IC)

ZF-IC is based on the properties of the QR factorization of $\mathbf{H}$. Recall from matrix theorem Section 1.9.2-(v) that any $m \times n$ matrix ($m \geq n$) can be decomposed into a product of two matrices, $\mathbf{QV}$, where $\mathbf{Q}$ is unitary and dimensioned $m \times m$, and $\mathbf{V}$ is upper triangular and dimensioned $m \times n$. It follows that the QR factorization of $\mathbf{H}$ is given by the following:

$$\mathbf{H} = \mathbf{QV}, \tag{8.20}$$

where $\mathbf{Q}$ is dimensioned $N_r \times N_r$ and $\mathbf{V}$ is $N_r \times N_t$. In ZF-IC, the received matrix is premultiplied by $\mathbf{Q}^H$ rather than $\mathbf{H}^+$. If, as before, we denote the matrix that results from the premultiplication step by $\tilde{\mathbf{R}}$, it follows that

$$\begin{aligned} \tilde{\mathbf{R}} &\triangleq \mathbf{Q}^H\mathbf{R} & (8.21)\\ &= \sqrt{\rho}\mathbf{Q}^H\mathbf{HS} + \mathbf{Q}^H\mathbf{Z} \\ &= \sqrt{\rho}\mathbf{Q}^H\mathbf{QVS} + \mathbf{Q}^H\mathbf{Z} \\ &= \sqrt{\rho}\mathbf{VS} + \tilde{\mathbf{Z}}, & (8.22) \end{aligned}$$

where the fourth line follows from the fact that $\mathbf{Q}$ is unitary and $\tilde{\mathbf{Z}} \triangleq \mathbf{Q}^H\mathbf{Z}$. To help see the significance of Eq. 8.22, it is helpful to write it out in expanded matrix format form as follows:

$$\tilde{\mathbf{R}} = \sqrt{\rho}\begin{pmatrix} v_{1,1} & v_{1,2} & \cdots & v_{1,N_t} \\ 0 & v_{2,2} & \cdots & v_{2,N_t} \\ \vdots & \vdots & \ddots & \vdots \\ 0 & 0 & \cdots & v_{N_t,N_t} \\ \vdots & \vdots & \ddots & \vdots \\ 0 & 0 & \cdots & 0 \end{pmatrix}\begin{pmatrix} s_1(1) & \cdots & s_1(p) \\ \vdots & \ddots & \vdots \\ s_{N_t}(1) & \cdots & s_{N_t}(p) \end{pmatrix} + \begin{pmatrix} z_1(1) & \cdots & z_1(p) \\ \vdots & \ddots & \vdots \\ z_{N_r}(1) & \cdots & z_{N_r}(p) \end{pmatrix}, \tag{8.23}$$

where we have explicitly indicated that the bottom $N_r - N_t$ rows of $\mathbf{V}$ are all zeros, assuming that $N_r \geq N_t$. Note that if that is not the case, then the $\mathbf{V}$ matrix is truncated and $\tilde{\mathbf{R}}$ will have fewer than N_t rows, which will make it impossible to detect the data streams from all N_t antennas. This shows that it is necessary that $N_r \geq N_t$. It can be shown that this is a general requirement that exists for all of the spatial multiplexing schemes discussed in this chapter.

We now consider the ith row of $\tilde{\mathbf{R}}$, which we denote by the sequence $\{\tilde{r}_i(k): k = 1, \ldots, p\}$. It follows from Eq. 8.23 that

$$\begin{aligned} \tilde{r}_i(k) &= \sqrt{\rho}\sum_{j=1}^{N_t} v_{i,j}s_j(k) + \tilde{z}_i(k) \\ &= \sqrt{\rho}\left[v_{i,i}s_i(k) + \sum_{j=i+1}^{N_t} v_{i,j}s_j(k)\right] + \tilde{z}_i(k), \quad k = 1, \ldots, p \end{aligned} \tag{8.24}$$

where the second equality follows from the fact that $\mathbf{V}$ is upper triangular.

We now develop the ZF-IC algorithm for demultiplexing and demodulating the transmitted symbols, using Eq. 8.24 as our starting point.

Step 1 We begin by setting $i = N_t$. Under that assumption, Eq. 8.24 becomes

$$\tilde{r}_{N_t}(k) = \sqrt{\rho} v_{N_t,N_t} s_{N_t}(k) + \tilde{z}_{N_t}(k). \tag{8.25}$$

We now define a new quantity called $\hat{r}_{N_t}(k)$ as follows:

$$\hat{r}_{N_t}(k) \triangleq \frac{\tilde{r}_{N_t}(k)}{\sqrt{\rho}\, v_{N_t,N_t}} \tag{8.26}$$

$$= s_{N_t}(k) + \frac{\tilde{z}_{N_t}(k)}{\sqrt{\rho}\, v_{N_t,N_t}}. \quad k = 1, \ldots, p \tag{8.27}$$

Equations 8.26 and 8.27 show that for each value of k, $\hat{r}_{N_t}(k)$ is equal to the symbol transmitted from the N_tth antenna at time k plus a noise term, which is the same form as Eq. 6.22. As a result, we can perform maximum likelihood detection to obtain estimates of $s_{N_t}(k)$, which we denote by $\hat{s}_{N_t}(k)$. It follows from Eq. 6.26 that the maximum likelihood estimate of $s_{N_t}(k)$ is

$$\hat{s}_{N_t}(k) = \arg\min_{\{s_{N_t}(k)\}} |\hat{r}_{N_t}(k) - s_{N_t}(k)|^2. \quad k = 1, \ldots, p \tag{8.28}$$

Step 1, therefore, involves estimating the symbols transmitted by antenna N_t (i.e., layer N_t) during a frame of length p symbols.

Step 2 Next, we set $i = N_t - 1$. Under this assumption, Eq. 8.24 becomes

$$\tilde{r}_{N_t-1}(k) = \sqrt{\rho}\left[v_{N_t-1,N_t-1} s_{N_t-1}(k) + v_{N_t-1,N_t} s_{N_t}(k)\right] + \tilde{z}_{N_t-1}(k). \tag{8.29}$$

The first term in the right-hand side of Eq. 8.29 consists of a scaled version of the signal from layer $N_t - 1$, the second term is interference from layer N_t, and the last term is noise. Since we have estimates of the signals from layer N_t that we obtained in step 1, we can use those estimates to *cancel interference* from layer N_t. We do this by subtracting $\sqrt{\rho} v_{N_t-1,N_t} \hat{s}_{N_t}(k)$ from $\tilde{r}_{N_t}(k)$. We denote the resulting signal by $\tilde{\tilde{r}}_{N_t-1}(k)$ as follows:

$$\tilde{\tilde{r}}_{N_t-1}(k) \triangleq \tilde{r}_{N_t-1}(k) - \sqrt{\rho} v_{N_t-1,N_t} \hat{s}_{N_t}(k) \tag{8.30}$$

$$= \sqrt{\rho}\left[v_{N_t-1,N_t-1} s_{N_t-1}(k) + v_{N_t-1,N_t}\left(s_{N_t}(k) - \hat{s}_{N_t}(k)\right)\right] + \tilde{z}_{N_t-1}(k). \tag{8.31}$$

If we now divide $\tilde{\tilde{r}}_{N_t-1}(k)$ by $\sqrt{\rho} v_{N_t-1,N_t-1}$, we denote the resulting signal by $\hat{r}_{N_t-1}(k)$, which has the following form:

$$\hat{r}_{N_t-1}(k) \triangleq \frac{\tilde{\tilde{r}}_{N_t-1}(k)}{\sqrt{\rho} v_{N_t-1,N_t-1}} \tag{8.32}$$

$$= s_{N_t-1}(k) + \frac{v_{N_t-1,N_t}\left(s_{N_t}(k) - \hat{s}_{N_t}(k)\right)}{v_{N_t-1,N_t-1}} + \frac{\tilde{z}_{N_t-1}(k)}{\sqrt{\rho} v_{N_t-1,N_t-1}}. \tag{8.33}$$

Equations 8.32 and 8.33 show that $\hat{r}_{N_t-1}(k)$ is equal to the signal from layer $N_t - 1$, plus residual interference from layer N_t, plus a thermal noise term. However, since interference cancellation has been performed using the estimate of the signal from layer N_t, the impact of the interference from layer N_t is reduced. We can, therefore, perform maximum likelihood detection on $\hat{r}_{N_t-1}(k)$ to obtain estimates of the symbols from layer $N_t - 1$. Denoting these estimates by $\{\hat{s}_{N_t-1}(k)\}$, it follows that

$$\hat{s}_{N_t-1}(k) = \underset{\{s_{N_t-1}(k)\}}{\arg\min} |\hat{r}_{N_t-1}(k) - s_{N_t-1}(k)|^2. \ k = 1, \ldots, p \tag{8.34}$$

Step i From the foregoing discussion, it is straightforward to generalize this process. For the general ith layer, it follows from Eq. 8.24 that

$$\tilde{\tilde{r}}_i(k) \triangleq \tilde{r}_i(k) - \sqrt{\rho} \sum_{j=i+1}^{N_t} v_{i,j}\hat{s}_j(k) \tag{8.35}$$

$$= \sqrt{\rho}\left[v_{i,i}s_i(k) + \sum_{j=i+1}^{N_t} v_{i,j}\left(s_j(k) - \hat{s}_j(k)\right)\right] + \tilde{z}_i(k), \tag{8.36}$$

which implies that

$$\hat{r}_i(k) \triangleq \frac{\tilde{\tilde{r}}_i(k)}{\sqrt{\rho}v_{i,i}} \tag{8.37}$$

$$= s_i(k) + \underbrace{\frac{\sum_{j=i+1}^{N_t} v_{i,j}\left(s_j(k) - \hat{s}_j(k)\right)}{v_{i,i}}}_{\text{interference}} + \underbrace{\frac{\tilde{z}_i(k)}{\sqrt{\rho}v_{i,i}}}_{\text{noise}}. \tag{8.38}$$

We see from this expression that $\hat{r}_i(k)$ consists of the signal from layer i plus interference and thermal noise. Therefore, we can estimate the transmitted symbols from layer i using maximum likelihood detection as follows:

$$\hat{s}_i(k) = \underset{\{s_i(k)\}}{\arg\min} \ |\hat{r}_i(k) - s_i(k)|^2. \ k = 1, \ldots, p \tag{8.39}$$

In general, the interference term in Eq. 8.38 is observed to be smaller than it would otherwise be because the interference from layers $i+1$ to N_t is cancelled using estimates of the signals from those layers. Not only does Eq. 8.38 show *interference cancellation*, it also shows that there is no interference coming from layers lower than layer $i+1$. The total absence of such interference is called *interference suppression*. It is important to keep the difference between interference cancellation and interference suppression clear since they are distinct effects and the MIMO literature distinguishes between the two. We conclude from this that when detecting layer i using ZF-IC, interference from layers $1, \ldots, (i-1)$ is *suppressed* and interference from layers $(i+1), \ldots, N_t$ is *cancelled*. It should also be pointed out that the use of the term "zero forcing" in the context of ZF-IC refers to the interference suppression part of the detection process since interference suppression implies the total elimination (or zeroing out) of interference in select layers.

Table 8.2 ZF-IC decoding algorithm for H-BLAST and V-BLAST.

1. Perform a QR decomposition of $\mathbf{H}$ (i.e., $\mathbf{H} = \mathbf{QV}$).
2. Compute $\tilde{\mathbf{R}}$ using Eq. 8.21.
3. Compute $\hat{r}_{N_t}(k)$ using the N_tth row of $\tilde{\mathbf{R}}$ as defined by Eq. 8.26.
4. Estimate the transmitted symbols associated with layer N_t by performing maximum likelihood detection of $\hat{r}_{N_t}(k)$ using Eq. 8.28.
5. Let $i = N_t - 1$.
6. While $i \geq 1$:
   (a) cancel interference in layer i using Eq. 8.35;
   (b) compute $\hat{r}_i(k)$ using Eq. 8.37;
   (c) estimate the transmitted symbols associated with layer i by performing maximum likelihood detection of $\hat{r}_i(k)$ using Eq. 8.39;
   (d) $i = i - 1$;
   end of while loop.

Table 8.2 summarizes the ZF-IC algorithm for use with H-BLAST and V-BLAST systems.

8.3.3 Linear minimum mean square detection (LMMSE)

8.3.3.1 The LMMSE criterion

The LMMSE detection method is based on the assumption that each transmitted signal can be estimated at the receiver by linearly combining weighted versions of the received signals. For example, the estimate of the symbol transmitted from antenna 1 at time k is given by $\hat{s}_1(k) = w_{1,1}r_1(k) + w_{1,2}r_2(k) + \cdots + w_{1,N_r}r_{N_r}(k)$, where the parameters $w_{1,1}, w_{1,2}, \ldots, w_{1,N_r}$ are complex numbers that are chosen to minimize the error between the true and estimated values of the transmitted symbols. Similarly, we could write expressions for the estimates of the symbols from the other transmitters as follows:

$$\begin{array}{ccccccccc}
\hat{s}_1(k) & = & w_{1,1}r_1(k) & + & w_{1,2}r_2(k) & + & \ldots & + & w_{1,N_r}r_{N_r}(k) \\
\hat{s}_2(k) & = & w_{2,1}r_1(k) & + & w_{2,2}r_2(k) & + & \ldots & + & w_{2,N_r}r_{N_r}(k) \\
\vdots & & \vdots & & \vdots & & & & \vdots \\
\hat{s}_{N_t}(k) & = & w_{N_t,1}r_1(k) & + & w_{N_t,2}r_2(k) & + & \ldots & + & w_{N_t,N_r}r_{N_r}(k).
\end{array} \tag{8.40}$$

If we now define the following matrix and vectors:

$$\mathbf{W} \triangleq \begin{pmatrix} w_{1,1} & w_{1,2} & \ldots & w_{1,N_r} \\ w_{2,1} & w_{2,2} & \ldots & w_{2,N_r} \\ \vdots & \vdots & \vdots & \vdots \\ w_{N_t,1} & w_{N_t,2} & \ldots & w_{N_t,N_r} \end{pmatrix} \tag{8.41}$$

and

$$\mathbf{r}(k) \triangleq \begin{pmatrix} r_1(k) \\ r_2(k) \\ \vdots \\ r_{N_r}(k) \end{pmatrix} \qquad \hat{\mathbf{s}}(k) \triangleq \begin{pmatrix} \hat{s}_1(k) \\ \hat{s}_2(k) \\ \vdots \\ \hat{s}_{N_t} \end{pmatrix}, \tag{8.42}$$

we can write Eq. 8.40 in matrix format as

$$\hat{\mathbf{s}}(k) = \mathbf{W}\mathbf{r}(k). \tag{8.43}$$

This expression, of course, can be generalized to p symbols, which results in the following expression for the estimated symbol matrix:

$$\hat{\mathbf{S}} = \mathbf{W}\mathbf{R}, \tag{8.44}$$

where $\mathbf{R}$ is defined in Eq. 8.6 and $\hat{\mathbf{S}}$ has the form

$$\hat{\mathbf{S}} = \begin{pmatrix} \hat{s}_1(1) & \hat{s}_1(2) & \dots & \hat{s}_1(p) \\ \hat{s}_2(1) & \hat{s}_2(2) & \dots & \hat{s}_2(p) \\ \vdots & \vdots & \ddots & \vdots \\ \hat{s}_{N_t}(1) & \hat{s}_{N_t}(2) & \dots & \hat{s}_{N_t}(p) \end{pmatrix}. \tag{8.45}$$

In LMMSE detection, we seek the matrix $\mathbf{W}$ that minimizes the mean square error between $\mathbf{S}$ and $\hat{\mathbf{S}}$. We denote this optimum matrix by $\mathbf{W}_o$ and define it as follows:

$$\begin{aligned} \mathbf{W}_o &\triangleq \underset{\{\mathbf{W}\}}{\arg\min}\ \mathbb{E}\left\{\|\mathbf{S} - \hat{\mathbf{S}}\|_F^2\right\} \\ &= \underset{\{\mathbf{W}\}}{\arg\min}\ \mathbb{E}\left\{\|\mathbf{S} - \mathbf{W}\mathbf{R}\|_F^2\right\}. \end{aligned} \tag{8.46}$$

To solve Eq. 8.46, we start by recalling that the Frobenius norm of a matrix is equal to the trace of the product of the matrix with its Hermitian. Therefore,

$$\begin{aligned} \mathbf{W}_o &= \underset{\{\mathbf{W}\}}{\arg\min}\ \mathbb{E}\left\{\mathrm{Tr}\left[(\mathbf{S} - \mathbf{W}\mathbf{R})(\mathbf{S} - \mathbf{W}\mathbf{R})^H\right]\right\} \\ &= \underset{\{\mathbf{W}\}}{\arg\min}\ \mathbb{E}\left\{\mathrm{Tr}\left[\mathbf{S}\mathbf{S}^H - \mathbf{S}\mathbf{R}^H\mathbf{W}^H - \mathbf{W}\mathbf{R}\mathbf{S}^H + \mathbf{W}\mathbf{R}\mathbf{R}^H\mathbf{W}^H\right]\right\}. \end{aligned} \tag{8.47}$$

To find the value of $\mathbf{W}$ that minimizes the right side of this equation, we appeal to the properties of matrix differentiation presented in [36] and summarized in matrix theorem Section 1.9.2-(y). The optimum value of $\mathbf{W}$ is obtained by taking the partial derivative with respect to $\mathbf{W}$, setting the result to zero, and solving the resulting equation for $\mathbf{W}$. This results in the following:

$$\mathbb{E}\left\{\frac{\partial \mathrm{Tr}\left[\mathbf{S}\mathbf{S}^H\right]}{\partial \mathbf{W}} - \frac{\partial \mathrm{Tr}\left[\mathbf{S}\mathbf{R}^H\mathbf{W}^H\right]}{\partial \mathbf{W}} - \frac{\partial \mathrm{Tr}\left[\mathbf{W}\mathbf{R}\mathbf{S}^H\right]}{\partial \mathbf{W}} + \frac{\partial \mathrm{Tr}\left[\mathbf{W}\mathbf{R}\mathbf{R}^H\mathbf{W}^H\right]}{\partial \mathbf{W}}\right\} = 0. \tag{8.48}$$

Since the numerator of the first term is not a function of **W**, its derivative is zero. The other terms are easily evaluated using matrix theorem Section 1.9.2-(y), which allows us to simplify Eq. 8.48 to the following:

$$\mathbb{E}\left\{\mathbf{R}\mathbf{R}^H\mathbf{W}^H\right\} = \mathbb{E}\left\{\mathbf{R}\mathbf{S}^H\right\}. \tag{8.49}$$

After substituting $\sqrt{\rho}\,\mathbf{HS} + \mathbf{Z}$ for **R** and simplifying, Eq. 8.49 reduces to

$$\mathbb{E}\left\{\rho\mathbf{HSS}^H\mathbf{H}^H + \mathbf{ZZ}^H\right\}\mathbf{W}^H = \mathbb{E}\left\{\sqrt{\rho}\,\mathbf{HSS}^H\right\}. \tag{8.50}$$

To apply the expectation operation in this equation, we assume that **H** is fixed over the frame length of p modulation symbols and that the transmitted signals and noise are uncorrelated. Under the latter assumption, it follows that $\mathbb{E}\left\{\mathbf{SS}^H\right\} = \sigma_s^2\mathbf{I}_{N_t}$ and $\mathbb{E}\left\{\mathbf{ZZ}^H\right\} = \sigma_z^2\mathbf{I}_{N_r}$, where σ_s^2 and σ_z^2 denote the variances of the transmitted signals and noise, respectively. Furthermore, our normalization implies that $\sigma_s^2 = 1/N_t$ and $\sigma_z^2 = 1$. Using these assumptions, Eq. 8.50 simplifies to

$$\left(\frac{\rho}{N_t}\mathbf{HH}^H + \mathbf{I}_{N_r}\right)\mathbf{W}^H = \frac{\sqrt{\rho}}{N_t}\mathbf{H}. \tag{8.51}$$

Solving this equation for **W** gives the following expression for $\mathbf{W}_o$:

$$\mathbf{W}_o = \frac{\sqrt{\rho}}{N_t}\mathbf{H}^H\left(\frac{\rho}{N_t}\mathbf{HH}^H + \mathbf{I}_{N_r}\right)^{-1}. \tag{8.52}$$

It can be shown that $\mathbf{W}_o$ can alternatively be expressed as

$$\mathbf{W}_o = \frac{\sqrt{\rho}}{N_t}\left(\frac{\rho}{N_t}\mathbf{H}^H\mathbf{H} + \mathbf{I}_{N_t}\right)^{-1}\mathbf{H}^H. \tag{8.53}$$

The equivalence of the two expressions in Eqs. 8.52 and 8.53 can be proven theoretically and easily demonstrated empirically in Matlab using arbitrary matrices for **H**. We use the second version later in the chapter when we discuss performance results for H-BLAST and V-BLAST.

Both the LMMSE and LMMSE-IC decoding techniques are based on the use of $\mathbf{W}_o$. In the next subsection, we describe how $\mathbf{W}_o$ is used in the LMMSE technique.

8.3.3.2 The LMMSE detection algorithm

The LMMSE detection scheme is similar to ZF detection except that the received matrix is premultiplied by $\mathbf{W}_o$ instead of $\mathbf{H}^+$. Denoting the resulting matrix by $\tilde{\mathbf{R}}$ as before, it follows that

$$\tilde{\mathbf{R}} \triangleq \mathbf{W}_o\mathbf{R} \tag{8.54}$$

$$= \sqrt{\rho}\,\mathbf{W}_o\mathbf{HS} + \mathbf{W}_o\mathbf{Z} \tag{8.55}$$

$$= \sqrt{\rho}\,\mathbf{W}_o\mathbf{HS} + \tilde{\mathbf{Z}}, \tag{8.56}$$

Table 8.3 LMMSE decoding algorithm for H-BLAST and V-BLAST.

1. Compute $\mathbf{W}_o$ using Eq. 8.52.
2. Compute $\tilde{\mathbf{R}}$ using Eq. 8.54.
3. Set $j = 1$.
4. While $j \leq N_t$:
   (a) Compute $\{\hat{r}_j(k)\}$ by normalizing the elements of $\tilde{\mathbf{R}}$ using Eq. 8.58;
   (b) Estimate the transmitted symbols $\{\hat{s}_j(k)\}$ using Eq. 8.60.
   end

where $\tilde{\mathbf{Z}} \triangleq \mathbf{W}_o\mathbf{Z}$. It follows that the kth element of the jth row of $\tilde{\mathbf{R}}$ is

$$\begin{aligned}\tilde{r}_j(k) &= \sqrt{\rho}\sum_{m=1}^{N_r} w_{j,m}\sum_{n=1}^{N_t} h_{m,n}s_n(k) + \tilde{z}_j(k)\\ &= \sqrt{\rho}\sum_{m=1}^{N_r} w_{j,m}\left(h_{m,j}s_j(k) + \sum_{n\neq j}^{N_t} h_{m,n}s_n(k)\right) + \tilde{z}_j(k)\\ &= \sqrt{\rho}\left(\sum_{m=1}^{N_r} w_{j,m}h_{m,j}\right)s_j(k) + \sqrt{\rho}\sum_{m=1}^{N_r} w_{j,m}\sum_{n\neq j}^{N_t} h_{m,n}s_n(k) + \tilde{z}_j(k).\end{aligned} \tag{8.57}$$

We now define a normalized version of $\tilde{\mathbf{R}}$, which we denote by $\hat{\mathbf{R}}$. The elements of this new array, which are denoted by $\{\tilde{r}_j(k) \mid j = 1, \ldots, N_t; k = 1, \ldots, p\}$, are defined as follows:

$$\hat{r}_j(k) \triangleq \frac{\tilde{r}_j(k)}{\sqrt{\rho}\sum_{m=1}^{N_r} w_{j,m}h_{m,j}}, \tag{8.58}$$

which is equivalent to

$$\hat{r}_j(k) = \underbrace{s_j(k)}_{\text{signal}} + \underbrace{\frac{\sum_{m=1}^{N_r} w_{j,m}\sum_{n\neq j}^{N_t} h_{m,n}s_n(k)}{\sum_{m=1}^{N_r} w_{j,m}h_{m,j}}}_{\text{interference}} + \underbrace{\frac{\tilde{z}_j(k)}{\sqrt{\rho}\sum_{m=1}^{N_r} w_{j,m}h_{m,j}}}_{\text{noise}}. \tag{8.59}$$

This equation shows that $\hat{r}_j(k)$ is equal to the signal from the jth transmitter plus interference and noise. It follows that the maximum likelihood estimate of $s_j(k)$ is

$$\hat{s}_j(k) = \underset{\{s_j(k)\}}{\arg\min}\ |\hat{r}_j(k) - s_j(k)|^2 \quad j = 1, \ldots, N_t.\ \ k = 1, \ldots, p \tag{8.60}$$

The steps outlined in this section constitute the LMMSE detection algorithm, which we summarize in Table 8.3.

This algorithm is similar to ZF detection; however, unlike ZF, LMMSE detection does not suffer from noise amplification. This follows from the differences in the functional forms of $\mathbf{H}^+$ and $\mathbf{W}_o$. It is straightforward to show that the noise power associated with LMMSE (i.e., $\sum_{j,k}\mathbb{E}\left\{|\tilde{z}_j(k)|^2\right\}$) is given by

$$P_{\tilde{Z}} = \left(\frac{p\sigma_z^2\rho}{N_t^2}\right)\mathrm{Tr}\left[\mathbf{H}^H\left(\frac{\rho}{N_t}\mathbf{H}\mathbf{H}^H + \mathbf{I}_{N_r}\right)^{-1}\left(\frac{\rho}{N_t}\mathbf{H}\mathbf{H}^H + \mathbf{I}_{N_r}\right)^{-H}\mathbf{H}\right]. \tag{8.61}$$

Recall that $(\frac{\rho}{N_t}\mathbf{H}\mathbf{H}^H + \mathbf{I}_{N_r})^{-1} \propto (\det(\frac{\rho}{N_t}\mathbf{H}\mathbf{H}^H + \mathbf{I}_{N_r}))^{-1}$. Since the determinant of a square matrix is equal to the product of its eigenvalues, it follows from matrix theorem Section 1.9.2-(z) that

$$\det(\frac{\rho}{N_t}\mathbf{H}\mathbf{H}^H + \mathbf{I}_{N_r}) = \prod_{i=1}^{r}(\frac{\rho}{N_t}\lambda_i + 1), \tag{8.62}$$

where $\{\lambda_i\}$ and r denote the eigenvalues and rank of $\mathbf{H}\mathbf{H}^H$, respectively. This result implies that even if one or more of the eigenvalues of $\mathbf{H}\mathbf{H}^H$ is zero (i.e., if $r < N_r$), the determinant of $(\frac{\rho}{N_t}\mathbf{H}\mathbf{H}^H + \mathbf{I}_{N_r})$ will not be zero, and, accordingly, $P_{\tilde{Z}}$ for LMMSE will not grow without limit as it did with ZF. This is an advantage that LMMSE has over ZF detection.

Despite this advantage, LMMSE detection appears to suffer from considerable interference effects, based on examination of Eq. 8.59. While it is true that LMMSE is degraded by interference effects, it should be kept in mind that the impact is not as large as might first appear from a casual examination of the interference term in Eq. 8.59, due to the fact that the elements of $\mathbf{W}_o$ are specifically chosen to minimize interference. This ameliorating effect on interference plus the lack of noise amplification results in LMMSE detection having better performance than ZF detection, as shown later in the chapter.

8.3.4 LMMSE with interference cancellation (LMMSE-IC)

The fourth detection method that we describe for use with H-BLAST and V-BLAST spatial multiplexing is LMMSE-IC. Although $\mathbf{W}_o$ is chosen to minimize interference, it does not eliminate it completely, so there is room for improvement. The LMMSE-IC detection scheme attempts to cancel the residual interference still remaining after premultiplying the received matrix by $\mathbf{W}_o$.

8.3.4.1 The MMSE-IC concept

Before describing the MMSE-IC algorithm in detail, we begin by explaining the underlying concept of MMSE-IC, which is similar to that of ZF-IC, but which differs in its mathematical particulars. The basic concept starts by considering the N_tth row of $\tilde{\mathbf{R}}$, which is normalized using Eq. 8.58 to obtain $\{\hat{r}_{N_t}(k), k = 1, \ldots, p\}$. The maximum likelihood estimates of the symbols from antenna N_t are computed using Eq. 8.60 for $j = N_t$.

The resulting estimates could be used to cancel interference from layer N_t when decoding the next layer down – layer $(N_t - 1)$. If we were to do that, we would set $j = N_t - 1$ in Eq. 8.59, but subtract $\hat{s}_{N_t}(k)$ from the $n = N_t$ term in the interference portion of Eq. 8.59. Denoting this modified version of $\hat{r}_{N_t-1}(k)$ by $\hat{\hat{r}}_{N_t-1}(k)$, it follows that

$$\hat{\hat{r}}_{N_t-1}(k) = s_{N_t-1}(k) + \frac{\sum_{m=1}^{N_r} w_{N_t-1,m}\left[\sum_{n=1}^{N_t-2} h_{m,n}s_n(k) + h_{m,N_t}\left(s_{N_t}(k) - \hat{s}_{N_t}(k)\right)\right]}{\sum_{m=1}^{N_r} w_{N_t-1,m}h_{m,N_t-1}} + \hat{z}_{N_t-1}(k), \tag{8.63}$$

where, for notational convenience, $\hat{z}_{N_t-1}(k)$ denotes the noise term in Eq. 8.59 for $j = N_t - 1$. This shows that if our estimates of the symbols from layer N_t are perfect, the interference from that layer is eliminated completely. Even if the estimates are not perfect, the interference tends to be reduced and our ability to decode layer $N_t - 1$ is enhanced.

Clearly, the next step would be to estimate the symbols from antenna $N_t - 1$ by applying maximum likelihood detection to $\{\hat{\hat{r}}_{N_t-1}(k) \mid k = 1, \ldots, p\}$. This process could then be continued for all of the layers, resulting in increasingly better estimates with each successive layer due to there being increasingly greater interference cancellation as we work our way from layer N_t to layer 1.

This describes the effect that we would like to be able to achieve, but we need to find a way to duplicate this process using only those parameters that are available to the receiver, namely, $\mathbf{R}$, $\mathbf{H}$, and $\mathbf{W}_o$ (which is based on $\mathbf{H}$). In the next subsection, we describe how this is accomplished.

8.3.4.2 The LMMSE-IC algorithm

We begin by defining a matrix, $\mathbf{P}_j$, as follows:

$$\mathbf{P}_j \triangleq \begin{pmatrix} h_{1,j} \\ h_{2,j} \\ \vdots \\ h_{N_r,j} \end{pmatrix} \begin{pmatrix} \hat{s}_j(1) & \hat{s}_j(2) & \ldots & \hat{s}_j(p) \end{pmatrix} \tag{8.64}$$

$$= \begin{pmatrix} h_{1,j}\hat{s}_j(1) & \ldots & h_{1,j}\hat{s}_j(p) \\ \vdots & \ddots & \vdots \\ h_{N_r,j}\hat{s}_j(1) & \ldots & h_{N_r,j}\hat{s}_j(p) \end{pmatrix}. \tag{8.65}$$

Next, we define another new matrix, $\mathbf{R}^{(N_t-1)}$, as

$$\mathbf{R}^{(N_t-1)} \triangleq \mathbf{R}^{(N_t)} - \sqrt{\rho}\,\mathbf{P}_{N_t}, \tag{8.66}$$

where $\mathbf{R}^{(N_t)} \triangleq \mathbf{R}$ (i.e., $\mathbf{R}^{(N_t)}$ is defined to be the received signal matrix defined in Eq. 8.6 before any interference cancellation takes place). We also make the following definition:

$$\tilde{\mathbf{R}}^{(N_t-1)} \triangleq \mathbf{W}_o\mathbf{R}^{(N_t-1)}. \tag{8.67}$$

Denote the $(N_t - 1)$th row of $\tilde{\mathbf{R}}^{(N_t-1)}$ by $\tilde{\mathbf{r}}_{N_t-1}^{(N_t-1)}(k)$. It is straightforward (but tedious) to show (see Appendix E for details) that

$$\tilde{\mathbf{r}}_{N_t-1}^{(N_t-1)}(k) = \sqrt{\rho}\left(\sum_{m=1}^{N_r} w_{N_t-1,m}h_{m,N_t-1}\right)s_{N_t-1}(k)$$
$$+\sqrt{\rho}\sum_{m=1}^{N_r} w_{N_t-1,m}\left[\sum_{n=1}^{N_t-2} h_{m,n}s_n(k) + h_{m,N_t}\left(s_{N_t}(k) - \hat{s}_{N_t}(k)\right)\right]$$
$$+\tilde{z}_{N_t-1}(k). \qquad k = 1,\ldots,p. \tag{8.68}$$

It follows that

$$\hat{\mathbf{r}}_{N_t-1}^{(N_t-1)}(k) \triangleq \frac{\tilde{\mathbf{r}}_{N_t-1}^{(N_t-1)}(k)}{\sqrt{\rho}\left(\sum_{m=1}^{N_r} w_{N_t-1,m}h_{m,N_t-1}\right)} \tag{8.69}$$

is equivalent to Eq. 8.63. We conclude from this observation that performing the operation defined by Eq. 8.66 and then normalizing the $(N_t - 1)$th row of $\mathbf{R}^{(N_t-1)}$ according to Eq. 8.69 is equivalent to canceling interference from layer N_t when attempting to detect the symbols in layer $N_t - 1$.

This result suggests an iterative process that involves repeatedly canceling interference from successively lower layers by iteratively applying the type of operation defined in Eq. 8.66 to all layers. In general, therefore, we define $\mathbf{R}^{(j-1)}$ as follows:

$$\mathbf{R}^{(j-1)} \triangleq \mathbf{R}^{(j)} - \sqrt{\rho}\,\mathbf{P}_j. \tag{8.70}$$

The basic steps of the LMMSE-IC algorithm are, therefore: a) start with the receive matrix and then premultiply by $\mathbf{W}_o$ to get $\tilde{\mathbf{R}}$; b) estimate the transmitted symbols from antenna N_t by applying maximum likelihood decoding to row N_t of the normalized version of $\tilde{\mathbf{R}}$; c) repeat this process by iteratively applying Eq. 8.70, followed by maximum likelihood detection in successively lower layers.

These are the fundamental steps in the LMMSE-IC algorithm; however, there is one additional issue that needs to be considered to complete the algorithm. Since interference from layer j is removed (or at least significantly reduced) each time Eq. 8.70 is applied, $\mathbf{H}$ and $\mathbf{W}_o$ should be modified after each application of that formula in such a way that the contribution from the jth transmitter is rendered negligible. This can be accomplished by ignoring the jth column of $\mathbf{H}$ after each application of Eq. 8.70, and recomputing $\mathbf{W}_o$ based on this truncated version of $\mathbf{H}$. The process of removing columns in $\mathbf{H}$ after canceling interference from a layer is an integral part of the LMMSE-IC algorithm.

Table 8.4 summarizes the details of the LMMSE-IC algorithm.

8.3.5 BLAST performance results

In this section, we present performance results of H-BLAST and V-BLAST spatial multiplexing systems using the four decoding methods just described. These results are based on computer simulations assuming BPSK modulation and Rayleigh fading. The first set of results is shown in Figure 8.5, which plots the bit error probability versus the signal-to-noise ratio assuming $N_t = N_r = 5$. This plot shows that LMMSE-IC

Table 8.4 LMMSE-IC decoding algorithm for H-BLAST and V-BLAST.

1. Set $j = N_t$.
2. Set $\mathbf{R}^{(N_t)} = \mathbf{R}$.
3. Compute $\mathbf{W}_o$ based on the entire $\mathbf{H}$ array using Eq. 8.52.
4. While $j \geq 1$:
   (a) Compute $\tilde{\mathbf{R}} = \mathbf{W}_0\mathbf{R}$;
   (b) Use the elements of the jth row of $\tilde{\mathbf{R}}$ to compute the set of values $\{\hat{r}_j(k) \mid k = 1, \ldots, p\}$, which are defined by Eq. 8.58;
   (c) Estimate the transmitted symbols from transmit antenna j by performing maximum likelihood detection on the elements $\{\hat{r}_j(k) \mid k = 1, \ldots, p\}$ using Eq. 8.60;
   (d) Compute $\mathbf{P}_j$ using Eq. 8.64;
   (e) Set $\mathbf{R} = \mathbf{R}^{(j-1)}$ using Eq. 8.70;
   (f) Update $\mathbf{H}$ by removing the jth column;
   (g) Re-compute $\mathbf{W}_o$ with the updated $\mathbf{H}$ using Eq. 8.52;
   (h) Set $j = j - 1$
   end of while loop.

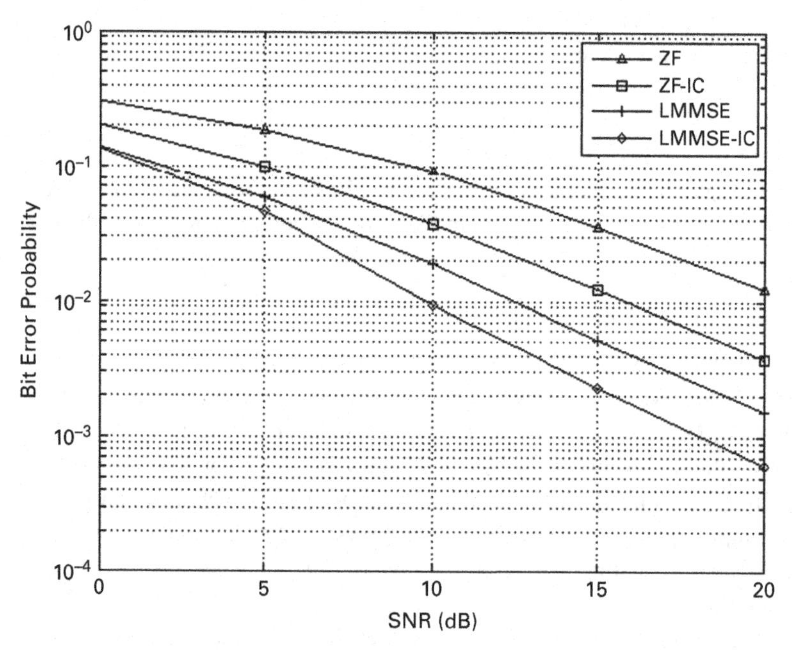

Figure 8.5 Bit error probability comparison for ZF and LMMSE detection schemes. This plot assumes uncoded BPSK modulation and $N_t = N_r = 5$.

performs best, LMMSE second best, ZF-IC third best, and ZF by itself performs the worst. This is partially explained by the fact that, as we discussed, ZF suffers from noise amplification and LMMSE does not. Since the computational complexity of the ZF and LMMSE schemes is not dramatically different, these results show that LMMSE decoding is recommended over ZF as a general rule.

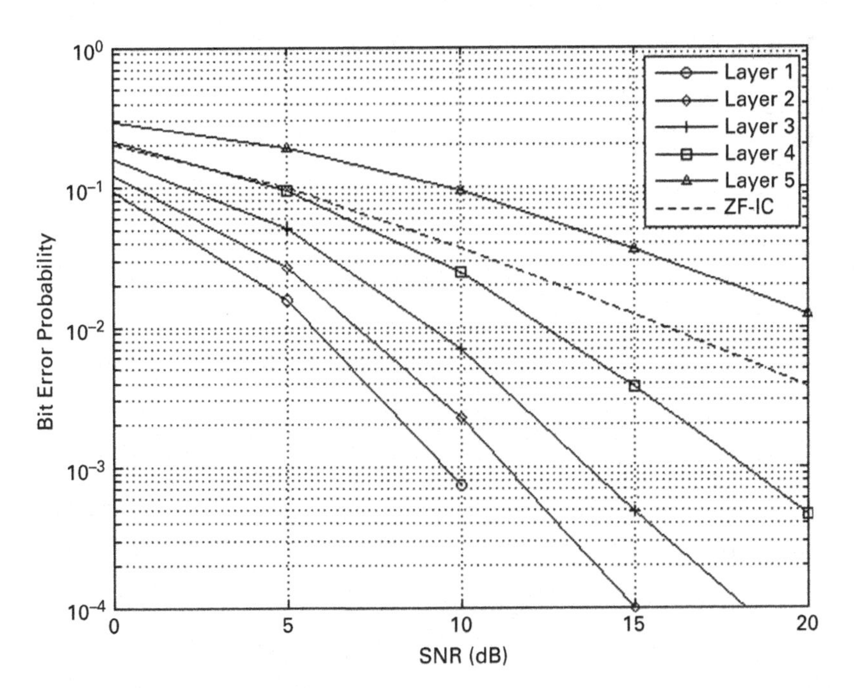

Figure 8.6 Bit error rate performance for individual layers using ZF-IC detection. This plot assumes uncoded BPSK modulation and $N_t = N_r = 5$.

The second set of results shows how the performance and spatial diversity order in spatial multiplexing vary with the layer being decoded. Figure 8.6 shows the bit error probability plotted versus SNR for different layers based on ZF-IC decoding. For the purpose of this plot, each curve is generated by assuming perfect interference cancellation in layers above the layer associated with each curve. That is, in Eq. 8.38, for each layer, i, we assume the interference term is zero prior to performing maximum likelihhood detection. The results show two key trends. First, we observe that performance improves as we go from layer N_t to layer 1. Since these results ignore interference effects, the differences in performance are totally due to differences in the noise term in Eq. 8.38 as the layer number changes (i.e., as i changes). Since $\tilde{z}_i(k)$ is independent of layer number because $\mathbf{Q}$ is a unitary matrix, it follows that the layer dependence of the noise term in Eq. 8.38 is due to the layer dependence of $v_{i,i}$. The Matlab simulation that was used to generate Figure 8.6 also computed the variance of v_{ii} as a function of i. The results are shown in Table 8.5.

These results show that as the layer number varies from 1 to $N_t = 5$, the noise term increases, which results in worse performance, consistent with the trend in Figure 8.6.

The second trend shown in Figure 8.6 is that the spatial diversity increases as the layer goes from $N_t = 5$ to 1. The overall performance, indicated by the dashed curve, has the same asymptotic slope as layer 5, which is the worst layer. This is a common property of many spatial multiplexing decoding techniques – the overall diversity order is approximately the same as the diversity order of the worst performing layer (i.e.,

Table 8.5 Dependence of the variance of $v_{i,i}$ on i for ZF-IC.

Layer	$\mathbb{E}\{\|v_{i,i}\|\}$
1	5.0
2	4.0
3	3.0
4	2.0
5	1.0

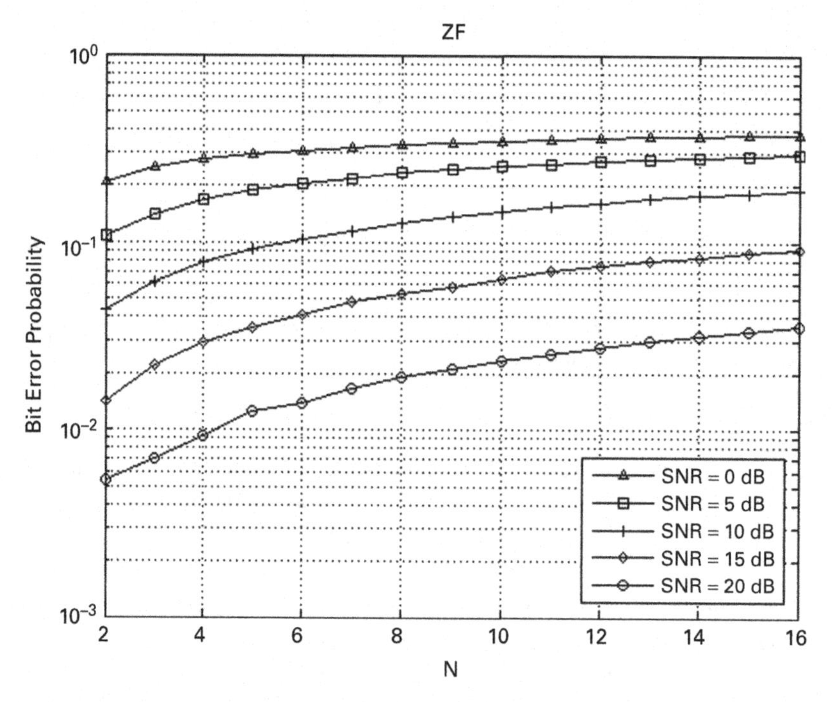

Figure 8.7 Bit error rate performance versus number of antennas for an $N \times N$ MIMO system using ZF detection for uncoded BPSK modulation.

layer N_t). We have more to say about the layer dependence of the diversity order when we discuss MGSTC.

The third set of results is summarized in Figures 8.7–8.10, which show the bit error probability versus the number of antennas for the four BLAST detection schemes. Each plot assumes that $N_t = N_r$, and displays results for $\rho = 0, 5, 10, 15,$ and 20 dB. These results show that the bit error probability for ZF and ZF-IC increases as the number of antennas and, hence, as the throughput increases. In contrast, the opposite is true with LMMSE and LMMSE-IC. In those detection schemes, the bit error probability decreases as the number of antennas increases when the SNR becomes large. This is another advantage of LMMSE over ZF, and a further reason to use LMMSE detection instead of either of the ZF methods.

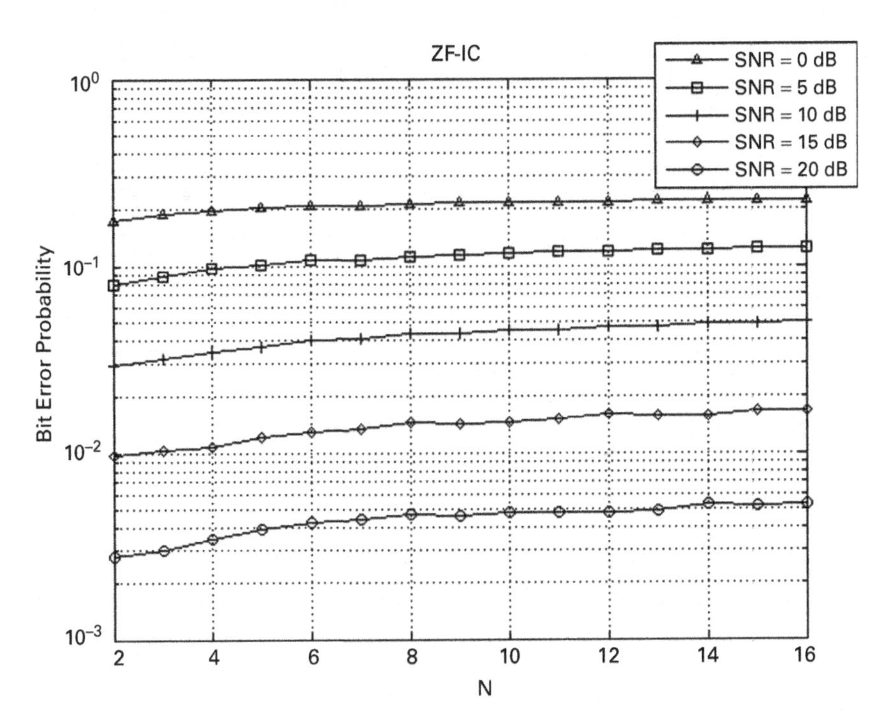

Figure 8.8 Bit error rate performance versus number of antennas for an $N \times N$ MIMO system using ZF-IC detection for uncoded BPSK modulation.

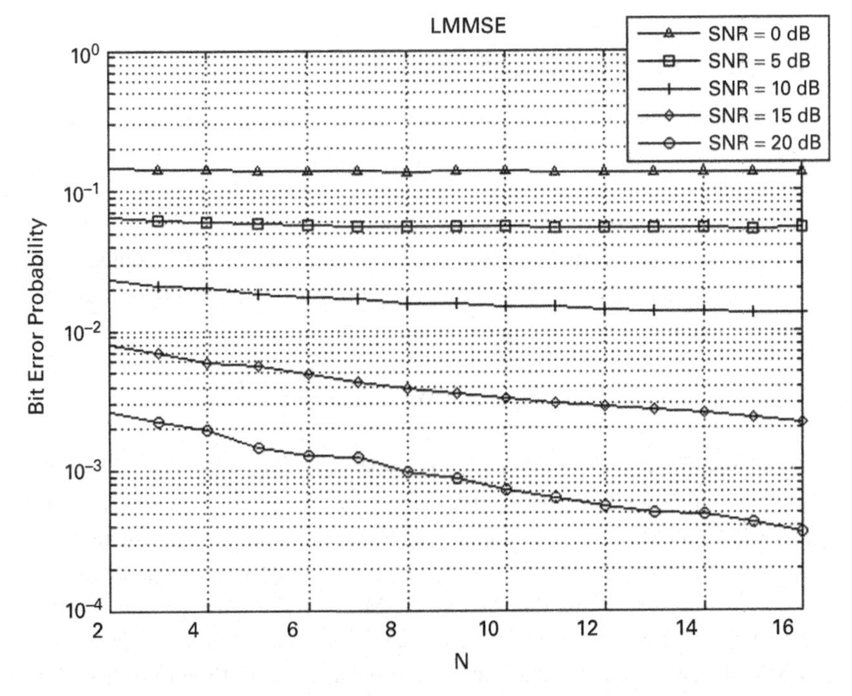

Figure 8.9 Bit error rate performance versus number of antennas for an $N \times N$ MIMO system using LMMSE detection for uncoded BPSK modulation.

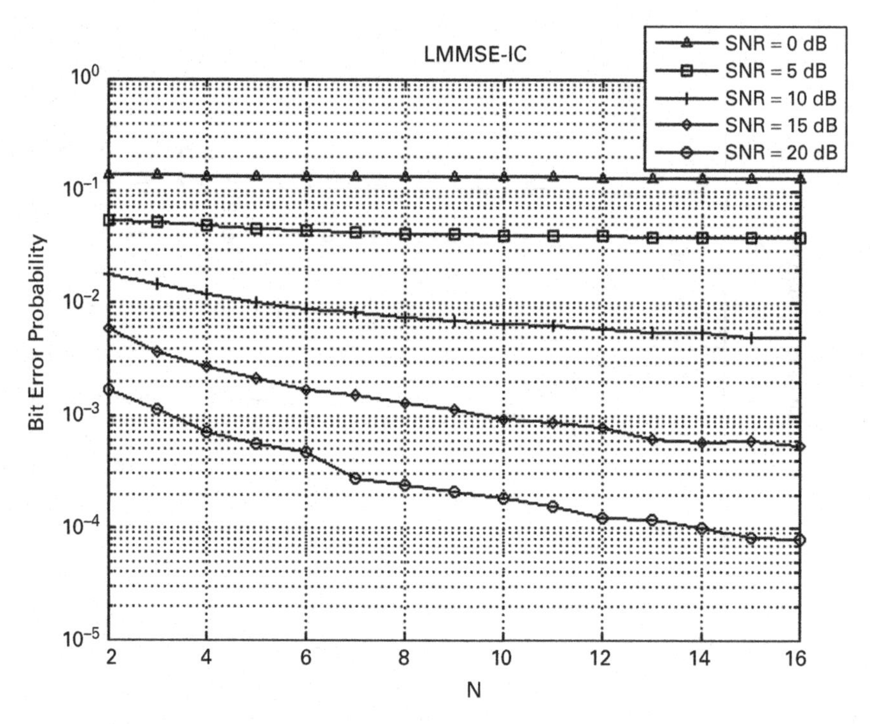

Figure 8.10 Bit error rate performance versus number of antennas for an $N \times N$ MIMO system using LMMSE-IC detection for uncoded BPSK modulation.

8.3.6 Comparison of ZF and LMMSE at large SNR

As we have just discussed, ZF suffers from noise amplification and LMMSE decoding does not, which, in part, explains why LMMSE performance is better than the performance associated with ZF. On closer examination, however, there are additional interesting and subtle complexities at work that are are seldom discussed in the literature when comparing these two decoding schemes. To help appreciate these effects, consider what happens to $\mathbf{W}_o$ as the signal-to-noise ratio becomes large. It follows from Eq. 8.53 that

$$\lim_{\rho \to \infty} \mathbf{W}_o = \frac{\left(\mathbf{H}^H\mathbf{H}\right)^{-1}\mathbf{H}^H}{\sqrt{\rho}}. \tag{8.71}$$

By comparing Eq. 8.71 with 8.12, it is evident that premultiplying $\mathbf{R}$ by $\mathbf{H}^+$ is equivalent to premultiplying by $\mathbf{W}_o$ in the limit of large ρ, since $\mathbf{W}_o$ is equivalent to $\mathbf{H}^+$ in the high-SNR limit to within a multiplicative scale factor. Since any such scale factor affects the signal and noise portions of $\mathbf{R}$ equally, it does not affect the performance and we would expect by this reasoning that ZF and LMMSE should have the same performance when ρ is large. An examination of Figure 8.5, however, shows no evidence that the ZF and LMMSE curves are tending to converge at large signal-to-noise ratios. This interesting and seemingly contradictory result is addressed in detail by Jian, Varanasi, and Li in [45], where they demonstrate this counterintuitive result theoretically. In that paper, they show that the SNRs of each of the decoupled data streams after LMMSE detection are given by

$$\rho_{\text{lmmse},n} = \rho_{\text{zf},n} + \eta_{\text{snr},n}, \qquad 1 \leq n \leq N_t, \tag{8.72}$$

where $\rho_{\text{lmmse},n}$ and $\rho_{\text{zf},n}$ denote the signal-to-noise ratios associated with the nth data stream (i.e., layer) after LMMSE and ZF demultiplexing, respectively, and $\eta_{\text{snr},n}$ is a *non-decreasing* function of the SNR. This result shows that even as the SNR becomes large, the SNR associated with LMMSE is larger than that of ZF, which results in LMMSE having lower bit error probabilities than ZF even when ρ becomes large. The details of this analysis are beyond the scope of this book, but interested readers are encouraged to consult [45] for further details.

8.4 Multi-group space-time coded modulation (MGSTC)

The decoding techniques that have been considered up to this point in our discussion achieve spatial multiplexing, but because they do not employ space-time coding, they provide little in the way of spatial diversity (e.g., see Figure 8.5). Since the publication of the BLAST techniques, researchers have attempted to find ways to achieve both spatial diversity and spatial multiplexing at the same time. One of the first examples of such a concept was published in 1999 by Tarokh *et al.* [92], and is called multi-group space-time coding (MGSTC). This section describes MGSTC, beginning with a description of the MGSTC encoder, followed by discussions of the MGSTC decoding algorithm, its spatial diversity properties, and its performance in Rayleigh fading.

8.4.1 The MGSTC encoder structure

Figure 8.11 shows the MGSTC encoder architecture. In MGSTC, a block of B bits enters the encoder, passes through a serial-to-parallel converter that splits the sequence of B bits into q groups, each with $B_1, B_2, \ldots, B_q$ bits, respectively. Each group of bits is then fed to a space-time encoder, which, in turn, generates $n_j, j = 1, \ldots, q$ output bits. Each of the space-time codes is called a *component code* and, in general, these codes can be either STBCs or STTCs, and can differ from one another within the same encoder. From the diagram it is clear that

$$\sum_{j=1}^{q} B_j = B \qquad \text{and} \qquad \sum_{j=1}^{q} n_j = N_t. \tag{8.73}$$

It is evident from the nature of the MGSTC encoder that the received signals from each component code encounter interference from all the other codes. Just as the BLAST schemes suffer from inter-layer interference, MGSTC systems suffer from inter-group inteference. This is one of the fundamental differences between the BLAST schemes and MGSTCs. In MGSTC, interference cancellation is employed, but it is performed on a group-by-group basis instead of on a layer-by-layer basis. The following subsection describes how interference cancellation is performed in MGSTC decoding.

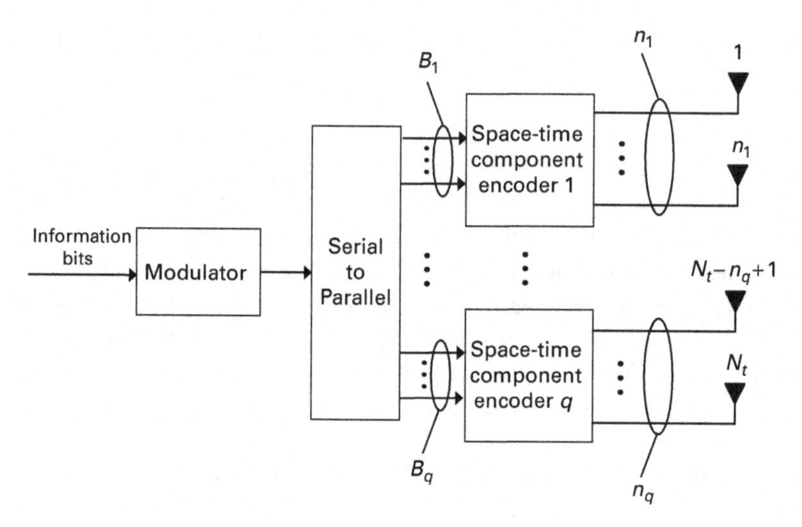

Figure 8.11 MGSTC encoder.

8.4.2 Nomenclature

We start by introducing some new nomenclature that will prove useful in describing the MGSTC decoding algorithm. We denote the outputs of the component codes by $\mathbf{S}_{C_1}$, $\mathbf{S}_{C_2}, \ldots, \mathbf{S}_{C_q}$, where the transmitted signal associated with code 1 is given by

$$\mathbf{S}_{C_1} \triangleq \begin{pmatrix} s_1(1) & \ldots & s_1(p) \\ \vdots & \ddots & \vdots \\ s_{n_1}(1) & \ldots & s_{n_1}(p) \end{pmatrix}. \tag{8.74}$$

Similarly, the output from component code 2 is

$$\mathbf{S}_{C_2} \triangleq \begin{pmatrix} s_{n_1+1}(1) & \ldots & s_{n_1+1}(p) \\ \vdots & \ddots & \vdots \\ s_{n_1+n_2}(1) & \ldots & s_{n_1+n_2}(p) \end{pmatrix}. \tag{8.75}$$

In general, the entire transmit signal matrix, $\mathbf{S}$, is given by

$$\mathbf{S} \triangleq \begin{pmatrix} s_1(1) & \ldots & s_1(p) \\ \vdots & \ddots & \vdots \\ s_{n_1}(1) & \ldots & s_{n_1}(p) \\ \vdots & \ddots & \vdots \\ s_{n_1+n_2}(1) & \ldots & s_{n_1+n_2}(p) \\ \vdots & \ddots & \vdots \\ s_{N_t}(1) & \ldots & s_{N_t}(p) \end{pmatrix} = \begin{pmatrix} \mathbf{S}_{C_1} \\ \mathbf{S}_{C_2} \\ \vdots \\ \mathbf{S}_{C_q} \end{pmatrix}. \tag{8.76}$$

It is also convenient to decompose the channel matrix into components as follows:

$$\mathbf{H} = \begin{pmatrix} h_{1,1} & \dots & h_{1,n_1} & \dots & h_{1,n_1+n_2} & \dots & h_{1,N_t} \\ \vdots & \ddots & \vdots & \ddots & \vdots & \ddots & \vdots \\ h_{N_r,1} & \dots & h_{N_r,n_1} & \dots & h_{N_r,n_1+n_2} & \dots & h_{N_r,N_t} \end{pmatrix} \tag{8.77}$$

$$= \begin{pmatrix} \mathbf{H}_{C_1} & \mathbf{H}_{C_2} & \dots & \mathbf{H}_{C_q} \end{pmatrix}, \tag{8.78}$$

where $\mathbf{H}_{C_i}$ is dimensioned $N_r \times n_i$.

We also find it convenient to define portions of the channel matrix where the parts associated with various component codes are excised. For example, consider the channel matrix with the first n_1 columns removed. We denote that as follows:

$$\mathbf{H}_{C-C_1} = \begin{pmatrix} h_{1,n_1+1} & \dots & h_{1,N_t} \\ \vdots & \ddots & \vdots \\ h_{N_r,n_1+1} & \dots & h_{N_r,N_t} \end{pmatrix}$$

$$= \begin{pmatrix} \mathbf{H}_{C_2} & \mathbf{H}_{C_3} & \dots & \mathbf{H}_{C_q} \end{pmatrix}. \tag{8.79}$$

More generally, we denote the truncated channel matrix where the first $n_1 + \cdots + n_i$ columns are omitted by

$$\mathbf{H}_{C-C_{1,i}} = \begin{pmatrix} \mathbf{H}_{C_{i+1}} & \mathbf{H}_{C_{i+2}} & \dots & \mathbf{H}_{C_q} \end{pmatrix}. \tag{8.80}$$

Finally, it proves useful in subsequent discussions to keep in mind the dimensions of the various matrices defined above. The following is a summary of the dimensions of key matrices:

1. $\mathbf{H}_{C_i}$ is dimensioned $[N_r \times n_i]$;
2. $\mathbf{H}_{C-C_1}$ is dimensioned $[N_r \times (N_t - n_1)]$;
3. $\mathbf{H}_{C-C_{1,i}}$ is dimensioned $\left[N_r \times (N_t - \sum_{k=1}^{i} n_k)\right]$.

8.4.3 MGSTC decoding

This section describes the decoding algorithm for multi-group space-time coding. We begin by considering how to decode space-time component code 1, which we denote by $\mathcal{C}_1$.

8.4.3.1 Decoding $\mathcal{C}_1$

Since code $\mathcal{C}_1$ is transmitted using n_1 transmit antennas, it follows that the transmitted signals from the remaining $N_t - n_1$ antennas constitute interference when decoding $\mathcal{C}_1$. In order to suppress this interference (i.e., eliminate it by forcing it to zero), it is necessary that

$$N_r \geq N_t - n_1 + 1. \tag{8.81}$$

It should be noted that this requirement is completely analogous to the requirement when performing ZF-IC. In that case, we were attempting to detect a single data stream and the other $N_t - 1$ streams appear as interference. In ZF-IC, therefore, n_1 is effectively equal to 1, so Eq. 8.81 reduces to $N_r \geq N_t$. Given the assumption in Eq. 8.81 and the fact that $\mathbf{H}_{C-C_1}$ is $[N_r \times n_i]$, it follows that

$$\text{rank}(\mathbf{H}_{C-C_1}) \leq N_t - n_1. \tag{8.82}$$

Next, we appeal to the concept of the *null space* of a matrix, which is defined in Section 1.9.1-(n). That definition states that the null space of any $m \times n$ matrix, $\mathbf{A}$, is equal to the set of $[1 \times m]$ vectors such that when $\mathbf{A}$ is multiplied by any of those vectors, the result is the all-zeros vector. We denote the null space of $\mathbf{A}$ by $N(\mathbf{A})$. It follows that the null space of $\mathbf{H}_{C-C_1}$ is expressed mathematically as follows:

$$N(\mathbf{H}_{C-C_1}) = \left\{\mathbf{x} \,|\, \mathbf{x}\mathbf{H}_{C-C_1} = \mathbf{0}\right\}, \tag{8.83}$$

where $\mathbf{x}$ is $[1 \times N_r]$ and $\mathbf{0}$ is $[1 \times (N_t - n_1)]$. Since the null space of an array is a vector space, it has a dimension that is equal to the minimum number of basis vectors that span that space. We can learn something about the dimension of $N(\mathbf{H}_{C-C_1})$ by appealing to the *rank plus nullity theorem* described in Section 1.9.2-(w), which, when applied to $N(\mathbf{H}_{C-C_1})$ has the following form:

$$\dim[N(\mathbf{H}_{C-C_1})] + \text{rank}[\mathbf{H}_{C-C_1}] = N_r. \tag{8.84}$$

It therefore follows from Eq. 8.82 that

$$\dim[N(\mathbf{H}_{C-C_1})] \geq N_r - (N_t - n_1). \tag{8.85}$$

This, in turn, means that there are at least $N_r - (N_t - n_1)$ basis vectors in $N(\mathbf{H}_{C-C_1})$ and that it is always possible to find a set of exactly $N_r - (N_t - n_1)$ orthonormal vectors (not necessarily unique)[2] that span the null set. Denote these $[1 \times N_r]$ orthonormal vectors by $\{\mathbf{v}_i\}$ and define the following matrix:

$$\mathbf{\Theta}_{C-C_1} \triangleq \begin{pmatrix} \mathbf{v}_1 \\ \mathbf{v}_2 \\ \vdots \\ \mathbf{v}_{N_r-(N_t-n_1)} \end{pmatrix}. \tag{8.86}$$

Thus, $\mathbf{\Theta}_{C-C_1}$ is dimensioned $[(N_r - N_t + n_1) \times N_r]$.[3]

Because the rows of $\mathbf{\Theta}_{C-C_1}$ are orthonormal, it follows that

$$\mathbf{\Theta}_{C-C1}\mathbf{\Theta}_{C-C_1}^H = \mathbf{I}_{N_r-(N_t-n_1)}. \tag{8.87}$$

Furthermore, since each row of $\mathbf{\Theta}_{C-C_1}$ is in $N(\mathbf{H}_{C-C_1})$, it is clear from Eq. 8.83 that

$$\mathbf{\Theta}_{C-C_1}\mathbf{H}_{C-C_1} = \mathbf{0}_{[(N_r-N_t+n_1)\times(N_t-n_1)]}. \tag{8.88}$$

[2] This means that there are possibly multiple sets of orthonormal vectors, but the vectors in each set are unique.

[3] A set of orthonormal null vectors that make up $\mathbf{\Theta}$ can be obtained using the built-in Matlab function called `null.m`.

Define $\tilde{\mathbf{R}}_{C_1}$ to be the result of premultiplying the received signal matrix, $\mathbf{R}$, by $\boldsymbol{\Theta}_{C-C_1}$. Thus,

$$\begin{aligned}
\tilde{\mathbf{R}}_{C_1} &\triangleq \boldsymbol{\Theta}_{C-C_1}\mathbf{R} \\
&= \sqrt{\rho}\,\boldsymbol{\Theta}_{C-C_1}\mathbf{H}\mathbf{S} + \boldsymbol{\Theta}_{C-C_1}\mathbf{Z} \\
&= \sqrt{\rho}\,\boldsymbol{\Theta}_{C-C_1}\begin{pmatrix} \mathbf{H}_{C_1} & \mathbf{H}_{C-C_1} \end{pmatrix}\begin{pmatrix} \mathbf{S}_{C_1} \\ \mathbf{S}_{C-C_1} \end{pmatrix} + \tilde{\mathbf{Z}}_{C-C_1} \\
&= \sqrt{\rho}\begin{pmatrix} \boldsymbol{\Theta}_{C-C_1}\mathbf{H}_{C_1} & \boldsymbol{\Theta}_{C-C_1}\mathbf{H}_{C-C_1} \end{pmatrix}\begin{pmatrix} \mathbf{S}_{C_1} \\ \mathbf{S}_{C-C_1} \end{pmatrix} + \tilde{\mathbf{Z}}_{C-C_1} \\
&= \sqrt{\rho}\left(\boldsymbol{\Theta}_{C-C_1}\mathbf{H}_{C_1}\mathbf{S}_{C_1} + \mathbf{0}\mathbf{S}_{C-C_1}\right) + \tilde{\mathbf{Z}}_{C-C_1} \\
&= \sqrt{\rho}\boldsymbol{\Theta}_{C-C_1}\mathbf{H}_{C_1}\mathbf{S}_{C_1} + \tilde{\mathbf{Z}}_{C-C_1},
\end{aligned} \tag{8.89}$$

where $\tilde{\mathbf{Z}}_{C-C_1} \triangleq \boldsymbol{\Theta}_{C-C_1}\mathbf{Z}$ and the fifth equality follows from Eq. 8.88.

At this point it should be noted that although $\mathbf{R}$ has energy from all the transmit antennas arriving at each receive antenna, $\tilde{\mathbf{R}}_{C_1}$ only has energy associated with code $\mathcal{C}_1$. Thus, by premultiplying $\mathbf{R}$ by $\boldsymbol{\Theta}_{C-C_1}$, the interference from codes $\mathcal{C}_2$ to $\mathcal{C}_q$ (i.e., from antennas $(n_1+1), \ldots, N_t$) has been *suppressed* (i.e., zeroed out).

Since the receiver is assumed to have knowledge of the channel matrix, we assume that $\boldsymbol{\Theta}_{C-C_1}$ and $\mathbf{H}_{C_1}$ are known, so the receiver can compute the *effective channel* matrix for decoding $\mathcal{C}_1$, which we denote by $\mathbf{H}_1^{(\text{eff})}$. Thus, from Eq. 8.89 it follows that

$$\mathbf{H}_1^{(\text{eff})} \triangleq \boldsymbol{\Theta}_{C-C_1}\mathbf{H}_{C_1}, \tag{8.90}$$

which has dimension $[(N_r - N_t + n_1) \times n_1]$. Therefore, we can write $\tilde{\mathbf{R}}_{C_1}$ as

$$\tilde{\mathbf{R}}_{C_1} = \sqrt{\rho}\mathbf{H}_1^{(\text{eff})}\mathbf{S}_{C_1} + \tilde{\mathbf{Z}}_{C-C_1}. \tag{8.91}$$

Since $\mathbf{H}_1^{(\text{eff})}$ is assumed to be known, the receiver can compute the maximum likelihood estimate of the transmitted code block $\mathbf{S}_{C_1}$ associated with the first space-time component code from $\tilde{\mathbf{R}}_{C_1}$. We denote this estimate by $\hat{\mathbf{S}}_{C_1}$, which is computed as follows:

$$\hat{\mathbf{S}}_{C_1} = \underset{\{\hat{\mathbf{S}}_{C_1}\}}{\arg\min} \|\tilde{\mathbf{R}}_{C_1} - \sqrt{\rho}\mathbf{H}_1^{(\text{eff})}\mathbf{S}_{C_1}\|_F^2. \tag{8.92}$$

If $\mathbf{S}_{C_1}$ encodes k modulation symbols from an M-ary alphabet, then $\mathbf{S}_{C_1}$ can take on k^M different values when performing the arg min operation.

8.4.3.2 Decoding $\mathcal{C}_2$

After $\hat{\mathbf{S}}_{C_1}$ is obtained, it can be used to *cancel interference* associated with $\mathcal{C}_1$ similarly to the way that interference cancellation is performed in ZF-IC and LMMSE-IC. Interference from $\mathcal{C}_1$ is canceled by subtracting $\sqrt{\rho}\mathbf{H}_{C_1}\hat{\mathbf{S}}_{C_1}$ from $\mathbf{R}$ as follows:

$$\mathbf{R}_{C_1} \triangleq \mathbf{R} - \sqrt{\rho}\,\mathbf{H}_{C_1}\hat{\mathbf{S}}_{C_1}. \tag{8.93}$$

This new matrix, $\mathbf{R}_{C_1}$, can now be used as the starting point for estimating the second component code (i.e., $\hat{\mathbf{S}}_{C_2}$). We do this by premultiplying $\mathbf{R}_{C_1}$ by a new matrix comprising vectors that span the null space of $\mathbf{H}_{C-C_{1,2}}$ instead of $\mathbf{H}_{C-C_1}$. We conclude from the rank plus nullity theorem that the dimension of $\mathbf{H}_{C-C_{1,2}}$ is greater than or equal to $N_r - N_t + (n_1 + n_2)$, so the number of orthonormal spanning vectors in $N(\mathbf{H}_{C-C_{1,2}})$ equals $N_r - N_t + (n_1 + n_2)$. As before, these vectors define the rows in the new matrix that we denote by $\boldsymbol{\Theta}_{C-C_{1,2}}$, which is dimensioned $[(N_r - N_t + (n_1 + n_2) \times N_r]$. Premultiplying $\mathbf{R}_{C_1}$ by $\boldsymbol{\Theta}_{C-C_{1,2}}$ results in

$$\tilde{\mathbf{R}}_{C_2} \triangleq \boldsymbol{\Theta}_{C-C_{1,2}} \mathbf{R}_{C_1}. \tag{8.94}$$

It is straightforward to show by plugging Eq. 8.93 into this equation that

$$\tilde{\mathbf{R}}_{C_2} = \sqrt{\rho}\, \boldsymbol{\Theta}_{C-C_{1,2}} \mathbf{H}_{C_2} \mathbf{S}_{C_2} + \sqrt{\rho} \boldsymbol{\Theta}_{C-C_{1,2}} \mathbf{H}_{C_1} (\mathbf{S}_{C_1} - \hat{\mathbf{S}}_{C_1}) + \boldsymbol{\Theta}_{C-C_{1,2}} \mathbf{Z}. \tag{8.95}$$

This equation demonstrates that by performing the operations defined in Eqs. 8.93 and 8.94, the resulting matrix consists of a term with just the C_2 code block (first term) plus a residual interference term (second term) and, finally, a thermal noise term. From this equation we observe that we can interpret the term that premultiplies $\mathbf{S}_{C_2}$ as an effective channel matrix, which we denote by $\mathbf{H}_2^{(\text{eff})}$. Thus,

$$\mathbf{H}_2^{(\text{eff})} \triangleq \boldsymbol{\Theta}_{C-C_{1,2}} \mathbf{H}_{C_2}, \tag{8.96}$$

from which it follows that $\mathbf{H}_2^{(\text{eff})}$ is dimensioned $[(N_r - N_t + (n_1 + n_2)) \times n_2]$. Since $\mathbf{H}_2^{(\text{eff})}$ is known, the receiver can compute the ML estimate of $\mathbf{S}_{C_2}$ as follows:

$$\hat{\mathbf{S}}_{C_2} = \arg \min_{\{\hat{\mathbf{S}}_{C_2}\}} \|\tilde{\mathbf{R}}_{C_2} - \sqrt{\rho} \mathbf{H}_2^{(\text{eff})} \mathbf{S}_{C_2}\|_F^2.$$

8.4.3.3 Decoding C_i

From the foregoing discussion, it is clear how to generalize to an arbitrary code group. In general, the ith code word, $\mathbf{S}_i$, is estimated by first canceling interference from component code $i-1$ as follows:

$$\mathbf{R}_{C_{i-1}} \triangleq \mathbf{R}_{C_{i-2}} - \sqrt{\rho}\, \mathbf{H}_{C_{i-1}} \hat{S}_{C_{i-1}}. \tag{8.97}$$

Next, we compute the set of $N_r - N_t + \sum_{i=1}^{i} n_i$ orthonormal basis vectors that span $N(\mathbf{H}_{C-C_{1,i}})$, which can be done using the Matlab function `null.m`. These vectors are then used to construct $\boldsymbol{\Theta}_{C-C_{1,i}}$ by assigning each vector to a row in $\boldsymbol{\Theta}_{C-C_{1,i}}$. This allows us to compute $\tilde{\mathbf{R}}_{C_i}$ as follows:

$$\tilde{\mathbf{R}}_{C_i} \triangleq \boldsymbol{\Theta}_{C-C_{1,i}} \mathbf{R}_{C_{i-1}}, \tag{8.98}$$

which can be shown to be equal to the following:

$$\tilde{\mathbf{R}}_{C_i} = \sqrt{\rho}\, \boldsymbol{\Theta}_{C-C_{1,i}} \mathbf{H}_{C_i} \mathbf{S}_{C_i} + \sqrt{\rho} \sum_{k=1}^{i-1} \boldsymbol{\Theta}_{C-C_{1,i}} \mathbf{H}_{C_k} \left(\mathbf{S}_{C_k} - \hat{\mathbf{S}}_{C_k} \right) + \boldsymbol{\Theta}_{C-C_{1,i}} \mathbf{Z}. \tag{8.99}$$

This shows that in the general case it is possible to perform ML estimation of $\mathbf{S}_{C_i}$ using $\tilde{\mathbf{R}}_{C_i}$ as follows:

$$\hat{\mathbf{S}}_{C_i} = \arg \min_{\{\hat{\mathbf{S}}_{C_i}\}} \|\tilde{\mathbf{R}}_{C_i} - \sqrt{\rho} \mathbf{H}_i^{(\text{eff})} \mathbf{S}_{C_i}\|_F^2.$$

Table 8.6 MGSTC decoding algorithm.

1. Set $i = 1$.
2. Compute $\mathbf{R} = \sqrt{\rho}\,\mathbf{HS} + \mathbf{Z}$.
3. Compute $\mathbf{H}_{C-C_1}$ by deleting the first n_1 columns of $\mathbf{H}$.
4. Compute $\mathbf{\Theta}_{C-C_1}$ from $\mathbf{H}_{C-C_1}$ (the Matlab function `null.m` can be used for this as follows:
   a) `temp = null(transpose(`$\mathbf{H}_{C-C1}$`)`; $\mathbf{\Theta}_{C-C1}$ `= transpose(temp);`).
5. Compute $\tilde{\mathbf{R}}_{C_1} = \mathbf{\Theta}_{C-C_1}\mathbf{R}$.
6. Compute $\mathbf{H}_1^{(\text{eff})} = \mathbf{\Theta}_{C-C_1}\mathbf{H}_{C_1}$.
7. Solve for $\hat{\mathbf{S}}_{C_1}$ by performing ML detection on $\tilde{\mathbf{R}}_{C_1}$.
8. Define $\mathbf{R}_0 = \mathbf{R}$.
9. Set $i = 2$.
10. While $i \leq (q-1)$:
    (a) Cancel interference from $\mathcal{C}_1, \ldots, \mathcal{C}_{i-1}$ by computing $\mathbf{R}_{i-1} = \mathbf{R}_{i-2} - \sqrt{\rho}\mathbf{H}_{C_{i-1}}\hat{\mathbf{S}}_{C_{i-1}}$;
    (b) Compute $\mathbf{H}_{C-C_{1,i}}$ by deleting the first $(n_1 + \cdots + n_i)$ columns of $\mathbf{H}$;
    (c) Compute $\mathbf{\Theta}_{C-C_{1,i}}$ from $\mathbf{H}_{C-C_{1,i}}$;
    (d) Compute $\tilde{\mathbf{R}}_{C_i} = \mathbf{\Theta}_{C-C_{1,i}}\mathbf{R}_{i-1}$;
    (e) Compute $\mathbf{H}_i^{(\text{eff})} = \mathbf{\Theta}_{C-C_{1,i}}\mathbf{H}_{C_i}$;
    (f) Solve for $\hat{\mathbf{S}}_{C_i}$ by performing ML detection on $\tilde{\mathbf{R}}_{C_i}$;
    (g) Set $i = i + 1$;
    end of while loop.
11. Cancel interference from $\mathcal{C}_1, \ldots, \mathcal{C}_{q-1}$ by computing $\mathbf{R}_{q-1} = \mathbf{R}_{q-2} - \sqrt{\rho}\mathbf{H}_{C_{q-1}}\hat{\mathbf{S}}_{C_{q-1}}$.
12. Compute $\mathbf{H}_q^{(\text{eff})} = \mathbf{H}_{C_q}$.
13. Solve for $\hat{\mathbf{S}}_{C_q}$ by performing ML detection on $\mathbf{R}_{q-1}$.

In general,

$$\mathbf{H}_i^{(\text{eff})} = \mathbf{\Theta}_{C-C_{1,i}}\mathbf{H}_{C_i}, \tag{8.100}$$

which is dimensioned $[(N_r - N_t + \sum_{k=1}^{i} n_k) \times n_i]$.

8.4.3.4 The MGSTC decoding algorithm

This section combines the findings in the previous three sections into a general decoding algorithm for MGSTC. A summary of the MGSTC decoding algorithm is given in Table 8.6.

8.4.4 Group-dependent diversity

As we have seen, in MGSTC each group is decoded by applying the maximum likelihood decoding algorithm to $\tilde{\mathbf{R}}_{C_i}, i = 1, \ldots, q$. We have shown that each group, in turn, has an effective channel matrix associated with it, $\mathbf{H}_i^{(\text{eff})}$, that is dimensioned $[(N_r - N_t + \sum_{k=1}^{i} n_k) \times n_i]$. Therefore, when decoding the ith group, the effective number of transmitters is n_i and the effective number of receivers is $(N_r - N_t + \sum_{k=1}^{i} n_k)$.

Table 8.7 Diversity order associated with each group when decoding MGSTC.

Group number	Diversity order
1	$n_1\,(N_r - N_t + n_1)$
2	$n_2\,(N_r - N_t + n_1 + n_2)$
$\vdots$	$\vdots$
i	$n_i(N_r - N_t + \sum_{k=1}^{i} n_k)$
$\vdots$	$\vdots$
q	$n_q N_r$

It follows that the diversity order associated with group i is $n_i(N_r - N_t + \sum_{k=1}^{i} n_k)$, $i = 1, \ldots, q$. We summarize this result in Table 8.7

These results show that the diversity increases as we go from group 1 to group q. In general, however, the overall probability of bit error is dominated by the worst group (i.e., the group with the smallest diversity order). To see why this is so, denote the diversity of the ith group by N_{d_i}. The overall probability of bit error is equal to the probability that group 1 has a bit error **or** that group 2 has an error **or** etc.. Denoting the probability of there being a bit error associated with group i by P_{b_i}, it follows that the total bit error probability, P_b, is given by

$$P_b = P_{b_1} + P_{b_2} + \cdots + P_{b_q}. \tag{8.101}$$

Therefore, when ρ is large,

$$\begin{aligned} P_b &= k_1 \rho^{-N_{d_1}} + k_2 \rho^{-N_{d_2}} + \cdots + k_q \rho^{-N_{d_q}} \\ &\approx \kappa \rho^{-\min\{N_{d_i}\}}, \end{aligned} \tag{8.102}$$

where κ denotes the coefficient associated with the smallest diversity order.

In the next section, we present simulated bit error results for MGSTC and see that the bit error performance obeys this behavior at large SNR.

8.4.5 MGSTC performance results

In this section, we show an example of the performance of MGSTC in Rayleigh fading and compare those results with the corresponding performance of the four V-BLAST and H-BLAST demultiplexing schemes described earlier. For the purpose of this comparison, we assume that $N_t = N_r = 6$ and that the MGSTC encoder employs three Alamouti space-time component codes as illustrated in Figure 8.12.

The performance results are shown in Figure 8.13. In this figure, the bit error probability is plotted versus the signal-to-noise ratio, ρ, for ZF, ZF-IC, LMMSE, LMMSE-IC, and MGSTC assuming uncoded BPSK with $N_t = N_r = 6$ in Rayleigh fading. The results in this plot are based on computer simulations in which there were 2000 Monte

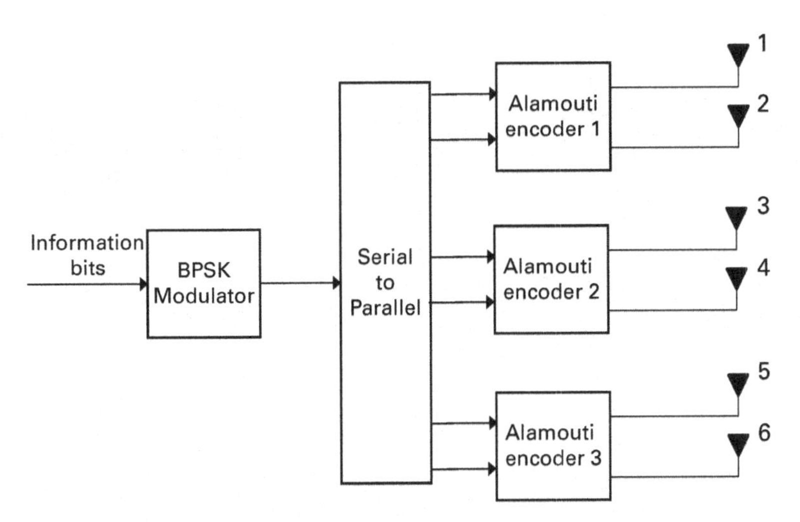

Figure 8.12 Encoder used to generate the MGSTC results in this section.

Carlo iterations, each of which involved randomly choosing a different 6×6 channel matrix. Each of these 2000 iterations, in turn, involved generating a different receive matrix, $\mathbf{R}$, that was dimensioned $[N_r \times 2000]$. Each of the 2000 receive vectors in $\mathbf{R}$ was computed using randomly chosen noise terms; thus, the simulation models quasi-static Rayleigh fading where the channel is assumed to be fixed for 2000 bit periods.

The results show that MGSTC achieves significantly lower bit errors than any of the other techniques. This should not be surprising, however, since MGSTC employs space-time block coding that provides diversity gain, which the other techniques do not provide. Since $N_t = N_r$ and $n_1 = 2$ in this example, Table 8.7 shows that the minimum diversity order, which is associated with group number 1, is equal to $n_1^2 = 4$. Therefore, based on Eq. 8.102, we would expect the slope of the bit error rate curve to approach $-\min\{N_{d_i}\} = -4$ as the SNR becomes large for the MGSTC curve. An examination of the MGSTC curve in Figure 8.13 confirms that the slope does, indeed, approach -4 at large ρ.

Problems

8.1 Derive Eq. 8.11.

8.2 Prove that the ZF receiver produces spatially colored noise. (Hint: spatially white noise means that the noise covariance matrix is diagonal.)

8.3 Derive Eq. 8.61.

8.4 Prove that Eq. 8.53 is equivalent to Eq. 8.52.

8.5 Prove Eq. 8.88.

8.6 Prove Eq. 8.95.

8.7 Write Matlab functions that simulate ZF, ZF-IC, LMMSE, and LMMSE-IC decoding for uncoded M-PSK modulation. Allow the user to specify the values of N_t, N_r, the

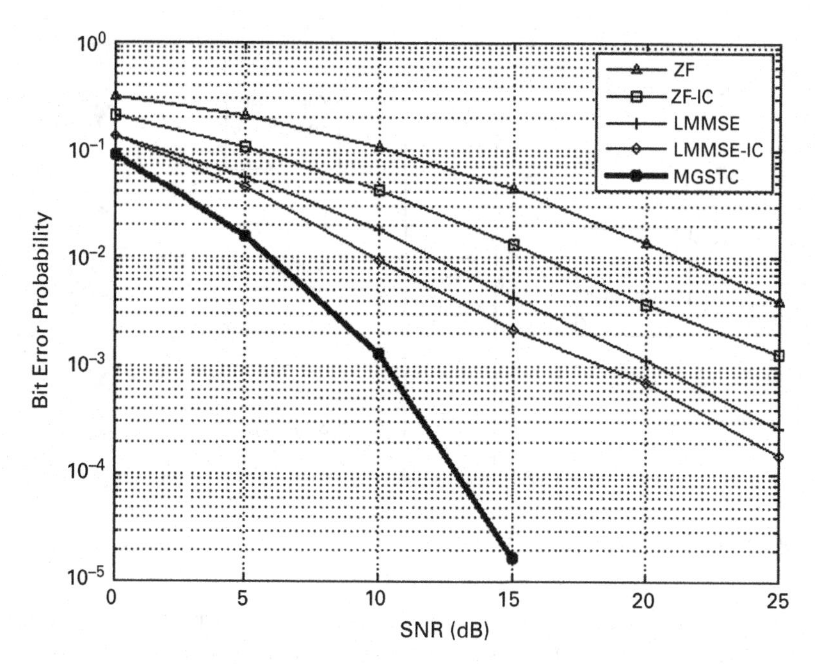

Figure 8.13 Comparison of V-BLAST and MGSTC bit error probability performance. This plot assumes uncoded BPSK modulation and $N_t = N_r = 6$.

PSK order, M, and the number of iterations. Run these functions and duplicate the results in Figure 8.5.

8.8 Write a Matlab function that simulates the performance of MGSTC with q Alamouti component codes, assuming uncoded M-PSK modulation. Allow the user to specify the values of q, N_t, N_r, the PSK order, M, and the number of iterations. Run this function and duplicate the MGSTC curve in Figure 8.13.

9 Broadband MIMO

Thus far, the focus in this book has been on flat fading channels; that is, we have assumed that the signal bandwidth is less than the coherence bandwidth of the channel. This assumption underlies all our discussions on space-time coding and spatial multiplexing since we have assumed throughout that the elements of the channel matrix consist of simple complex gain values. In practice, however, most modern wireless communication systems that use MIMO techniques are broadband systems; that is, they operate at high data rates – too high for the flat fading assumption to be valid. In most practical systems, the channel is frequency selective, not flat. Nevertheless, as we explain in this chapter, the concepts and techniques that have been described up to this point can still be applied. The purpose of this chapter is to explain how practical MIMO systems operate in frequency-selective environments.

9.1 Flat and frequency-selective fading

In Chapter 4 we showed that when fading is flat, the channel can be modeled as a complex gain applied to each symbol, and this has allowed us to use the following system equation when analyzing MIMO performance:

$$\mathbf{r} = \sqrt{\rho}\,\mathbf{H}\mathbf{s} + \mathbf{z}. \tag{9.1}$$

This equation is valid if the modulation symbol rate, R_s (i.e., baud rate), is less than the coherence bandwidth, B_c, of the channel. If, however, $R_s > B_c$, the channel is said to be frequency selective and the MIMO system equation becomes

$$r_i(n) = \sqrt{\rho}\sum_{j=1}^{N_t}\sum_{l=0}^{L-1} h_{i,j}^{(l)} s_j(n-l) + z_i(k), \qquad \begin{array}{l} i = 1,\ldots,N_r \\ n = 1,\ldots,N_f \end{array} \tag{9.2}$$

where N_f is the number of modulation symbols in a frame, $h_{i,j}^{(l)}$ denotes the channel impulse response during a frame, and $r_i(n)$ denotes the received signal at receive antenna i at time index n. The parameter L denotes the number of taps in the channel response, which is equal to the ratio of the RMS delay spread to the modulation symbol period (i.e., $L = \sigma_\tau/T_s$). It is important to note that Eq. 9.2 models quasi-static fading in which the channel is assumed to be fixed over a single frame. For this reason, the channel is not a function of the time index n.

Table 9.1 Delay spread values and maximum data rates for which the flat fading assumption is valid for selective environments (based on delay spread values listed in [61]).

Environment	Frequency (MHz)	RMS delay spread (σ_τ) (ns)	Max baud rate (ksym/s)
Urban (New York)	910	1300	154
Urban (San Francisco)	892	20 000	10
Typical suburban	910	300	660
Extreme suburban	910	2000	100
Indoor (office bldg.)	1500	25	8000
Indoor (office bldg.)	850	270	740
Indoor (office bldg.)	1900	90	2200

Table 9.2 Representative maximum data rates for selected wireless standards.

Standard	Maximum supported data rates (Mbps)
802.11n (WiFi)	6 to 600
Mobile WiMAX (802.16e-2005)	1 to 75
LTE (3GPP Rel. 8)	5 to 300
LTE-Advanced (3GPP Rel. 10)	1 (mobile); 1000 (fixed)

To see that frequency-selective conditions normally dominate with modern wireless systems, consider Table 9.1, which lists typical delay spread values for a variety of environments. The last column lists the maximum baud rate for which the flat fading condition is met (i.e., where $R_s = B_{c,50} = 1/(5\sigma_\tau)$). These results show that in outdoor environments the maximum data rate for which flat fading is a valid assumption is less than 1 Msym/sec. For comparison, consider Table 9.2, which lists maximum data rates that are supported for several commercial wireless standards. These results show that most commercial wireless technologies employ data rates that exceed typical outdoor coherence bandwidths and, therefore, experience frequency-selective fading.

One of the primary side effects of frequency-selective fading is inter-symbol interference (ISI), which is due to the symbol period being shorter than the delay spread. When this occurs, each symbol is spread out in time, resulting in its energy overlapping the energy in subsequent symbols. ISI can result in significantly degraded performance if not properly mitigated. In the next section, we discuss the two fundamental approaches that can be used to cope with frequency-selective fading.

9.2 Strategies for coping with frequency-selective fading

In general, frequency-selective fading can be dealt with either by exploiting it or by combating it. In practice, most MIMO systems attempt to combat frequency-selective fading effects using orthogonal frequency division multiplexing (OFDM), which is our primary focus in this chapter. For completeness, we begin this section by including a brief discussion of techniques that exploit frequency selectivity.

9.2.1 Exploiting frequency-selective fading

In frequency-selective fading, the energy from each transmitted symbol arrives at each receiver over multiple "paths" that span a time period that is longer than the symbol period itself. Although this leads to ISI, it also means that if each "path"[1] fades independently of the others, frequency selectivity potentially provides greater inherent diversity than flat fading. Since the effective number of resolvable paths in a frequency-selective channel is equal to $L = \sigma_\tau / T_s$, it follows that each transmitted symbol arrives at the receiver over $N_t N_r L$ diversity paths. Denoting the number of independent diversity channels by N_d as before, we conclude

$$\max\{N_d\} = \begin{cases} N_t N_r, & \text{flat channels,} \\ N_t N_r L, & \text{frequency-selective channels.} \end{cases} \tag{9.3}$$

These results imply, perhaps counterintuitively, that the capacity of a MIMO system operating in frequency-selective fading is greater than if it operates in flat fading. This issue has been analyzed in detail by Zhang, Duman, and Kurtas in [85]. In that paper, they use a simulation-based method to compute the achievable information rates for MIMO systems operating in frequency-selective fading. Figure 9.1, which is from this paper, shows the maximum information rate plotted versus SNR for a 2×2 MIMO system in Rayleigh fading for different levels of frequency selectivity as measured by the value of L. The three bottom curves assume BPSK modulation, so their asymptotic information rate is 2 bps/Hz (1 bps/Hz for each of the two transmitters). The top curve, which shows the true capacity of the system (i.e., no modulation constraint), is included as a reference. These results show that the maximum theoretical information rate increases as the frequency selectivity (i.e., L) increases, although the increase is relatively modest (less than about 5 %) and the improvement appears to reach a point of diminishing returns when $L = 2$.

Theoretical results such as those shown in Figure 9.1 have prompted research to find space-time coding techniques that are capable of exploiting the inherent diversity associated with frequency-selective fading. The following is a list of examples of such research: In [24], El Gamal *et al.* describe a framework for designing algebraic codes for MIMO in frequency-selective fading; in [86], Zhang, Xia, and Ching present space-time codes for use in MIMO frequency-selective channels that can achieve full diversity equal to $N_t N_r L$; and in [64], Schober, Gerstacker, and Lampe investigate space-time block codes over frequency-selective channels and show that they are capable of exploiting frequency selectivity to achieve enhanced diversity gain.

[1] The word *path* in this context is placed in quotes to indicate that we are talking about effective paths, not necessarily physical paths. In general, the smallest resolvable delay resolution is equal to the symbol period. Within that resolution, however, there are likely to be many physical scattered components that cannot be resolved independently.

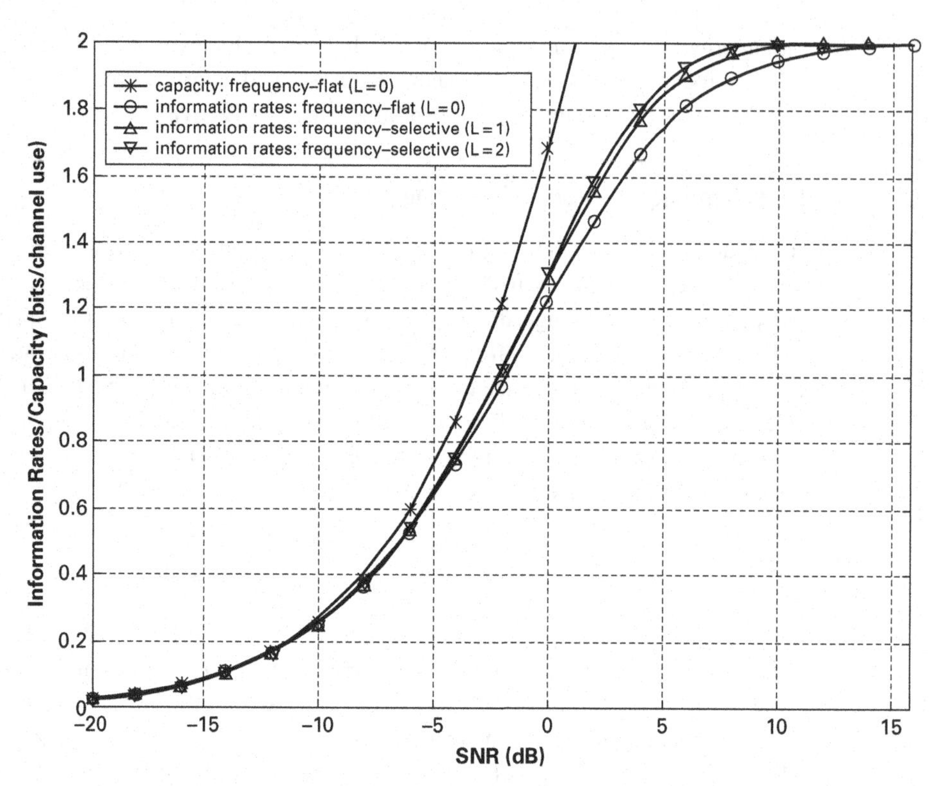

Figure 9.1 Maximum information rates in Rayleigh frequency-selective fading for a 2×2 MIMO system using BPSK modulation (from [85]).

9.2.2 Combating frequency-selective fading

Methods for combating frequency-selective fading fall into one of two categories: *single carrier techniques* or *multiple carrier techniques*, which we discuss in the following two subsections.

9.2.2.1 Single carrier techniques

As the name suggests, in single carrier methods, all of the data from each transmit antenna are modulated using a single carrier, but filtering, called equalization, is used to counteract ISI. Figure 9.2 shows a high-level block diagram of a MIMO system that employs equalization. The details of the equalizer block can take many forms, but two common classes of equalization are linear equalization and decision feedback equalization. Regardless of the details of the equalizer, however, the key aspect of a single channel approach is that it uses equalization to mitigate ISI and all the transmitters use one channel.

The primary advantage of using equalization to combat frequency selectivity is that the peak-to-average power ratio (PAPR) of the transmitted signal is relatively low. As a consequence, the power amplifiers in the transmit antenna RF chains can be operated relatively close to their cutoff power levels, which enables amplifiers to be used that are

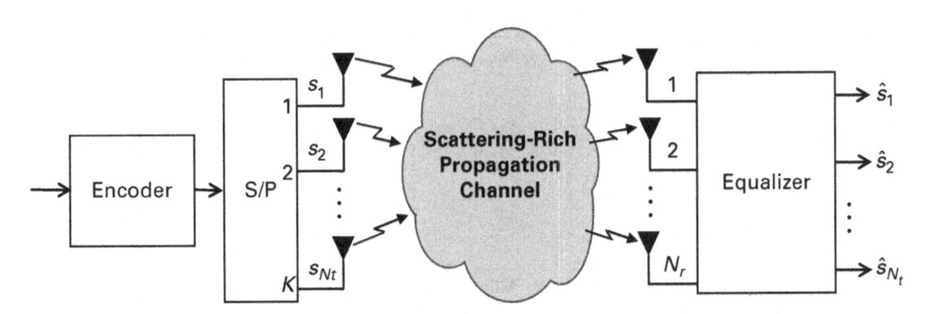

Figure 9.2 High-level diagram of a single channel MIMO system that uses equalization to combat frequency-selective multipath.

less expensive and consume less power than amplifiers that need to be operated at large backoff values. The chief disadvantage of the single channel approach, however, is that the processing in the equalizer can be computationally complex. In general, the complexity increases exponentially with the modulation order, the number of transmitters, and the ISI length (i.e., the value of L). It can be shown that even for binary modulation, high computational complexity can occur. In [74], van Nee and Prasad show that OFDM is computationally less demanding and, therefore, is preferred over single channel systems with equalization when $R_s\sigma_\tau > 1$. For this reason, most MIMO systems do not employ a single carrier approach for mitigating frequency-selective multipath.

9.2.2.2 Multiple carrier techniques

The idea behind multiple carrier techniques is to split a broadband signal that experiences frequency-selective fading and its attendant ISI into narrow sub-bands so that each sub-band experiences flat fading. Once that is done, the MIMO techniques that have been discussed up to this point can be applied to the individual sub-bands. The objective in such an approach is to make the bandwidths of the sub-bands less than the coherence bandwidth of the channel. As examples, in IEEE 802.11n, the sub-band bandwidth is 312.5 kHz and in LTE it is 15 kHz. These values, in turn, have been chosen to ensure that they are less than the coherence bandwidths anticipated to be experienced in the environments in which these standards are deployed. For example, 802.11n devices typically are deployed in indoor settings, which have delay spreads between roughly 25 and 270 ns according to Table 9.1, corresponding to 740 kHz $\leq B_{c,50} \leq$ 8 MHz. In contrast, LTE is designed to operate in larger cellular environments, which, according to Table 9.1, have delay delay spread values ranging from 300 to 20 000 ns, corresponding to 10 kHz $\leq B_{c,50} \leq$ 666 kHz. It should be noted that the value of 20 000 ns (which corresponds to $B_{c,50} =$ 10 kHz) represents an extreme case and that the maximum delay spread assumed in the extended typical urban model (ETU) used by LTE for development and testing purposes is only 5000 ns [25], which corresponds to $B_{c,50} =$ 40 kHz. Under these less extreme assumptions, we see that LTE's sub-band bandwidth of 15 kHz is well within the flat fading regime.

Figure 9.3 depicts a generic multiple carrier MIMO communication system. As illustrated, separate multiple carrier processors (depicted by the boxes labeled "MC") are

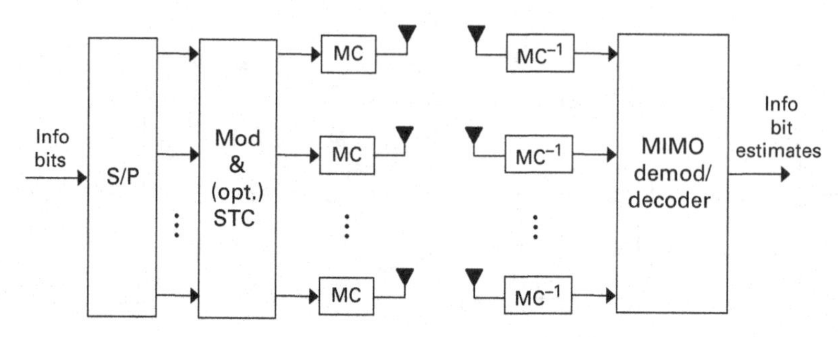

Figure 9.3 A generic multiple carrier MIMO system.

used on each transmitter. These boxes perform frequency division multiplexing on each transmitted data stream by splitting the broadband signal into sub-bands and then adding them together before being transmitted over the frequency-selective channel. At the receiving end, each received signal is passed through a complementary processing block (labeled "MC^{-1}") that performs frequency demultiplexing, which translates each of the sub-bands back to baseband before performing performing MIMO receiver processing on the resulting signal.

Figure 9.4 illustrates the operations performed by the MC and MC^{-1} blocks. The key property of the MC block is that it takes a broadband signal that is vulnerable to ISI and divides it into N sub-streams that have signaling rates that are a factor of N less than the input, resulting in sub-streams that are far less vulnerable to ISI than the original input stream. Each of those sub-streams are then mixed to sub-bands (also called *subcarriers*) before being summed together prior to transmission. The MC^{-1} part of the figure mixes each of the sub-bands down to baseband, samples the resulting signal, and then passes the sampled sequences through a parallel-to-serial converter before being processed by the MIMO demodulator and decoding stages.

Example 9.1 Consider a multiple carrier system operating in an office building environment where the RMS delay spread is 50 ns and the total symbol rate (i.e., baud rate) is 40 Msym/s. What is the minimum number of carriers needed to ensure that each sub-band experiences flat fading?

Answer A reasonable criterion for ensuring flat fading is that the bandwidth of a channel be less than $B_{c,50}$. In practice, we may want the channel bandwidth to be smaller than this, but this is a reasonable upper bound, which results in a minimum bound for the number of required carriers. Since $\sigma_\tau = 50$ ns, it follows that $B_{c,50} = 1/(5\sigma_\tau) = 4$ MHz. If the total bandwidth is twice the baud rate (a typical assumption), then the number of sub-band carriers is 40 Msym/s $\div$ 4 Msym/s = 10 carriers.

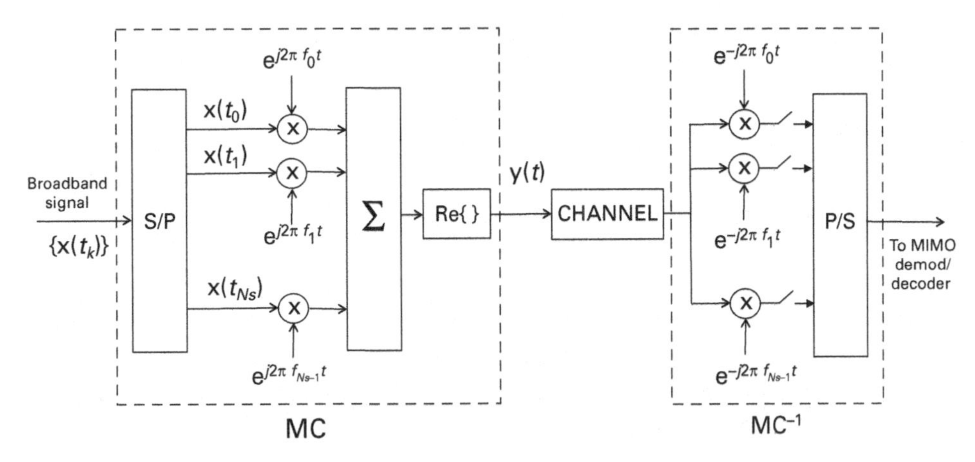

Figure 9.4 The operations performed by the "MC" and "MC^{-1}" blocks in Figure 9.3.

OFDM is a specific form of multiple carrier transmission that is both spectrally and computationally efficient. In the next section, we briefly review the basics of conventional OFDM before analyzing its use in MIMO applications.

9.3 Conventional OFDM

OFDM is a multiple carrier transmission scheme where the frequency spacing between the carriers, Δf_{OFDM}, is related to the OFDM symbol duration, T_{OFDM}, as follows:

$$\Delta f_{\mathrm{OFDM}} = \frac{1}{T_{\mathrm{OFDM}}}. \tag{9.4}$$

In contrast, early multiple carrier systems that were developed before OFDM used conventional frequency division multiplexing techniques in which the carriers were typically spaced twice as far apart or more than they are in OFDM. This is one of the advantages of OFDM – it is spectrally efficient.

The second major advantage of OFDM is that it can be implemented in a computationally efficient manner. To see why this is so, we express the transmitted signal, $y(t)$, in Figure 9.4 mathematically. It follows from the figure that

$$y(t) = \mathfrak{Re}\left\{\sum_{k=0}^{N_s-1} x(k)\mathrm{e}^{\mathrm{j}2\pi(f_c+k/T_{\mathrm{OFDM}})t}\right\}, \; 0 \le t \le T_{\mathrm{OFDM}} \tag{9.5}$$

where $T_{\mathrm{OFDM}} \triangleq N_s T_s$ is the total length of the input signal that is processed in this equation. In this expression, we assume that there are N_s carriers with center frequencies equal to $\{f_c + k/T_{\mathrm{OFDM}} \mid k = 0, \ldots, N_s - 1\}$, and that the symbols to be transmitted are denoted by $\{x(k) \triangleq x(t_k) = x(kT_s) \mid k = 0, \ldots, N_s - 1\}$. Equation 9.5 can be expressed as $y(t) = \mathfrak{Re}\left\{s'(t)\mathrm{e}^{\mathrm{j}2\pi f_c t}\right\}$, where $s'(t)$ is defined to be the equivalent complex baseband form of $y(t)$. Thus,

$$s'(t) = \sum_{k=0}^{N_s-1} x(k)e^{j2\pi kt/T_{\mathrm{OFDM}}}.\ 0 \le t \le T_{\mathrm{OFDM}} \tag{9.6}$$

The discrete time version of this equation is obtained by letting $t = nT_s$ and $T_{\mathrm{OFDM}} = N_sT_s$. Substituting these expressions for t and T_{OFDM} results in the following discrete version of an OFDM signal:

$$s'(n) = \sum_{k=0}^{N_s-1} x(k)e^{j2\pi kn/N_s}. \quad n = 0, \ldots, N_s - 1 \tag{9.7}$$

The signal in Eq. 9.7 is often referred to as an OFDM symbol, so it is important to distinguish between OFDM symbols and modulation symbols, which are used to generate the OFDM symbol.

Equation 9.7 shows that an OFDM symbol is obtained by taking the inverse discrete Fourier transform (IDFT) of a block of N_s modulation symbols at the transmitter. At the receiver, the forward discrete Fourier transform (DFT) is performed on the received signal to demodulate the transmitted symbols. In practice, the IDFT and DFT are implemented using fast Fourier transform (FFT) algorithms, which is the reason why OFDM is a computationally efficient technique.

Although each OFDM subcarrier is narrowband (i.e., does not experience ISI), the bandwidth of the OFDM symbol is greater than the coherence bandwidth in a frequency-selective channel. This implies that each OFDM symbol is still vulnerable to delay spread induced ISI. To mitigate the effects of this ISI, guard intervals are inserted between OFDM symbols so that any smearing of the OFDM symbol due to time dispersion will not interfere with subsequent OFDM symbols. The guard interval duration, T_{CP}, is chosen such that

$$T_{CP} \ge \sigma_\tau. \tag{9.8}$$

In principle, the OFDM guard interval could be a silent period during which no signal is transmitted; however, in practice this is not normally done. Instead, the guard interval period is usually filled with what is called a *cyclic extension* or *cyclic prefix* of the OFDM symbol. Figure 9.5 illustrates the concept. In this figure, the symbols that comprise the OFDM symbol are shown on the right part of the diagram and the cyclic prefix guard interval is shown on the left. The cyclic prefix consists of symbols that have been right-hand circularly shifted beginning with symbol $(N_s - 1)$ through symbol $(N_s - N_{CP})$; where, N_{CP} denotes the number of symbols in the guard interval (i.e., cyclic extension) given by

$$N_{CP} = \frac{T_{CP}}{T_s} \approx \frac{\sigma_\tau}{T_s} = L, \tag{9.9}$$

where we have assumed that $T_{CP} = \sigma_\tau$. When the cyclic extension is included as part of the OFDM symbol, we call the resulting symbol the *augmented OFDM symbol* (AOFDM), which we denote by $s(n)$. It follows that

$$s(n) = \begin{cases} s'(n + N_s - N_{CP}), & n = 0, \ldots, N_{CP} - 1 \\ s'(n - N_{CP}). & n = N_{CP}, \ldots, N_S + N_{CP} - 1 \end{cases} \tag{9.10}$$

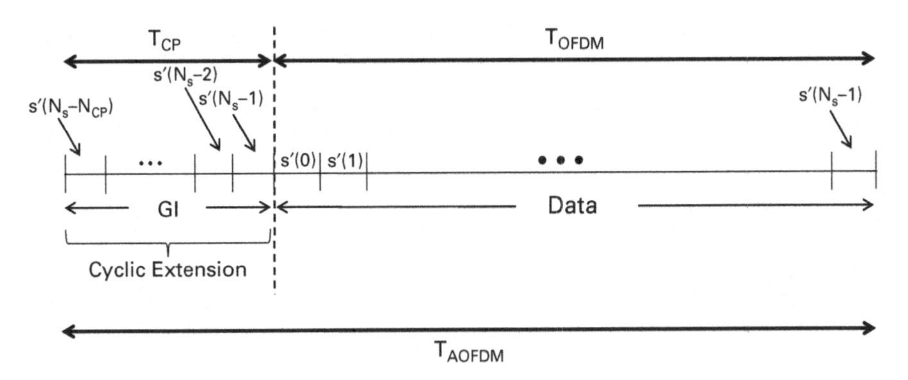

Figure 9.5 An augmented OFDM symbol with a cyclic prefix extension added.

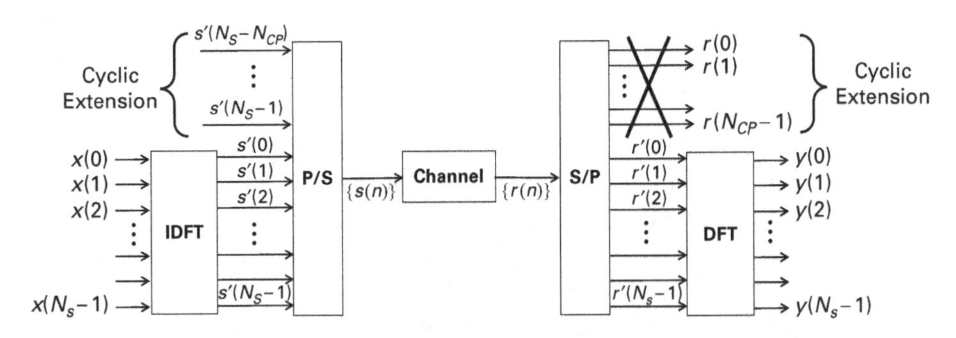

Figure 9.6 Discrete time diagram illustrating the transmission and reception of an *augmented* OFDM symbol.

Figure 9.6 is the block diagram of a full end-to-end OFDM system showing the application of the IDFT at the transmitter and the DFT operation at the receiver. In addition, N_{CP} cyclic prefix symbols are shown being added to the beginning of each augmented OFDM symbol at the transmitter and being removed at the receiver prior to the DFT processing. The combination of the IDFT and P/S blocks define the operations performed in each "MC" block in Figure 9.3 and the S/P and DFT blocks constitute the operations performed by the "MC^{-1}" blocks in that figure. In the next section, we analyze these operations when used in a MIMO configuration.

9.4 MIMO OFDM

In the previous section, we reviewed conventional SISO OFDM and showed that it can be implemented using an IFFT at the transmitter and an FFT at the receiver. The generalization to systems with multiple antennas is straightforward. Figure 9.7 illustrates how OFDM is implemented conceptually in a MIMO communication system. As indicated in the figure, the inputs to the DFTs and the outputs from the DFTs are interpreted as being in the frequency domain, whereas the transmitted and received signals are in the time domain. The IDFT blocks shown in the figure are each assumed to include the addition

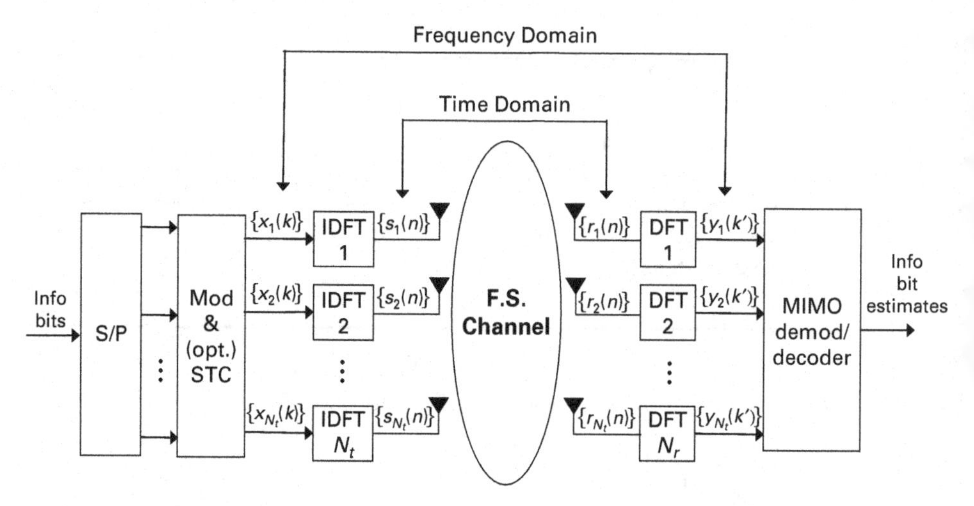

Figure 9.7 Implementation of OFDM in a MIMO system.

of the cyclic extension as shown in Figure 9.6 plus the parallel-to-serial converter. The DFT blocks are assumed to include serial-to-parallel conversion followed by deletion of the cyclic prefix and the application of DFT processing on the remaining received symbols. The indices associated with the various signals shown in the figure match the nomenclature we use in the analysis that follows.

We now analyze the operations shown in Figure 9.7 and demonstrate mathematically that by applying the IDFT and DFT functions as illustrated, we are able to convert a single frequency-selective channel into multiple flat fading channels, each of which can have any of the MIMO techniques that have been described earlier applied to it.

We begin by writing an expression for the transmitted sequence of symbols from transmitter j, $\{s_j(n)\}$ shown in Figure 9.7. Using Eq. 9.10, it follows that

$$s_j(n) = \begin{cases} s_j'(n + N_s - N_{CP}), & n = 0, \ldots, N_{CP} - 1 \\ s_j'(n - N_{CP}), & n = N_{CP}, \ldots, N_S + N_{CP} - 1 \end{cases} \tag{9.11}$$

which, when combined with Eq. 9.7, implies that

$$s_j(n) = \begin{cases} \sum_{k=0}^{N_s-1} x(k) \mathrm{e}^{+\mathrm{j}2\pi k(n+N_s-N_{CP})/N_s}, & n = 0, \ldots, N_{CP} - 1 \\ \\ \sum_{k=0}^{N_s-1} x(k) \mathrm{e}^{+\mathrm{j}2\pi k(n-N_{CP})/N_s}. & n = N_{CP}, \ldots, N_S + N_{CP} - 1 \end{cases} \tag{9.12}$$

The received signal at receive antenna i at time index n, which we denote by $r_i(n)$, is given by

$$r_i(n) = \sqrt{\rho} \sum_{j=1}^{N_t} \sum_{l=0}^{L-1} h_{i,j}^{(l)} s_j(n - l) + z_i(n), \quad n = 0, \ldots, N_s + N_{CP} - 1 \tag{9.13}$$

where $z_i(n)$ denotes the thermal noise at receiver i.

At the receiver, the cyclic prefix is removed from the received sequence, $\{r_i(n)\}$, which results in the truncated received sequence, $\mathbf{r}'$, given by

$$\mathbf{r}' \triangleq [r_i(N_{CP}), r_i(N_{CP}+1), \ldots, r_i(N_s + N_{CP} - 1)]^T. \tag{9.14}$$

Thus,

$$r_i'(n) = \begin{cases} r_i(n + N_{CP}), & n = 0, \ldots, N_s - 1 \\ 0. & \text{otherwise} \end{cases} \tag{9.15}$$

The receiver then performs the DFT operation on $\mathbf{r}'$, resulting in

$$\begin{aligned}
y_i(k') &= \sum_{n=0}^{N_s-1} r_i'(n) \mathrm{e}^{-\mathrm{j}2\pi k' n/N_s} \\
&= \sum_{n=0}^{N_s-1} r_i(n + N_{CP}) \mathrm{e}^{-\mathrm{j}2\pi k' n/N_s} \\
&= \sqrt{\rho} \sum_{n=0}^{N_s-1} \sum_{j=1}^{N_t} \sum_{l=0}^{L-1} h_{i,j}^{(l)} s_j(n + N_{CP} - l) \mathrm{e}^{-\mathrm{j}2\pi k' n/N_s} \\
&\quad + \sum_{n=0}^{N_s-1} z_i(n + N_{CP}) \mathrm{e}^{-\mathrm{j}2\pi k' n/N_s} \\
&= \sqrt{\rho} \sum_{n=0}^{N_s-1} \sum_{j=1}^{N_t} \sum_{l=0}^{L-1} h_{i,j}^{(l)} \left(\sum_{k=0}^{N_s-1} x_j(k) \mathrm{e}^{+\mathrm{j}2\pi k(n-l)/N_s} \right) \mathrm{e}^{-\mathrm{j}2\pi k' n/N_s} + Z_i(k') \\
&= \sqrt{\rho} \sum_{j,l,k} x_j(k) h_{i,j}^{(l)} \mathrm{e}^{-\mathrm{j}2\pi k l/N_s} \sum_{n=0}^{N_s-1} \mathrm{e}^{+\mathrm{j}2\pi n(k-k')/N_s} + Z_i(k'), \\
&\qquad k' = 0, \ldots, N_s - 1
\end{aligned} \tag{9.16}$$

where the frequency index at the receiver is denoted by k' to distinguish it from the transmitter, $Z_i(k') \triangleq \sum_{n=0}^{N_s-1} z_i(n + N_{CP}) \mathrm{e}^{-\mathrm{j}2\pi n k'/N_s}$, and the fourth step follows from the fact that $N_{cp} = L$, so $n + N_{cp} - l \geq 0$; therefore we can use the second line in Eq. 9.12.

Next, we make use of the fact that as long as N_s is an even integer (which we assume to be the case in this analysis), it follows that $\sum_{n=0}^{N_s-1} \mathrm{e}^{\mathrm{j}2\pi n(k-k')/N_s} = N_s \delta(k - k')$. As a result, Eq. 9.16 can be simplified as follows:

$$\begin{aligned}
y_i(k') &= N_s \sqrt{\rho} \sum_{j,l,k} x_j(k) h_{i,j}^{(l)} \mathrm{e}^{-\mathrm{j}2\pi k l/N_s} \delta(k - k') + Z_i(k') \\
&= N_s \sqrt{\rho} \sum_{j,l} x_j(k') h_{i,j}^{(l)} \mathrm{e}^{-\mathrm{j}2\pi k' l/N_s} + Z_i(k') \\
&= N_s \sqrt{\rho} \sum_{j=0}^{N_t} x_j(k') \sum_{l=0}^{L-1} h_{i,j}^{(l)} \mathrm{e}^{-\mathrm{j}2\pi k' l/N_s} + Z_i(k') \\
&= N_s \sqrt{\rho} \sum_{j=0}^{N_t} x_j(k') \mathcal{H}_{ij}(k') + Z_i(k'), \qquad k' = 0, \ldots, N_s - 1
\end{aligned} \tag{9.17}$$

where $\mathcal{H}_{ij}(k') \triangleq \sum_{l=0}^{L-1} h_{i,j}^{(l)} e^{-j2\pi k' l/N_s}$ is the frequency response of the channel between transmitter j and receiver i. It should be noted that, by definition, interpreting $\mathcal{H}_{ij}(k')$ to be the frequency response implies that $\sum_{l=0}^{L-1} h_{i,j}^{(l)} e^{-j2\pi k' l/N_s}$ is the DFT of the impulse response of the channel. However, since we have truncated the summation in this expression to $(L-1)$ instead of (N_s-1), we are implicitly assuming that: a) $N_s \geq L$; and b) $\mathcal{H}_{ij}(k') = 0$ for $l \geq L$.

Since $\mathcal{H}_{ij}(k')$ is a complex constant for each frequency index k', Eq. 9.17 shows that the combination of performing IFFT operations at the transmitter and FFT operations at the receiver transforms a single frequency-selective MIMO channel into N_s flat MIMO channels, which was our goal.[2]

It should be noted that Eq. 9.17 can also be expressed in the following matrix format:

$$\mathbf{Y}(k') = N_s\sqrt{\rho}\,\mathcal{H}(k')\mathbf{X}(k') + \mathbf{Z}(k'), \quad k' = 0, \ldots, N_s - 1 \tag{9.18}$$

where

$$\mathbf{Y}(k') \triangleq \begin{pmatrix} y_1(k') \\ y_2(k') \\ \vdots \\ y_{N_r}(k') \end{pmatrix}, \qquad \mathcal{H}(k') \triangleq \begin{pmatrix} \mathcal{H}_{1,1}(k') & \ldots & \mathcal{H}_{1,N_t}(k') \\ \vdots & \ddots & \vdots \\ \mathcal{H}_{N_r,1}(k') & \ldots & \mathcal{H}_{N_r,N_t}(k') \end{pmatrix},$$

and

$$\mathbf{X}(k') \triangleq \begin{pmatrix} x_1(k') \\ x_2(k') \\ \vdots \\ x_{N_t}(k') \end{pmatrix}, \qquad \mathbf{Z}(k') \triangleq \begin{pmatrix} Z_1(k') \\ Z_2(k') \\ \vdots \\ Z_{N_r}(k') \end{pmatrix}.$$

The following example illustrates how to compute the values of the basic parameters in an OFDM MIMO communication system.

Example 9.2 Design an OFDM MIMO communication system to achieve a data rate of $R_b = 40$ Mbps over a bandwidth of $W = 10$ MHz. Assume $N_t = N_r = 2$, $\sigma_\tau = 20\,\mu$s, and $B_d = 10$ Hz. Compute the OFDM symbol duration (T_{OFDM}), number of subcarriers (N_s), cyclic prefix duration (T_{CP}), and the required modulation order (M). Assume a BLAST-type architecture with no conventional coding.

Answer

1. Since we need to make sure the OFDM carrier spacing, Δf_{OFDM}, is less than or equal to the coherence bandwidth of the channel, we start by computing the coherence bandwidth as follows:

[2] To be clear, when we speak of a "frequency-selective MIMO channel" or "flat MIMO channel", we mean an entire $N_r \times N_t$ channel matrix; thus, by performing the IDFT and DFT operations, we have shown that the resulting effective channel can be represented by a standard $N_r \times N_t$ channel matrix with elements consisting of complex values.

$$B_{c,50} = \frac{1}{5\sigma_\tau} = \frac{1}{5 \times 20 \times 10^{-6}} = 10 \text{ kHz}. \tag{9.19}$$

2. According to Eq. 9.8, $T_{CP} \geq \sigma_\tau$. For the purposes of this design, we assume that

$$T_{CP} = \sigma_\tau = 20 \ \mu\text{s}. \tag{9.20}$$

3. To prevent the guard interval from introducing excessive overhead, we want $T_{\text{OFDM}} \gg T_{CP}$. For the purpose of this example, we set

$$T_{\text{OFDM}} = 10T_{CP} = 0.2 \text{ ms}. \tag{9.21}$$

4. Since $\Delta f_{\text{OFDM}} = 1/T_{\text{OFDM}}$, and we need to ensure that $\Delta f_{\text{OFDM}} \leq B_{c,50}$ to ensure flat fading in each sub-band, we need to make sure that our choice of T_{OFDM} meets this requirement. For $T_{\text{OFDM}} = 0.2$ ms, it follows that

$$\Delta f_{\text{OFDM}} = \frac{1}{T_{\text{OFDM}}} = \frac{1}{0.2 \times 10^{-3}} = 5\text{kHz}, \tag{9.22}$$

   which meets our requirement for flat subcarriers.
5. Next, we compute the size of our DFTs as follows:

$$N_s = \frac{W}{\Delta f_{\text{OFDM}}} = \frac{10 \times 10^6}{5 \times 10^3} = 2000. \tag{9.23}$$

6. To compute the required modulation order, M, we assume a BLAST-type spatial multiplexing scheme. In such a system, the total bit rate, R_b, is given by

$$R_b = N_t R_s \log_2 M \text{ bps}, \tag{9.24}$$

   where R_s is the modulation symbol (i.e., baud) rate. The bandwidth of each transmitted signal, W, is related to the baud rate by

$$W \simeq 2R_s = 10 \text{ MHz}, \tag{9.25}$$

   which implies that $R_s = 5$ Msym/sec. It follows that $R_b = 40 \times 10^6$ bps $= 2 \times 5 \times 10^6 \log M$, which implies that $M = 16$.
7. Lastly, we need to ensure that the channel can be assumed to be fixed during the duration of an OFDM symbol. That is, we need to ensure that $T_{\text{OFDM}} \leq T_{c,50}$. From Chapter 4 recall that

$$T_{c,50} = \frac{9}{16\pi B_d} = 17.9 \text{ ms}. \tag{9.26}$$

   Since $T_{\text{OFDM}} = 0.2$ ms $<<$ 17.9 ms, our design meets this requirement.

Although the set of parameter values that we have computed in this example is not the only permissible one (other sets of values would meet our requirements just a well), this example illustrates the interrelationship between the various parameters and the thought process that should be used when designing the OFDM parameters for a MIMO communication system.

9.5 OFDMA

Another concept that is closely related to OFDM is orthogonal frequency division multiple access (OFDMA). OFDMA is a multiple access scheme that is often used by communication systems that employ OFDM. Figure 9.8 illustrates the essential features of OFDMA. This figure depicts a segment of time and frequency that is to be shared by multiple OFDM users. Although the figure does not show it, since the users are assumed to employ OFDM modulation, the time dimension is assumed to be quantized into OFDM symbol segments and frequency is assumed to be divided into OFDM subcarrier bins. The different shade patterns in the figure represent three different users that are attempting to share the channel. The main point of the figure is that in OFDMA, users are able to dynamically share the time-frequency space of the channel by mapping their data symbols onto appropriate OFDM frequency bins. The assignment of subcarrier channels is usually performed by a centralized controller located at a base station in the case of a cellular wireless system.

Consider the left side of the figure. At that point in time, only Users 1 and 2 have data to transmit, so their data symbols are mapped to the OFDM bins that correspond to the frequency bands depicted in the figure. It should be noted that the subcarriers assigned to a given user do not have to be contiguous, which is why the bands occupied by Users 1 and 2 at the beginning are depicted as being non-contiguous. As time progresses, User 1 eventually does not need to use the channel and stops transmitting. At that point, the OFDM subcarriers that were allocated to him become available to other users. Shortly after User 1 stops transmitting, User 3 begins using some of the same (but not all) of the OFDM subcarrier bins that were previously assigned to User 1. This process continues as time progresses in the diagram.

It is important to note that OFDMA is intrinsically non-packet-based, in contrast, for example, with carrier sense multiple access (CSMA). In CSMA, each user monitors the channel for activity and when the channel becomes idle, users with data to transmit wait a random amount of time before transmitting a data packet. The randomized wait period (called a back-off period) is designed to minimize collisions, but it does not eliminate them entirely. In CSMA, each data packet utilizes the entire channel bandwidth for the

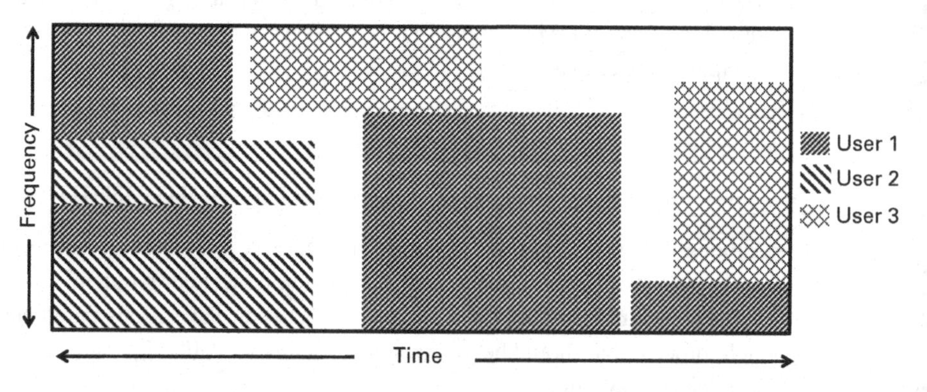

Figure 9.8 The basic concept of OFDMA.

duration of the packet. In contrast, OFDMA users are assigned OFDM subcarriers when they need them and only as many as they need. In OFDMA, packets are not needed and the efficiency is usually significantly higher than CSMA because there is no need for random back-off periods to avoid data collisions. CSMA is used by WiFi, but OFDMA is used by LTE and WiMAX. We have more to say about OFDMA when we describe how LTE uses MIMO in Chapter 11.

9.6 Space-frequency block coding (SFBC)

We conclude Chapter 9 with a discussion of space-frequency block coding (SFBC), which is a technique that is used to achieve spatial diversity in OFDM MIMO systems, including LTE and WiMAX. As its name suggests, SFBC is similar to STBC, except that in SFBC data symbols are encoded across space and frequency rather than across space and time. Figure 9.9 shows a simple block diagram of a generic SFBC system. At the transmitter, a block of N modulation symbols, $s_1, s_2, \ldots, s_N$, arrives at the SFBC encoder, which generates two or more encoded output streams using, in principle, any STBC encoding algorithm. The resulting output streams are then passed through IDFT blocks before being transmitted. Each IDFT block is assumed to consist of a serial-to-parallel converter, an IFFT block, and a parallel-to-serial converter. The encoded symbols on each of the SFBC outputs correspond to different frequency subcarriers that are transformed by the IDFT blocks into time-domain signals before being transmitted. At each receiver (only the ith receiver is shown in the figure), the received signal is processed by a DFT block and its output is fed to a SFBC decoder, which performs STBC decoding on the received signal. In general, the size of the input block, N, is equal to the IFFT size in each IDFT block.

To help solidify the concept of SFBC, it is helpful to consider a specific example. Figure 9.10 is a block diagram of a 2×1 communication system that uses SFBC with Alamouti coding. As illustrated, the SFBC encoder block is equivalent to a conventional Alamouti encoder. Each pair of input modulation symbols is encoded using the Alamouti encoding rule in Table 6.1. The dashed rectangular boxes depict successive code blocks, except that instead of interpreting these as space-time codewords, as we did in conventional Alamouti coding, these should be considered *space-frequency* code blocks,

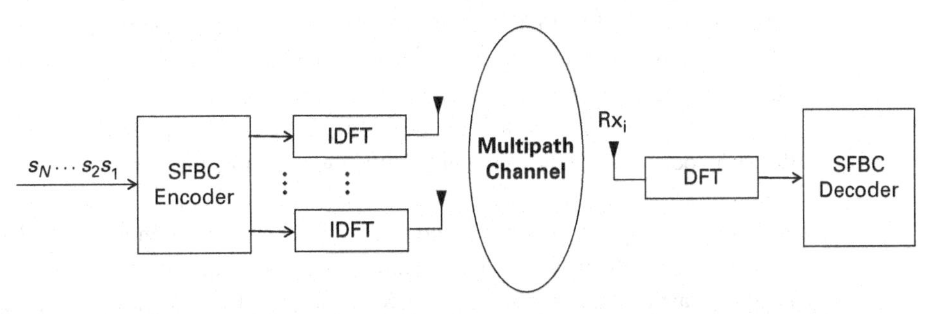

Figure 9.9 A SFBC communication system.

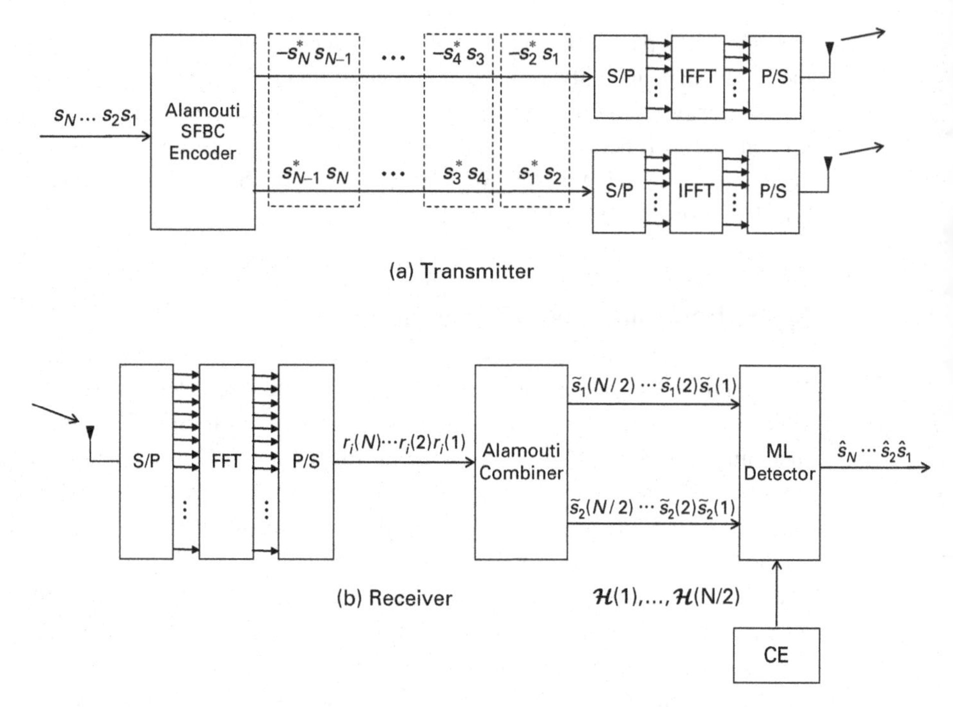

Figure 9.10 A 2 × 1 Alamouti-based SFBC system. Part (a) depicts the transmitter and part (b) depicts the receiver processing.

where the vertical dimension is associated with the spatial domain and the horizontal dimension corresponds to the frequency domain.

Each encoded stream is passed to an IDFT block, which is explicitly shown to consist of a serial-to-parallel conversion block, an IFFT block, followed, in turn, by a parallel-to-serial block. The IFFT is assumed to have a block size equal to N, so it takes $N/2$ Alamouti code blocks to perform one IFFT. Because of the IFFT blocks, the Alamouti encoded modulation symbols are interpreted to be in the frequency domain and the output of the of the IFFT is in the time domain. It is for this reason that when Alamouti encoding (or any STBC encoding) is used with OFDM, the code blocks are interpreted as space-frequency code blocks instead of space-time blocks.

The receiver in Figure 9.10 performs DFT processing on the received signal (denoted collectively by the S/P, FFT, and P/S blocks), and the resulting samples are applied to the Alamouti combiner block, which performs the combining operations defined by Eq. 6.10 on each successive pair of input samples. Each pair of inputs results in two output symbols from the combiner, which are denoted by $\tilde{s}_0$ and $\tilde{s}_1$, where the numbers in parentheses associated with these symbols in the figure denote the input pair count. Each pair of combiner outputs, in turn, is applied to a maximum-likelihood detector that generates a pair of estimates of the transmitted symbols using the logic in Eq. 6.30. To do that, it is necessary for the detector to have estimates of the frequency response of the channel for each pair of combiner outputs, which is provided by the channel estimator block, denoted by CE. The outputs of the CE block correspond to the $\mathcal{H}(k')$ terms in Eq. 9.18.

In this discussion, we have ignored some practical details in order to focus on the fundamental concept. For example, we have not shown the RF processing chain at the transmitter before the antenna or the receiver RF front end before the DFT processing. Nor have we explicitly shown where the cyclic prefix is appended and removed on each OFDM symbol. The purpose of this discussion was to define SFBC and describe the basic concept. In Chapter 11, we discuss how LTE uses SFBC with Alamouti coding.

Problems

9.1 Consider a 2×4 MIMO system that uses 8-PSK modulation and that transmits data at a rate equal to 24 Mbps. Assume that this system operates in a fading environment where the RMS delay spread is equal to 1000 ns. Now consider the performance of such a system. What is the maximum theoretical slope of the BER-vs-SNR curve for this system at large SNR values if the fading is Rayleigh and no conventional coding is used?

9.2 Consider a multicarrier SISO communication system with an information data rate of 50 Mbps that uses QPSK modulation and rate 1/2 conventional coding. Assume that the RMS delay spread is 100 ns. How many carriers are needed to ensure flat fading on each carrier? [Hint: Assume that flat fading occurs as long as the bandwidth of a signal is less than or equal to the 50th percentile coherence bandwidth of the channel.]

9.3 In our mathematical analysis of an OFDM MIMO system we relied on the following property:

$$\sum_{n=0}^{N_s-1} \mathrm{e}^{-\mathrm{j}2\pi n(k'-k)/N_s} = N_s \delta(k' - k).$$

Prove that this property holds when N_s is an even integer, for any integers n, k', and k.

9.4 Design a 5×5 V-BLAST OFDM MIMO communication system to achieve a data rate of 75 Mbps over a bandwidth of 10 MHz. Assume $\sigma_\tau = 50$ ns, and $B_d = 20$ Hz. Compute T_{OFDM}, N_s, T_{CP}, and the required modulation order M.

10 Channel estimation

Up to this point in the book, we have assumed that the receiver (and the transmitter in the case of eigenbeamforming) have perfect knowledge of the channel matrix. The theoretical performance results we have shown thus far have been based on that assumption. In practice, of course, the channel matrix must be estimated, and there is always some error associated with knowledge of the channel. This chapter introduces the basic concepts associated with channel estimation (CE) and presents results that illustrate how MIMO performance is affected by channel estimation errors.

10.1 Introduction

In general, there are two types of MIMO channel estimation methods: a) *training-based*, which uses known training symbols; and b) *blind-based* approaches, that perform CE without the benefit of known training symbols. In training-based CE, known training symbols are transmitted at certain prescribed times and frequencies that are known by the receiver. Since the receiver knows the training symbols, as well as when and where (i.e., at which frequencies) they are transmitted, it uses that information to estimate the gain and phase rotation imparted by the channel at each point in time and frequency based on the characteristics of the received training symbols. Although blind-based methods have higher bandwidth efficiencies because they do not use any resources for transmitting training symbols, they tend to have lower speed and poorer performance than training-based methods. For this reason, training-based CE is used more than blind-estimation, and it is the method we focus on in this chapter.

The placement of training symbols in time, frequency, and space (i.e., the transmit antenna's) dimensions is a key part of the design of a MIMO communication system. In general, training symbols should be spaced as far apart as possible to reduce training overhead, while still maintaining a required performance level. For example, in a high Doppler, fast fading environment, training symbols need to be placed relatively often in time. Similarly, in a highly frequency-selective channel, training symbols need to be placed close together in the frequency dimension. We give examples of the spacings used in WiFi, LTE, and WiMAX later in this chapter.

Another factor influencing the placement of training symbols is whether or not the training symbols are in the header or the body of a data packet. In a packet-based communication system, each packet has a header that performs synchronization and initial channel estimation, followed by the body of the packet that contains the data. A common strategy in such a system is to attempt to get an accurate estimate of the channel at the beginning of a packet and then to update that estimate during the body of the packet. This approach assumes that the channel only changes slightly during the duration of a packet, so only small updates are needed once the initial accurate estimate is generated during the header. Such an approach requires relatively dense training symbol placement in the header and relatively sparse placement in the body, along with appropriate interpolation to span the spaces between the relatively sparse training symbols. IEEE 802.11n, which is the MIMO extension to WiFi, is an example of a MIMO standard that employs training symbols in this way.

In the published literature on MIMO communications, there are several different terms used to describe training symbols. Some authors use the word *pilot* or *pilot tone* to refer to these symbols. In general, *training symbol* is the umbrella designation that refers to any symbol that is transmitted for the purpose of estimating the channel, and a pilot tone refers to a sequence of symbols that are transmitted one after the other with no gaps between them in the time dimension. We use the terms pilot and training symbol interchangeably in this chapter.

The following section describes training symbol allocation strategies in greater detail.

10.2 Pilot allocation strategies

In addition to the factors discussed above, another consideration that influences the distribution of training symbols is whether the channel is flat or frequency selective. In this section, we describe typical ways of distributing pilot symbols in narrowband and broadband channels, and we also talk about ways to avoid pilot interference, which is important when there are multiple transmit antennas.

10.2.1 Narrowband MIMO channels

As we showed in Chapter 9, most wireless systems operate in frequency-selective fading environments because their bandwidths are larger than the coherence bandwidths of typical channels. Nevertheless, there may be situations where flat fading occurs, so, for completeness, we begin by considering the allocation of training symbols in narrowband conditions.

Figure 10.1 shows a generic depiction of how training symbols (indicated by the shaded squares) might be allocated in a narrowband MIMO system. The figure assumes that the system is packet-based, where each packet has a preamble followed by data. In the preamble, each of the transmit antennas transmits N_p pilot symbols, which are used by the receiver to estimate the channel matrix at the beginning of the packet. If the packet length is shorter than the coherence time, quasi-static fading occurs, and the receiver

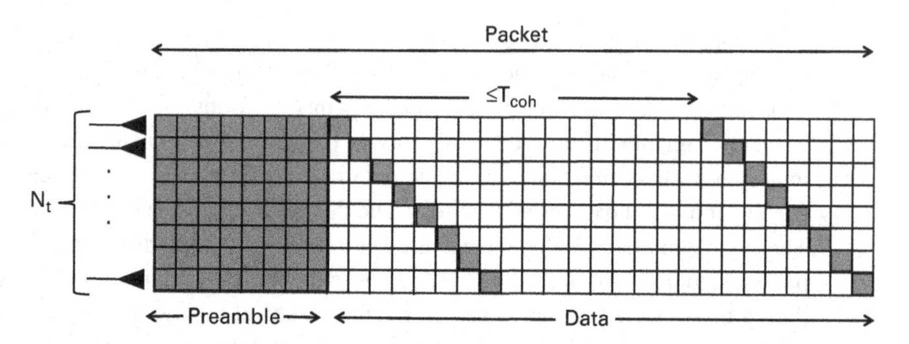

Figure 10.1 An example training symbol allocation for narrowband (i.e., flat fading) environments.

could use the estimate of the channel it computes during the preamble for the entire packet. If the packet is longer than the coherence time, it is necessary for the system to update the initial channel estimate at least once during each coherence time period in the body of the packet. One approach would be to transmit a block of N_p training symbols every T_{coh} seconds. Alternatively, an allocation strategy like that shown in Figure 10.1, which results in less overhead, could also be used. As usual, there is a tradeoff between accuracy and overhead inefficiency.

10.2.2 Broadband MIMO channels

Because OFDM is normally used in broadband MIMO applications (i.e., when the bandwidth of the MIMO signal is greater than the coherence bandwidth of the channel), it is necessary to consider not only how to allocate training symbols in space and time, as we did in flat fading, but also how to allocate them in the frequency dimension, since OFDM divides each transmitted signal into frequency subcarriers. Figure 10.2 depicts four generic time-frequency pilot allocation options for use with OFDM MIMO.

The allocation shown in part (a) depicts the case when entire OFDM symbols (indicated by the columns in each diagram) are used to transmit training symbols. This type of time-frequency allocation is useful in highly frequency-selective channels with moderate to slow fading. In contrast, part (b) depicts the allocation that could be used when the opposite is true – the frequency selectivity of the channel is moderate but the Doppler is relatively high, resulting in fast fading. Parts (c) and (d) depict allocation strategies that can be used when the channel has moderate frequency selectivity and Doppler. IEEE 802.11n uses the allocation in part (b) for the body of each packet, and LTE, LTE-advanced, and WiMAX use the allocation shown in part (d).

Actual MIMO systems may use a combination of the allocation options shown in Figure 10.2. For example, Figure 10.3 shows how 802.11n, LTE, and WiMAX allocate their pilots. As discussed above, 802.11n uses CSMA and is packet-based. Part (a) of the figure shows a single WiFi packet with a header at the beginning with the data

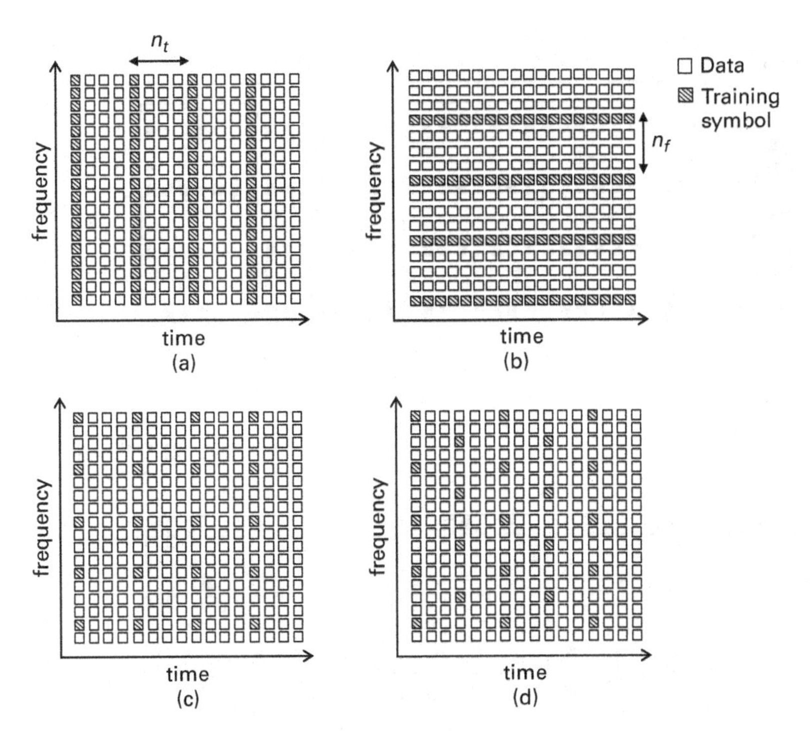

Figure 10.2 Four generic time-frequency training symbol allocations when OFDM is used.

portion following it. The header includes several consecutive OFDM symbols that are dedicated to transmitting training symbols in all of the frequency subcarriers. These OFDM symbols are used by the receiver to compute an accurate estimate of the channel matrix at the beginning of each packet. In the data payload, every OFDM symbol only uses four of its subcarriers to transmit training symbols. These symbols are used to update the original channel estimate and allow the receiver to track relatively small changes that occur to the channel over the duration of a packet. The training symbols in the data portion of the packet are called the pilot tones.

Part (b) of Figure 10.3 shows the pilot symbol allocation for LTE and WiMAX. As noted earlier, these systems employ OFDMA and, therefore, are not packet-based. As a result, the training symbols are distributed in both time and frequency similarly to part (d) in Figure 10.2. In the next subsection, we discuss how the pilot spacing is computed in both the time and frequency dimensions.

10.2.3 Designing pilot spacing

When designing a MIMO-OFDM system, pilot symbols need to be spaced close enough in time and frequency to capture time and frequency variations that are anticipated to occur in the multipath channel environment in which the system will operate. In general, the spacing in time between pilot symbols needs to be smaller than the coherence

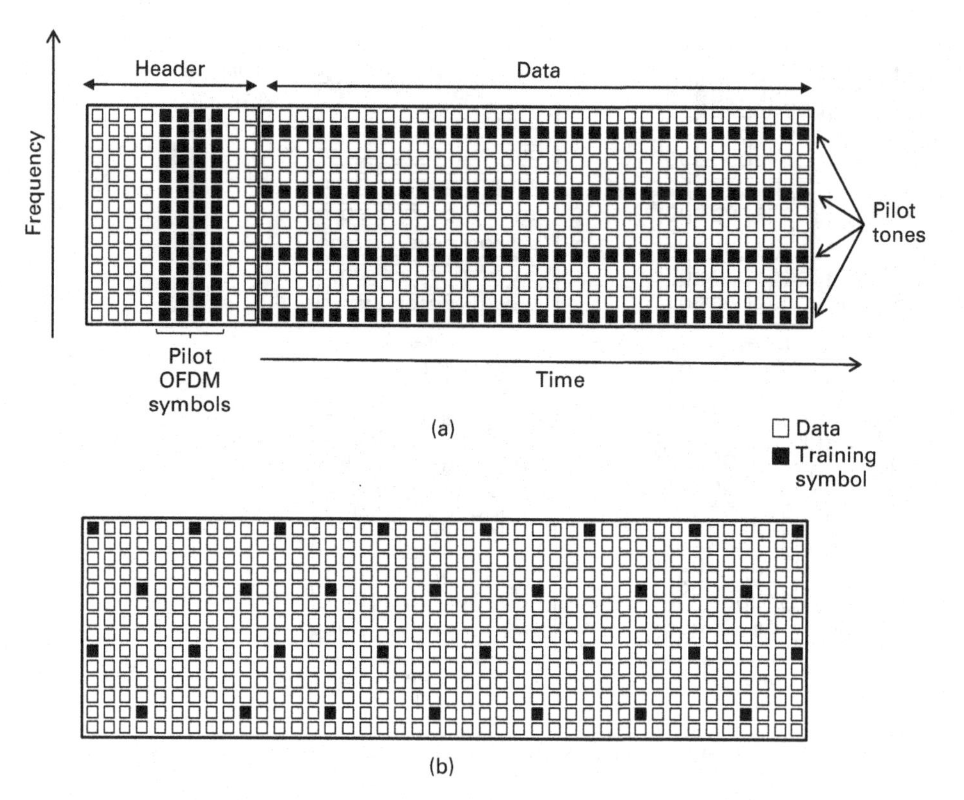

Figure 10.3 Pilot allocations used in actual wireless MIMO systems: (a) The type of allocation used by packet-based systems such as 802.11n; b) The type of allocation used by non-packet-based OFDMA systems such as LTE, LTE-advanced, and WiMAX.

time of the channel and the spacing in frequency needs to be less than the coherence bandwidth.

Let n_t denote the time spacing between pilots in units of OFDM symbol periods. A practical rule-of-thumb is that

$$n_t \leq \frac{T_{c,50}}{T_{\text{OFDM}}}, \tag{10.1}$$

where $T_{c,50}$ denotes the 50th percentile coherence time of the channel.

Similarly, let n_f denote the frequency spacing between pilots in units of OFDM subcarrier channels. Another useful rule-of-thumb is that

$$n_f \leq \frac{B_{c,50}}{\Delta f_{\text{OFDM}}} = B_{c,50} T_{\text{OFDM}}, \tag{10.2}$$

where $B_{c,50}$ and Δf_{OFDM} denote the 50th percentile channel coherence bandwidth and the OFDM subcarrier spacing, respectively. The following example presents a practical illustration of how the frequency and time spacings are computed.

Example 10.1 The LTE and LTE-advanced standards are based on the use of several multipath fading models. The model that has the largest delay spread is called the extended typical urban model (ETU) [25]. The RMS delay spread and Doppler bandwidths defined by the ETU are 5 μs and 600 Hz, respectively. The LTE and LTE-advanced standards specify that $T_{\text{OFDM}} = 1/15\,000$ seconds long. What are the required pilot symbol spacings in the time and frequency dimensions for LTE (i.e., the values of n_t and n_f in Figure 10.3 (b))?

Answer Since $\sigma_\tau = 5$ μs, it follows that $B_{c,50} = 1/(5\sigma_\tau) = 1/(5 \times 5 \times 10^{-6})$ Hz $=$ 40 kHz. Since $T_{\text{OFDM}} = 1/15\,000$ seconds, the OFDM subcarrier spacing is $\Delta f_{\text{OFDM}} =$ 15 kHz. It follows that $n_f = 40\,000/15\,000 = 2.7$. In practice, LTE uses $n_f = 3$.

To estimate the spacing in the time dimension, we use the specified Doppler spread. Since $B_d = 600$ Hz, it follows that $T_{c,50} = 9/(16\pi B_d) = 298.4$ μs. Therefore, $n_t =$ $298.4 \times 10^{-6} \times 15\,000 = 4.5$. In practice, LTE uses $n_t = 3$ and 4.

10.2.4 Spatial pilot allocation strategies

In addition to considering how to allocate training symbols in time and frequency, MIMO systems also need to address the question of how to allocate pilot symbols in the space dimension. This issue is important because in MIMO systems the received signal at each antenna is the superposition of the signals from all of the transmit antennas; therefore, the training symbols need to be transmitted without interfering with one another to accurately estimate the channel. In general, there are three fundamental ways to avoid pilot signal interference, which are illustrated in Figure 10.4.

Method (a) illustrates one approach that can be used, called temporal orthogonality. In this scheme, training symbols may be transmitted at the same frequency, but interference at the receiver is avoided by never transmitting training symbols from two or more transmit antennas at the same time. Method (b), called frequency orthogonality,

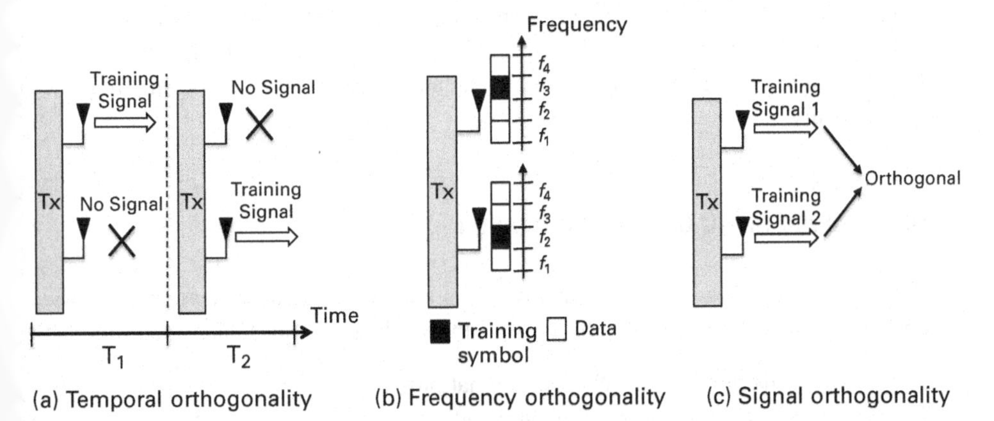

Figure 10.4 Spatial pilot allocation options to ensure pilot orthogonality [adapted from [3]].

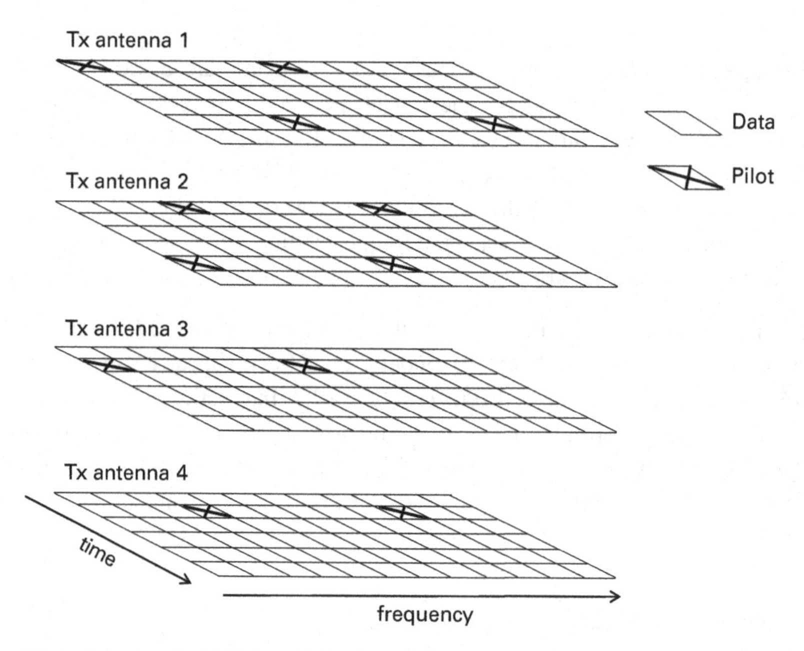

Figure 10.5 Pilot allocation for LTE downlink when there are four transmit antennas.

transmits training symbols at the same time, but avoids interference at the receiver by transmitting pilot symbols on different carrier frequencies. This approach is popular in a variety of wireless standards, and it is particularly well-suited for OFDM-based systems where each of the frequency cells shown in the figure would represent OFDM carriers. Method (c), called signal orthogonality, allows pilot symbols to be transmitted at the same time and frequency, but interference at the receiver is prevented by designing the sequences of pilot symbols to be orthogonal.

In practice, MIMO systems often use a combination of at least two of these. For example, LTE uses both temporal and frequency orthogonality. Figure 10.5 shows how LTE allocates pilots in space. This diagram depicts the LTE downlink transmission when there are four transmit antennas. The data symbols are denoted by the white cells and the pilot symbols by the cells marked with Xs. As illustrated, antennas 1 and 2 employ frequency orthogonality; however, antennas 2 and 3 use temporal and frequency orthogonality.

10.3 Narrowband MIMO channel estimation

In this section, we describe several ways of estimating a MIMO channel matrix when operating in a quasi-static flat fading environment. The techniques that we consider are:

1. Maximum likelihood (ML) channel estimation;
2. Least squares (LS) channel estimation; and
3. Linear minimum mean square error (LMMSE) channel estimation.

To derive expressions for the channel estimate based on these criteria, we start by assuming that each of the transmit antennas in a MIMO system transmits a sequence of p training symbols like the preamble portion shown in Figure 10.1. The resulting $N_t \times p$ transmitted pilot matrix, which is denoted by $\mathbf{S}_p$, and the resulting $N_r \times p$ receive matrix, $\mathbf{R}_p$, are given by

$$\mathbf{R}_p = \sqrt{\rho}\,\mathbf{H}\mathbf{S}_\mathbf{p} + \mathbf{Z}, \tag{10.3}$$

where $\sqrt{\rho}$ is the SNR, $\mathbf{H}$ is an $N_r \times N_t$ channel matrix, and $\mathbf{Z}$ is the $N_r \times p$ receive noise matrix. Under the quasi-static fading assumption, the channel is assumed to be fixed over the period of p modulation symbols.

We now consider each of the channel estimate criteria separately.

10.3.1 Maximum likelihood channel estimation

Since $\mathbf{R}_p$ is a random function of $\mathbf{H}$, by definition the likelihood function of $\mathbf{H}$ is given by $p(\mathbf{R}_p|\mathbf{H})$; where $p(\cdot)$ is a multivariate probability density function. It follows that the maximum likelihood (ML) estimate of $\mathbf{H}$ is given by

$$\hat{\mathbf{H}}_{ML} \triangleq \underset{\{H\}}{\arg\max}\; p(\mathbf{R}_p|\mathbf{H}). \tag{10.4}$$

It follows from Eq. 10.3 that if $\mathbf{Z}$ consists of independent Gaussian elements that are temporally and spatially white, the ML channel estimate becomes

$$\hat{\mathbf{H}}_{ML} = \underset{\{H\}}{\arg\min}\|\mathbf{R}_p - \sqrt{\rho}\mathbf{H}\mathbf{S}_\mathbf{p}\|_F^2. \tag{10.5}$$

Equation 10.5 can be solved by taking the derivative of $\|\mathbf{R}_p - \sqrt{\rho}\mathbf{H}\mathbf{S}_\mathbf{p}\|_F^2$ with respect to $\mathbf{H}$ and setting the resulting equation equal to 0, then solving that equation for $\mathbf{H}$. We do this as follows:

$$\begin{aligned}
\text{Let } C &\triangleq \|\mathbf{R}_p - \sqrt{\rho}\mathbf{H}\mathbf{S}_\mathbf{p}\|_F^2 \\
&= \mathrm{Tr}\{(\mathbf{R}_p - \sqrt{\rho}\mathbf{H}\mathbf{S}_\mathbf{p})(\mathbf{R}_p - \sqrt{\rho}\mathbf{H}\mathbf{S}_\mathbf{p})^H\} \\
&= \mathrm{Tr}\{\mathbf{R}_p\mathbf{R}_p^H - \sqrt{\rho}\mathbf{R}_p\mathbf{S}_p^H\mathbf{H}^H - \sqrt{\rho}\mathbf{H}\mathbf{S}_p\mathbf{R}_p^H + \rho\mathbf{H}\mathbf{S}_p\mathbf{S}_p^H\mathbf{H}^H\}.
\end{aligned} \tag{10.6}$$

Next, solve $\frac{\partial C}{\partial \mathbf{H}} = 0$. To do this, we use matrix theorem Section 1.9.2-(y), which states that for any two matrices, $\mathbf{A}$ and $\mathbf{B}$, $\partial(\mathrm{Tr}\,(\mathbf{AB}))/\partial\mathbf{B} = \mathbf{A}^T$. Therefore,

$$\begin{aligned}
\frac{\partial C}{\partial \mathbf{H}} &= -\sqrt{\rho}\frac{\partial}{\partial \mathbf{H}}\mathrm{Tr}\{\mathbf{H}\mathbf{S}_p\mathbf{R}_p^H\} + \rho\frac{\partial}{\partial \mathbf{H}}\mathrm{Tr}\{\mathbf{H}\mathbf{S}_p\mathbf{S}_p^H\mathbf{H}^H\} \\
&= -\sqrt{\rho}\left(\mathbf{S}_p\mathbf{R}_p^H\right)^T + \rho\left(\mathbf{S}_p\mathbf{S}_p^H\mathbf{H}^H\right)^T \\
&= -\sqrt{\rho}\left(\mathbf{R}_p^*\mathbf{S}_p^T\right) + \rho\left(\mathbf{H}^*\mathbf{S}_p^*\mathbf{S}_p^T\right) = 0.
\end{aligned} \tag{10.7}$$

It follows from Eq. 10.7 that

$$\begin{aligned}
\rho\mathbf{H}^*\mathbf{S}_p^*\mathbf{S}_p^T &= \sqrt{\rho}\mathbf{R}_p^*\mathbf{S}_p^T \\
\Longrightarrow \mathbf{H}^* &= \frac{1}{\sqrt{\rho}}\left(\mathbf{R}_p^*\mathbf{S}_p^T\right)\left(\mathbf{S}_p^*\mathbf{S}_p^T\right)^{-1}
\end{aligned}$$

$$\Longrightarrow \hat{\mathbf{H}}_{ML} = \frac{1}{\sqrt{\rho}} \left(\mathbf{R}_p^* \mathbf{S}_p^T\right)^* \left[\left(\mathbf{S}_p^* \mathbf{S}_p^T\right)^{-1}\right]^*$$

$$\Longrightarrow \hat{\mathbf{H}}_{ML} = \frac{1}{\sqrt{\rho}} \left(\mathbf{R}_p \mathbf{S}_p^H\right) \left[\left(\mathbf{S}_p^* \mathbf{S}_p^T\right)^*\right]^{-1}$$

$$\Longrightarrow \hat{\mathbf{H}}_{ML} = \frac{1}{\sqrt{\rho}} \left(\mathbf{R}_p \mathbf{S}_p^H\right) \left(\mathbf{S}_p \mathbf{S}_p^H\right)^{-1}. \tag{10.8}$$

It should be noted that the preceding sequence of operations has taken into account that: a) the trace operation is linear; and b) the transpose of a product of arrays is the product of the transposes of the individual arrays in reverse order. It has also taken into account that for any square array $\mathbf{A}$, $(\mathbf{A}^*)^{-1} = \left(\mathbf{A}^{-1}\right)^*$.

10.3.2 Least squares channel estimation

In least squares estimation, the estimate of $\mathbf{H}$ is the value that minimizes the squared error between the actual received signal, $\mathbf{R}_p$, and the estimated received signal, $\hat{\mathbf{R}}_p \triangleq \sqrt{\rho}\,\hat{\mathbf{H}}\mathbf{S}_p$. Thus,

$$\hat{\mathbf{H}}_{LS} = \underset{\{\hat{H}\}}{\arg\min}\ \|\hat{\mathbf{R}}_p - \mathbf{R}_p\|_F^2 \tag{10.9}$$

$$= \underset{\{\hat{H}\}}{\arg\min}\ \|\sqrt{\rho}\,\hat{\mathbf{H}}\mathbf{S}_p - \mathbf{R}_p\|_F^2. \tag{10.10}$$

To obtain $\hat{\mathbf{H}}_{LS}$, we compute the partial derivative of Eq. 10.10 with respect to $\hat{\mathbf{H}}$, set the resulting equation equal to zero, and then solve for $\hat{\mathbf{H}}$. These operations are straightforward to perform using the steps shown in the ML channel estimation derivation as a template. Doing so results in the following:

$$\hat{\mathbf{H}}_{LS} = \frac{1}{\sqrt{\rho}} \left(\mathbf{R}_p \mathbf{S}_p^H\right) \left(\mathbf{S}_p \mathbf{S}_p^H\right)^{-1}. \tag{10.11}$$

We note two observations. First, comparing Eqs. 10.8 and 10.11 shows the maximum likelihood and least squares channel estimates are equivalent. Secondly, we note that the ML and LS channel estimate involves the term $\left(\mathbf{S}_p \mathbf{S}_p^H\right)^{-1}$. It follows from matrix theorem Section 1.9.2-(k) that in order for this inverse to exist, $p \geq N_t$.

10.3.3 Linear minimum mean square channel estimation

The LMMSE channel estimate is the channel matrix that minimizes the mean square error between the true channel and the estimate of the channel, $\hat{\mathbf{H}}$, which is assumed, in the case of LMMSE, to be a linear superposition of the received signals. We express this assumption mathematically as follows:

$$\hat{h}_{i,j} = \sum_{k=1}^{P} r_i(k) w_{k,j}, \tag{10.12}$$

where $\hat{h}_{i,j}$ denotes the (i,j)th element of $\hat{\mathbf{H}}$, $r_i(k)$ is the (i,k)th element of the receive matrix, $\mathbf{R}_p$, and the $w_{k,j}$ values are complex weights chosen to minimize the mean square error between the true value of $\mathbf{H}$ and the estimate. The relationship in Eq. 10.12 can be expressed in the following matrix format:

$$\hat{\mathbf{H}} = \mathbf{R}_p \mathbf{W}, \tag{10.13}$$

where the (k,j)th element of $\mathbf{W}$ is $w_{k,j}$.[1] The LMMSE estimate of $\mathbf{H}$ is, therefore, expressed as

$$\begin{aligned}
\hat{\mathbf{H}}_{\text{LMMSE}} & \\
&\triangleq \arg\min_{\{\hat{\mathbf{H}}\}} \mathbb{E}\left\{\|\mathbf{H} - \hat{\mathbf{H}}\|_F^2\right\} \\
&= \arg\min_{\{\mathbf{W}\}} \mathbb{E}\left\{\| \mathbf{H} - \mathbf{R}_p \mathbf{W} \|_F^2\right\} \\
&= \arg\min_{\{\mathbf{W}\}} \mathbb{E}\left\{\text{Tr}\left[\left(\mathbf{H} - \sqrt{\rho}\mathbf{H}\mathbf{S}_p\mathbf{W} - \mathbf{Z}\mathbf{W}\right)^H \left(\mathbf{H} - \sqrt{\rho}\mathbf{H}\mathbf{S}_p\mathbf{W} - \mathbf{Z}\mathbf{W}\right)\right]\right\} \\
&= \arg\min_{\{\mathbf{W}\}} \text{Tr}\left[\mathbb{E}\left\{\left(\mathbf{H} - \sqrt{\rho}\mathbf{H}\mathbf{S}_p\mathbf{W} - \mathbf{Z}\mathbf{W}\right)^H \left(\mathbf{H} - \sqrt{\rho}\mathbf{H}\mathbf{S}_p\mathbf{W} - \mathbf{Z}\mathbf{W}\right)\right\}\right],
\end{aligned} \tag{10.14}$$

where the last equality is based on the linear nature of the trace operator, which allows the order of the expectation and trace operations to be reversed.

If we assume that $\mathbf{H}$ and $\mathbf{Z}$ consist of independent, complex Gaussian components with zero mean and unit variance, then $\mathbb{E}\left\{\mathbf{H}^H\mathbf{H}\right\} = N_r\mathbf{I}_{N_t}$, and $\mathbb{E}\left\{\mathbf{Z}^H\mathbf{Z}\right\} = N_r\mathbf{I}_p$. Using these relationships, Eq. 10.14 simplifies to

$$\begin{aligned}
\hat{\mathbf{H}}_{\text{LMMSE}} = \arg\min_{\{W\}} \text{Tr}\Big[&N_r\mathbf{I}_{N_t} - \sqrt{\rho}N_r\mathbf{S}_p\mathbf{W} - \sqrt{\rho}N_r\mathbf{W}^H\mathbf{S}_p^H \\
&+\rho N_r\mathbf{W}^H\mathbf{S}_p^H\mathbf{S}_p\mathbf{W} + N_r\mathbf{W}^H\mathbf{W}\Big].
\end{aligned} \tag{10.15}$$

If we now let

$$\begin{aligned}
C \triangleq \text{Tr}\Big[&N_r\mathbf{I}_{N_t} - \sqrt{\rho}N_r\mathbf{S}_p\mathbf{W} - \sqrt{\rho}N_r\mathbf{W}^H\mathbf{S}_p^H \\
&+ \rho N_r\mathbf{W}^H\mathbf{S}_p^H\mathbf{S}_p\mathbf{W} + N_r\mathbf{W}^H\mathbf{W}\Big],
\end{aligned} \tag{10.16}$$

[1] It will be noted that the LMMSE discussion in Section 8.3.3 places the **W** matrix on the left, but in this section it is mathematically more convenient to reverse the convention and place **W** on the right. Both conventions are acceptable.

it follows from matrix theorem Section 1.9.2-(y) that

$$\frac{\partial C}{\partial \mathbf{W}} = -\sqrt{\rho} N_r \mathbf{S}_p^T + \rho N_r \left(\mathbf{W}^H \mathbf{S}_p^H \mathbf{S}_p\right)^T + N_r \left(\mathbf{W}^H\right)^T . \tag{10.17}$$

Setting $\frac{\partial C}{\partial \mathbf{W}} = 0$ and solving for $\mathbf{W}$, which we denote by $\mathbf{W}_o$, yields

$$\mathbf{W}_o = \sqrt{\rho} \left(\mathbf{I}_p + \rho \mathbf{S}_p^H \mathbf{S}_p\right)^{-1} \mathbf{S}_p^H . \tag{10.18}$$

Since $\hat{\mathbf{H}}_{\mathrm{LMMSE}} = \mathbf{R}_p \mathbf{W}_o$, we conclude that

$$\hat{\mathbf{H}}_{\mathrm{LMMSE}} = \sqrt{\rho} \mathbf{R}_p \left(\mathbf{I}_p + \rho \mathbf{S}_p^H \mathbf{S}_p\right)^{-1} \mathbf{S}_p^H . \tag{10.19}$$

10.3.4 Choosing pilot signals

We now have expressions for the estimate of the channel matrix in a flat fading environment in terms of a known pilot matrix, $\mathbf{S}_p$, the received signal matrix, $\mathbf{R}_p$, and the signal-to-noise ratio. What properties should $\mathbf{S}_p$ have? In general, $\mathbf{S}_p$ should have the following two properties:

a) $p \geq N_t$; and
b) $\mathbf{S}_p \mathbf{S}_p^H = \frac{p}{N_t} \mathbf{I}_{N_t}$.

We have already seen that property (a) is required for the ML and LS methods in order to make it possible to compute $\left(\mathbf{S}_p \mathbf{S}_p^H\right)^{-1}$, which is needed in evaluating the ML and LS channel estimate. In [37], Hassibi and Hochwald examine the question of how to design pilot signals from an information theory perspective and draw the same conclusion. The second property states in a concise form that the rows of $\mathbf{S}_p$ should be orthogonal and that, in particular, the transmitted power from each transmit antenna during the pilot portion of the packet should be $1/N_t$, which is the same normalization that we have been using throughout this book. Property (b) is consistent with the conclusions in [37] and [10]. In general, the accuracy of the channel estimate improves as the value of p increases. As p increases, however, the overhead associated with channel estimation also increases, limiting the duty cycle for the data portion of the packet. Designing a system involves evaluating the tradeoff between channel estimation accuracy and the overhead associated with that process.

As an illustration, consider a MIMO system with $N_t = 2$ and $p = 4$. A possible choice for $\mathbf{S}_p$ in that case would be

$$\mathbf{S}_p = \frac{1}{\sqrt{2}} \begin{pmatrix} 1 & 1 & -1 & -1 \\ 1 & -1 & 1 & -1 \end{pmatrix} . \tag{10.20}$$

It is clear by inspection that this matrix has both properties (a) and (b) above. Since this pilot signal consists of only +1s and –1s, this is an example of a pilot matrix that could be used with BPSK modulation.

When $\mathbf{S}_p\mathbf{S}_p^H = \frac{p}{N_t}\mathbf{I}_{N_t}$, it follows from Eqs. 10.8 and 10.11 that

$$\hat{\mathbf{H}}_{ML} = \hat{\mathbf{H}}_{LS} = \mathbf{H} + \frac{N_t}{p\sqrt{\rho}}\mathbf{Z}\mathbf{S}_p^H. \tag{10.21}$$

This equation shows four important properties associated with the ML and LS channel estimates:

1. When the noise is zero-mean Gaussian, the ML and LS channel estimates are unbiased.
2. The error increases with the number of transmit antennas. This happens because for a fixed signal-to-noise ratio, the power from each antenna, which is used to estimate the channel, decreases as N_t increases.
3. The error increases as the signal-to-noise decreases, as expected.
4. The error increases as the length of the pilot training sequence decreases.

10.3.5 Narrowband CE performance

Now that we have expressions for the estimate of the channel in flat fading, we can use those results to study the sensitivity of MIMO performance to errors in the channel estimates. Figure 10.6 shows the results of a computer simulation that predicts the bit error probability associated with Alamouti coding when channel estimation is performed. Two sets of curves associated with $N_r = 1$ and $N_r = 2$ are shown. In each set, results are shown assuming perfect knowledge of the channel (the solid curve) as

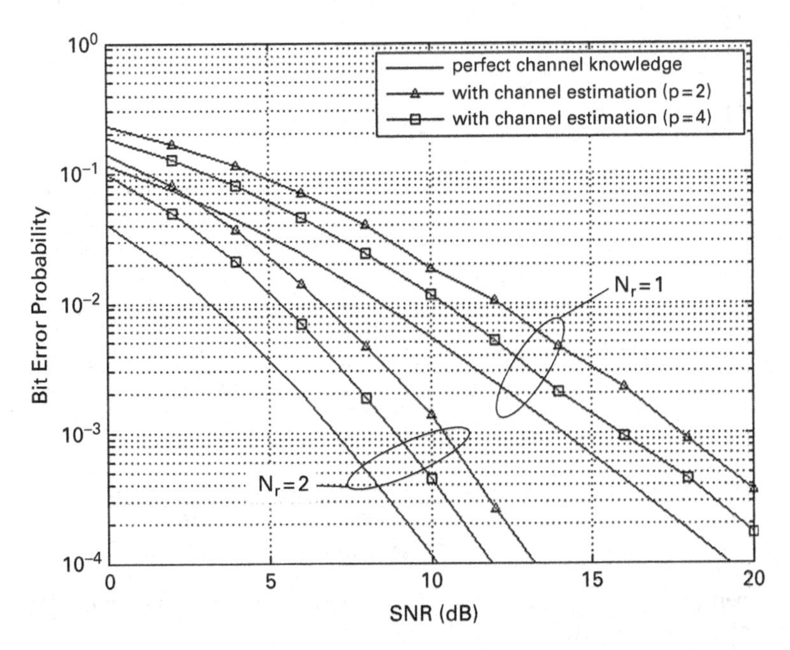

Figure 10.6 Alamouti performance showing the impact of channel estimation error on bit error probability.

well as predictions based on performing ML/LS channel estimation using pilot matrices with lengths equal to two and four symbols. The results in this plot assume Rayleigh fading, no conventional coding, BPSK modulation, and the following two pilot matrices for the $p = 2$ and $p = 4$ curves:

$$p = 2: \quad \mathbf{S}_p = \frac{1}{\sqrt{2}} \begin{pmatrix} 1 & 1 \\ 1 & -1 \end{pmatrix}, \tag{10.22}$$

$$p = 4: \quad \mathbf{S}_p = \frac{1}{\sqrt{2}} \begin{pmatrix} 1 & 1 & -1 & -1 \\ 1 & -1 & 1 & -1 \end{pmatrix}. \tag{10.23}$$

These results show that even with imperfect knowledge of the channel, the slopes and, hence, the diversity orders, are not impacted. We also note that even with the short pilot matrices assumed in this simulation, the performance degradation is relatively modest, ranging from about 1.75 to 3 dB for $p = 4$ and 2, respectively.

Figure 10.7 shows similar computer simulation results for a 2×2 MIMO system employing ZF-IC spatial multiplexing in Rayleigh fading. The pilot matrices defined in Eqs. 10.22 and 10.23 are also assumed in this figure. Again, we observe that the diversity order is unaffected by the channel estimation process and that the degradation in performance varies from about 2 to 3 dB for $p = 4$ and $p = 2$, respectively.

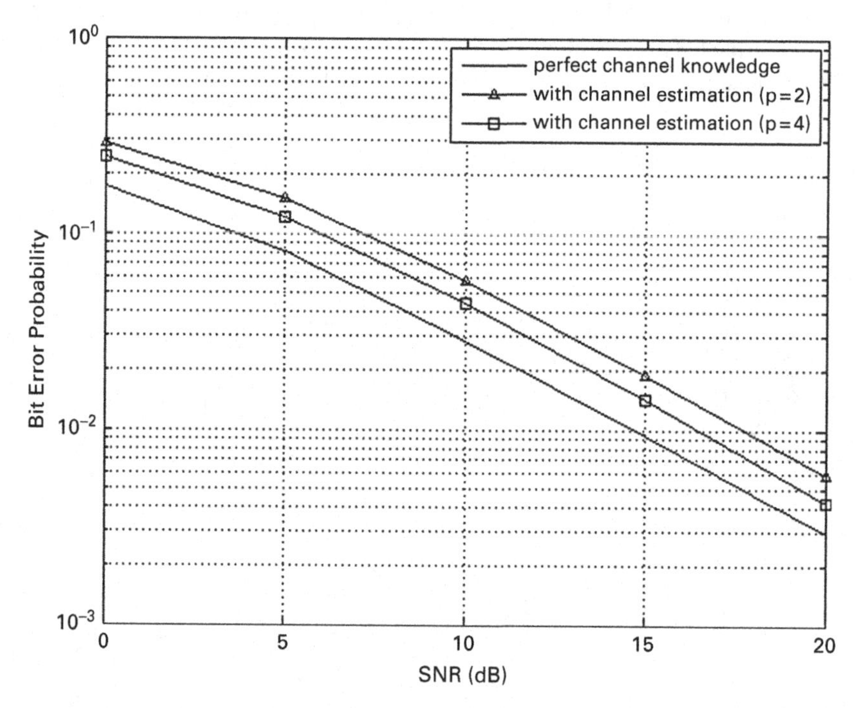

Figure 10.7 ZF-IC performance showing the impact of channel estimation error on bit error probability. These results assume that $N_t = N_r = 2$.

These results show that in flat fading pilot lengths of only a few symbols result in performance that is about only 2 to 3 dB worse than occurs with perfect channel knowledge. It is straightforward to show that such short pilot signals constitute extremely small overhead for environments of interest. As an illustration of this fact, consider a WiFi-like system that uses MIMO technology. We start by assuming that the packet length is designed to be $0.1T_{c,50}$ to ensure that the channel does not change significantly during the packet duration. If we assume that the Doppler shift between the access point and the mobile WiFi device is due to the motion of a pedestrian, we can appeal to the ITU pedestrian channel models, which define a pedestrian speed to be 3 km/h [25]. The Doppler shift associated with this speed for a carrier frequency of 2.4 GHz is $f_d = 6.6$ Hz. It follows that $T_{c,50} = 9/(16\pi f_d) = 27.1$ ms, which means that the packet length, T_{pkt}, is assumed to be 2.71 ms. If we denote the bit rate by R, then the fractional overhead per packet due to the pilot is given by $p/(T_{\text{pkt}}R)$. If we assume that $p = 4$ and $R = 10$ Mbps, it follows that the overhead is only 1.5%.

These results show that practical channel estimation techniques can achieve good performance with little overhead. In the next section, we discuss how to estimate the channel in a frequency-selective environment.

10.4 Broadband MIMO channel estimation

In some ways, broadband channel estimation is simpler than narrowband. The reason for this is, as we discussed earlier, pilot symbols are assumed to be transmitted in such a way that they do not interfere with one another (see Figure 10.5). As a result, when considering broadband channel estimation, it is only necessary to consider a single pair of antennas, which reduces the MIMO system to an effective SISO system for the purpose of analysis.

We assume that OFDM processing is used to mitigate the effects of frequency selectivity, as we discussed in the previous chapter. Under that assumption, we can depict the processing performed by any transmitter–receiver pair as shown in Figure 10.8. This figure is similar to Figure 9.6 except that it shows, for graphical simplicity, the insertion and removal of the cyclic prefix as being performed by functional blocks separate from the P/S and S/P blocks. As we discussed in Chapter 9, the signal going into the IDFT block ($\mathbf{x}$) and exiting the DFT block ($\mathbf{y}$) are interpreted to be in the frequency domain. Similarly, the signal exiting the IDFT block ($\mathbf{s}'$) and entering the DFT block ($\mathbf{r}'$) are interpreted to be time-domain signals. The channel between each pair of antennas can also be viewed in either the frequency or time domain, which we denote by $\mathcal{H}_{i,j}(k)$ and $h_{i,j}^{(l)}$, respectively.

When using OFDM processing on a frequency-selective channel, we have a choice of performing channel estimation in either the frequency or time dimensions. Because frequency domain processing is conceptually and mathematically simpler than time-domain processing, we focus our discussion in this book exclusively on

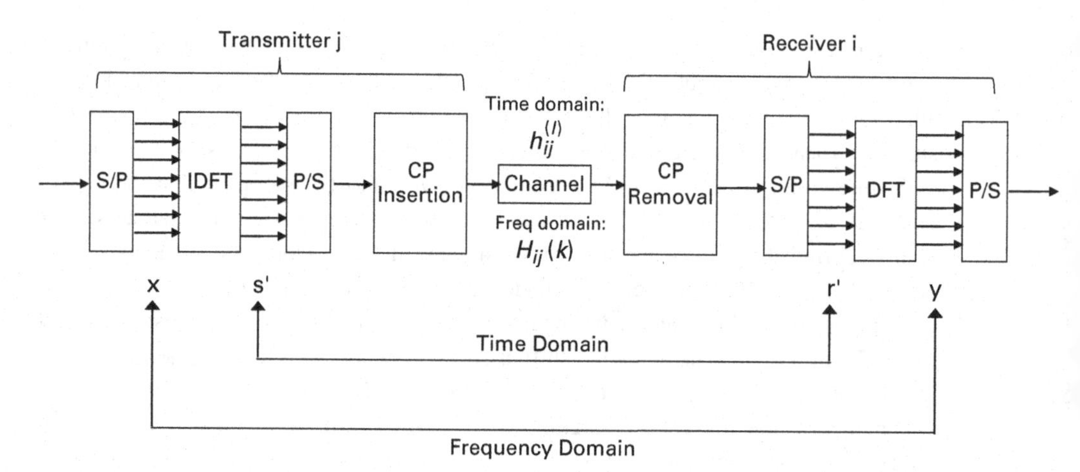

Figure 10.8 Processing steps used in an OFDM system between a transmitter–receiver pair.

frequency-domain channel estimation. Readers interested in time-domain channel estimation are referred to [27] and [3] for additional information.

10.4.1 Frequency-domain channel estimation

In Chapter 9 we showed that OFDM processing converts a frequency-selective channel into a series of flat fading channels. In the case of a single transmitter–receiver antenna pair, Eq. 9.18 simplifies to

$$y_i(k) = N_s\sqrt{\rho}\,\mathcal{H}_{ij}(k)x_j(k) + Z_i(k), \qquad \begin{array}{l} i = 1, \ldots, N_r \\ k = 0, \ldots, N_s - 1 \end{array} \tag{10.24}$$

where, for notational simplicity, we have dropped the prime on the frequency index, k. The estimate of the frequency response of the channel follows in a straightforward way from this equation.

We assume that the receiver is synchronized in both time and frequency with the transmitter and knows the time-frequency pattern where the training symbols are transmitted (e.g., it knows where black boxes are in Figure 10.3(b)). Denoting the pilot symbol emitted by transmitter j at frequency k by $p_j(k)$, it follows that the corresponding received signal is given by

$$y_i(k) = N_s\sqrt{\rho}\,\mathcal{H}_{ij}(k)p_j(k) + Z_i(k), \qquad k \in \mathbb{P} \tag{10.25}$$

where $\mathbb{P}$ denotes the set of frequency indices with pilot symbols for the current OFDM symbol.

We now use this equation along with the expressions we derived earlier for the LS, ML, and LMMSE narrowband channel estimates to find corresponding expressions for the estimates of the broadband channel frequency response. We start with Eq. 10.8 that gives the ML estimate assuming $\mathbf{R}_p = \sqrt{\rho}\,\mathbf{H}\mathbf{S}_p + \mathbf{Z}$. The ML estimate of the channel frequency response in Eq. 10.25 is obtained by replacing the terms in Eq. 10.8 as follows: $\sqrt{\rho} \rightarrow N_s\sqrt{\rho}$; $\mathbf{R}_p \rightarrow y_i(k)$; $\mathbf{S}_p \rightarrow p_j(k)$, and $\mathbf{S}_p^H \rightarrow p_j(k)^*$. It follows that

$$\hat{\mathcal{H}}_{ij}(k) = \frac{1}{N_s\sqrt{\rho}}\left(y_i(k)p_j^*(k)\right)\left(p_j(k)p_j^*\right)^{-1}, \tag{10.26}$$

which simplifies to

$$\textbf{LS/ML:} \quad \hat{\mathcal{H}}_{i,j}(k) = \frac{1}{N_s\sqrt{\rho}}\left(\frac{y_i(k)}{p_j(k)}\right) = \mathcal{H}_{i,j}(k) + V(k), \qquad k \in \mathbb{P} \tag{10.27}$$

where $V(k) \triangleq Z_i(k)/p_j(k)$ is the noise term.

Similarly, we can use the scalar version of Eq. 10.19 to obtain an expression for the LMMSE estimate of the channel frequency response by applying the substitutions defined above. The resulting LMMSE channel estimate is given by

$$\textbf{LMMSE:} \quad \hat{\mathbf{H}}_{i,j}(k) = \left(\frac{N_s\sqrt{\rho}\, y_i(k)}{1 + N_s^2\rho|p_j(k)|^2}\right)p_j^*(k). \qquad k \in \mathbb{P} \tag{10.28}$$

Equation 10.27 shows that the LS/ML channel estimate is an unbiased estimate as long as the noise term, $V(k)$, has a zero-mean symmetrical distribution. If we assume that the only noise source at each receiver is its thermal noise, and that it is zero-mean Gaussian, it follows that $V(k)$ is also zero-mean Gaussian and, therefore, under most practical conditions the LS/ML channel estimate will, indeed, be unbiased. The LMMSE channel estimate, however, is not, in general, unbiased. This can be seen by examining Eq. 10.28. It should be noted, however, that in the limit as the signal-to-noise ratio gets large, Eq. 10.28 becomes equivalent to the LS/ML estimate. Therefore, at large SNR, the LMMSE estimate is also unbiased.

Performing the simple calculations in either Eq. 10.27 or 10.28 enables the MIMO system to estimate the frequency response of the channel at the time-frequency points where there are training symbols. However, in order to perform MIMO processing, it is necessary to obtain channel estimates at all of the OFDM subcarrier frequencies in each OFDM symbol period. It follows that interpolations in both time and frequency are required when frequency-domain channel estimation is performed. The next section briefly discusses this topic.

10.4.2 Time-frequency interpolation

The subject of time-frequency interpolation in channel estimation has been an extensive area of research in the past and we do not attempt to delve into it in this book in detail. Instead, we provide a cursory outline of the general methods that have been proposed and list some references for readers that are interested in learning about these techniques in greater detail.

In general, interpolation schemes fall into one of the following three categories:

1. Simple linear interpolation;
2. Least squares interpolation; and
3. Wiener filtering interpolation.

Linear interpolation is the simplest method. There are two types of linear interpolation: one-dimensional (1-D) and two-dimensional (2-D). 1-D linear interpolation involves either linearly interpolating the channel estimates in the frequency dimension, then linearly interpolating in the time dimension, or performing the interpolation in the opposite order. References [15] and [41] describe 1-D interpolation in greater detail. 2-D interpolation, as its name suggests, performs interpolation in both the time and frequency dimensions at the same time.

Least squares interpolation is a generalization of linear interpolation and is based on modeling the channel frequency response as a weighted sum of basis functions. This technique also can be applied in 1-D and 2-D versions. For example, the 1-D version models $\mathcal{H}_{i,j}(k)$ as follows:

$$\mathcal{H}_{i,j}(k) = \sum_{i=0}^{N-1} \alpha_i \phi_i(k), \qquad (10.29)$$

where $\{\alpha_i\}$ are the weights and the $\{\phi_i(k)\}$ are the N basis functions at frequency index k. Similarly, the 2-D version has the following form:

$$\mathcal{H}_{i,j}(k,t) = \sum_{i=0}^{N-1} \alpha_i \phi_i(k,t), \qquad (10.30)$$

where t is the time index that specifies the OFDM symbol number. Many different options exist for the basis functions, including polynomials, Fourier series, discrete cosine transform functions, and discrete sine transform functions, to name only a few. The least squares interpolation method involves estimating the weighting coefficients in such a way as to minimize the mean square error between the true channel frequency response values and the interpolated values. A tutorial discussion of this technique is presented in [5].

Wiener filtering interpolation is the most complex of the three classes of schemes listed above, requiring greater computational complexity, but potentially offering greater accuracy. In addition to having estimates of the channel frequency response at certain time-frequency locations where there are training symbols, Wiener filtering also uses knowledge of the spaced-frequency and spaced-time correlation functions of the channel, which were defined in Chapter 4. Estimating these functions, of course, is an issue unto itself, but if they can be obtained, Wiener filtering has the potential of providing superior performance. Additional information on Wiener filtering can be found in [63].

Problems

10.1 Consider a MIMO system that operates in a flat fading environment and that uses $(N_t \times p)$ pilot matrices, $\mathbf{S}_p$, to estimate the channel. For each of the pilot matrices below, decide whether or not they are good candidates for performing LS/ML and LMMSE channel estimation. Explain why or why not.

(a)

$$\mathbf{S}_p = \frac{1}{2}\begin{pmatrix} 1 & 1 \\ 1 & 1 \\ 1 & 1 \\ 1 & -1 \end{pmatrix}; \tag{10.31}$$

(b)

$$\mathbf{S}_p = \frac{1}{2}\begin{pmatrix} 1 & 1 \\ 1 & -1 \\ 1 & 1 \\ 1 & -1 \end{pmatrix}; \tag{10.32}$$

(c)

$$\mathbf{S}_p = \frac{1}{\sqrt{2}}\begin{pmatrix} 1 & -1 & 1 & -1 \\ 1 & -1 & -1 & -1 \end{pmatrix}; \tag{10.33}$$

(d)

$$\mathbf{S}_p = \frac{1}{\sqrt{2}}\begin{pmatrix} 1 & -1 & 1 & -1 \\ 1 & -1 & -1 & 1 \end{pmatrix}. \tag{10.34}$$

10.2 Derive Eq. 10.11.

10.3 Write a Matlab function to reproduce Figure 10.7. Do this by modifying the Alamouti code from Problem 6.3. Call this new routine Alamouti_CE.m.

10.4 Write a Matlab function to reproduce Figure 10.8. Do this by modifying the Matlab ZF-IC program from Problem 8.7. Call this new routine ZF_IC_CE.m.

10.5 Consider an OFDMA MIMO system that operates at 2.4 GHz and that uses pilot symbols spaced in the time and frequency dimensions for channel estimation. Assume that this system uses the same OFDM symbol length as used in WiMAX, which is 102.4 μs [3].

a) Assume the RMS delay spread equals 20 μs and that the system is designed to operate in an environment defined by the ITU vehicular-A channel model [42], which assumes vehicular speeds up to 120 km/hr. Compute the pilot spacing in the frequency dimension in OFDM subcarrier units, and the pilot spacing in the time dimension in OFDM symbol period units. This represents an extreme cellular condition involving high speed (i.e., large Doppler) and large delay spread.

b) Now consider more typical cellular conditions. Compute the pilot spacing assuming that the RMS delay spread is only 1 μs and that pedestrian speeds of only 3 km/hr are involved.

11 Practical MIMO examples

Now that the basic concepts of MIMO communications have been introduced, we conclude the book with a chapter that describes how those concepts are applied in real-world wireless communications systems. For this purpose, we focus on two popular wireless families of standards, WiFi and LTE/LTE-advanced, which are used extensively for wireless LAN applications and conventional cellular communications, respectively. Although these communication standards are extremely complex, we focus primarily on the MIMO aspects of these technologies.

11.1 WiFi

WiFi is a virtually ubiquitous wireless technology that is designed to allow electronic devices to exchange data wirelessly at rates ranging from 1 Mbps to as much as 600 Mbps over distances of typically 10s to 100s of feet. The most common application of WiFi technology today is the use of wireless routers in homes and businesses for the purpose of enabling computers, cellular phones, and a growing list of personal wireless devices to connect to the Internet without cables. The version of WiFi that employs MIMO is defined by the IEEE 802.11n standard, which is the latest in a series of amendments to the original 802.11 standard that was developed in the late 1990s. The current version of the 802.11n standard at the time of this writing is called IEEE 802.11n-2009, which was published in October 2009. The information in this section is primarily gleaned from that standard [77]. In addition, Perahia and Stacey have written an excellent book on 802.11n [60] that is strongly recommended for those readers who want to delve into 802.11n in greater detail.

The following section provides a brief overview and history of IEEE 802.11n.

11.1.1 Overview of IEEE 802.11n

History

The first version of WiFi, which was simply called IEEE 802.11, was published in June 1997, and it supported a maximum data rate of only 2 Mbps using spread spectrum techniques. Subsequent to this, enhancements to the standard in September 1999 (802.11a) and again in June 2003 (802.11g) increased the maximum data rate from 2 Mbps to 54 Mbps through, among other techniques, the use of higher-order modulation

802.11 Protocol	Release	Freq. (GHz)	Bandwidth (MHz)	Data rate per stream (Mbps)	Allowable MIMO streams	Modulation
-	Jun 1997	2.4	20	1, 2	1	DSSS, FHSS
a	Sep 1999	5	20	6, 9, 12, 18, 24, 36, 48, 54	1	OFDM
		3.7				
b	Sep 1999	2.4	20	5.5, 11	1	DSSS
g	Jun 2003	2.4	20	6, 9, 12, 18, 24, 36, 48, 54	1	OFDM, DSSS
n	Oct 2009	2.4,5	20	7.2, 14.4, 21.7, 28.9, 43.3, 57.8, 65, 72.2	4	OFDM
			40	15, 30, 45, 60, 90, 120, 135, 150		

Figure 11.1 Comparison of IEEE 802.11 standards.

and the introduction of OFDM to combat multipath effects. Since 64-QAM requires higher SNR than the lower order modulation schemes that were in the earlier version of 802.11, the improvement in throughput was accomplished at the cost of shorter range and increased vulnerability to co-channel interference, which resulted in reduced total system capacity [74].

A recent enhancement to WiFi is the IEEE 802.11n amendment to the standard, which incorporates MIMO technology. The part of the 802.11n standard that employs MIMO is referred to in the standard as the *enhancement for higher throughput* (HT) mode. First published on 29 October 2009, IEEE 802.11n supports data rates up to 600 Mbps – more than a ten-fold increase over 802.11a/g, and a 300-fold increase over the original version of WiFi. IEEE 802.11n operates in the 2.4 and 5 GHz bands like 802.11a/g but, unlike these earlier standards, which only use a bandwidth equal to 20 MHz, 802.11n can operate with bandwidths equal to 20 and 40 MHz. In addition, 802.11n can use up to four antennas at the transmitter and receiver. Figure 11.1 summarizes some of the key differences between the various 802.11 standards.

MIMO techniques in 802.11n

IEEE 802.11n supports most of the basic types of MIMO techniques that we have discussed in this book. A summary of the MIMO schemes supported by 802.11n are listed below:

1. ***Antenna configurations*** – IEEE 802.11n supports up to four antennas at either the transmitter, receiver, or both, and all combinations thereof.

2. ***Spatial multiplexing*** – IEEE 802.11n supports spatial multiplexing. In this mode, a single high-speed stream of symbols is split and encoded into N_t parallel streams that are transmitted by up to four transmit antennas. At the receiver, any of the decoding techniques that we have discussed, such as ZF, ZF-IC, LMMSE, and LMMSE-IC can be used, or any other technique that is compatible with the transmitted signal. As has been pointed out previously, commercial wireless standards such as 802.11n only define the specifications for the transmitter; they do not explicitly state how the receiver is to operate or which algorithms should be used at the receiver. Vendors are free to use any method they choose as long as it works. We will see that 802.11n is capable of implementing V-BLAST, but not H-BLAST or D-BLAST.
3. ***Space-time block coding*** – IEEE 802.11n supports a variety of space-time block coding schemes based on the Alamouti method. Unlike spatial multiplexing, which requires that $N_r \geq N_t$, these STBC schemes are designed to mitigate the effects of multipath and can be used even when $N_t > N_r$. A common application is when an 802.11n network access point (AP) is transmitting to a handset client. In this case, the AP normally can support multiple antennas and transmitters but the handset may only have room (and battery power) for a single receiver. STBC decreases the effects of fading and, hence, decreases the error rate, which enables the handset to be located further away from the AP for a given error rate and average SNR. It should be emphasized that the current version of 802.11n does not support space-time block codes other than the Alamouti scheme.
4. ***Beamforming*** – 802.11n supports full eigenbeamforming as well as single-mode beamforming. As discussed earlier, eigenbeamforming enables the transmitter to transmit multiple data streams over the same channel; however, doing so requires that the transmitter has knowledge of the MIMO channel. IEEE 802.11n supports this capability, which is called *transmit beamforming* (TxBF).

The role of MIMO in 802.11n

As we have already mentioned, 802.11n is capable of supporting data rates up to 600 Mbps, compared with only 54 Mbps with 802.11a/g. How is this accomplished and what is the role of MIMO in achieving this enormous increase in throughput? The answer is that MIMO plays a significant role, but there are three other techniques that are also instrumental. Before describing how MIMO is implemented in 802.11n in further detail, we briefly discuss its role in the newest version of WiFi.

To help understand how 802.11n achieves such high data rates, we first consider 802.11a/g and see how it achieves its relatively modest maximum throughput of 54 Mbps. It does it by using 64-QAM modulation with a rate $r = 3/4$ error control code, combined with an augmented OFDM symbol length $T_{\text{AOFDM}} = 4.0\ \mu\text{s}$. Each augmented OFDM symbol is created using a 64-point IFFT filled with 48 data symbols. Thus, the total average data rate is $\left(\frac{48}{4\times10^{-6}}\right) \cdot \log_2(64) \cdot r = 54$ Mbps.

IEEE 802.11n also uses 64-QAM in its highest data rate mode, and employs a similar OFDM structure as that used in 802.11a/g, except that in order to achieve its maximum data rates, it uses a slightly shorter augmented OFDM symbol period equal to 3.6 μs, which is achieved using a guard interval of only 0.4 μs rather than the normal 0.8 μs

Table 11.1 Data rate improvement factors due to various enhancements in IEEE 802.11n compared with 802.11a/g.

Technology modification	Data rate improvement factor	Percent effect
MIMO	$\times$ 4	40.5
Doubling bandwidth (20 MHz $\rightarrow$ 40 MHz)	$\times$ 2.25	22.8
Shortening OFDM symbol	$\times$ 1.11	11.2
Increasing code rate	$\times$ 1.11	11.2

value. Secondly, rather than using a bandwidth of 20 MHz, which is what is used by 802.11a/g, 802.11n uses 40 MHz in its high data rate modes, which allows each OFDM symbol to carry about twice as many data symbols. The actual value is 108 symbols compared with 48 for 802.11a/g. A third difference has to do with the highest error control rate that is used. In 802.11a/g, $r = 3/4$, but in 802.11n, the maximum rate is $r = 5/6$. Lastly, 802.11n employs up to 4 $\times$ 4 MIMO, which supports a factor of 4 increase in throughput. Combining these effects together, we find that the maximum data rate for 802.11n is $\frac{108}{3.6\times 10^{-6}} \cdot \log_2(64) \cdot (5/6) \cdot 4 = 600$ Mbps.

Table 11.1 lists the data rate improvement factors associated with each of the four techniques that are used by 802.11n to achieve its maximum data rate. These results show that MIMO is by far the largest single factor and that it is nearly twice as large as the other three factors combined. Nevertheless, it is important to keep in mind that the enormous improvement in WiFi throughput with 802.11n is a result of multiple factors, with MIMO being only one – albeit the most significant.

11.1.2 802.11n packet structure

IEEE 802.11n uses CSMA to coordinate how users share the RF spectrum. As a result, 802.11n is a packet-based communication system. To help understand some of the key aspects of how 802.11n operates, it is useful to understand the basic structures of the 802.11n packets. There are three modes of operation in 802.11n called: *legacy mode*, *mixed mode*, and the *Greenfield mode*. The legacy mode is equivalent to 802.11a/g and is used in WiFi networks that consist of only legacy devices operating the older 802.11a/g protocol. The mixed mode supports both legacy and 802.11n functionality simultaneously, thus enabling heterogeneous networks that consist of legacy and newer 802.11n WiFi devices. In cases where only 802.11n devices operate, the so-called Greenfield mode can be used, which only supports the functionality of 802.11n. The legacy mode is designed to operate over a 20 MHz channel only; the HT modes (mixed and Greenfield modes) can operate in either 20 or 40 MHz channels.

In all three modes, each packet consists of a header, which is transmitted at 6 Mbps, followed by the data, which is transmitted in any one of a large number of modulation, coding, and data rate combinations called *modulation and coding schemes* (MCS). Figure 11.2 illustrates the packet structures for all three modes.

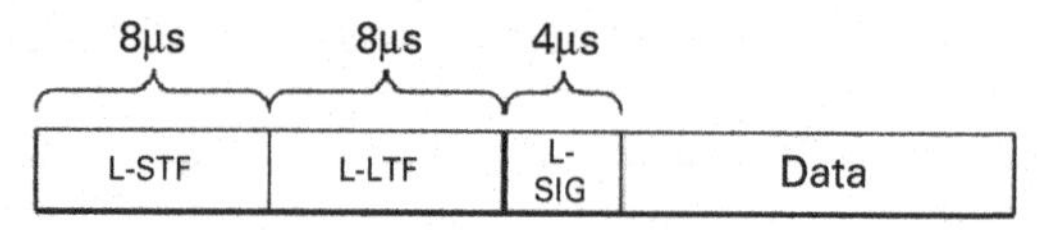

(a) Packet structure for legacy mode

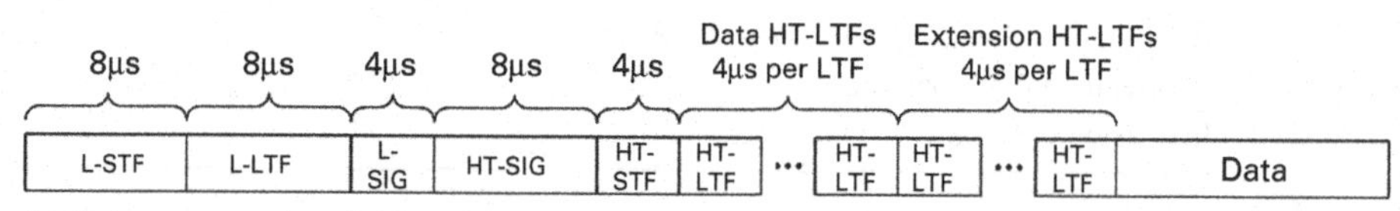

(b) Packet structure for mixed mode

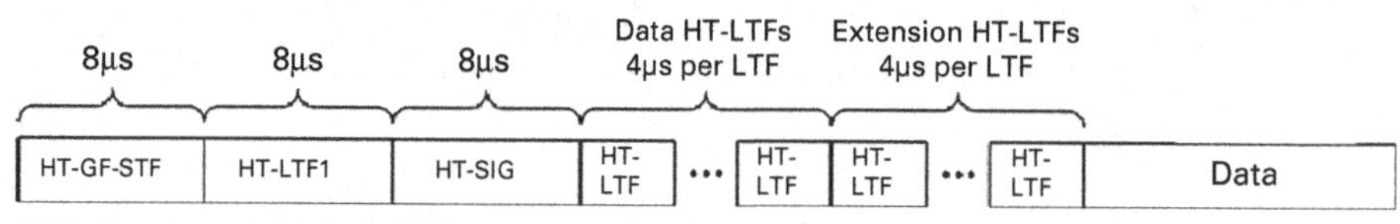

(c) Packet structure for Greenfield mode

Figure 11.2 Packet structures used in IEEE 802.11n (from [77]).

The simplest packet is the one used when operating in the legacy mode, which has the same format as the packets used in 802.11a/g. The L-STF field, which stands for the *legacy-short training field*, consists of two augmented OFDM symbols of 4 μs duration each, and these are used by the receiver to perform time and frequency synchronization and to adjust its AGC. After the two L-STF augmented OFDM symbols, there is a second field called L-LTF, which stands for *legacy-long training field*, which is used to perform fine timing adjustments after the initial coarse acquisition in the L-STF field. The third field, called L-SIG, which stands for *legacy signal field*, contains data that tell the receiver what modulation and coding will be used in the data portion of the packet. The L-SIG field consists of one augmented OFDM symbol.

The mixed mode packet structure illustrated in part (b) of the figure is more complex than the legacy-only packet because it must be capable of being both backwards compatible with 802.11a/g devices as well as being able to operate with newer 802.11n equipment. As a result, it needs to be capable of supporting all four combinations of links: legacy-to-legacy; legacy-to-HT; HT-to-legacy; and HT-to-HT. An examination of the structure of the preamble portion of the packet (the part before the data segment) shows that the first three fields are equivalent to the legacy preamble, which makes this packet compatible with 802.11a/g devices even if it is transmitted by an 802.11n-enabled device. Legacy users would, therefore, ignore the part of the preamble after the first three fields and then wait until the data portion begins.

In a mixed mode network, HT devices (i.e., devices that are 802.11n-enabled) also use the legacy preamble to perform acquisition. From this information, they are then able to demodulate the rest of the preamble that contains information of interest to HT users. The first field after the legacy preamble portion of the packet is the HT-SIG field, which stands for *HT-signal field*. This field consists of two augmented OFDM symbols that tell the HT receiver what the MCS is for the data, which bandwidth is being transmitted (i.e., 20 MHz or 40 MHz), which, if any, type of STBC will be used during the transmission of the data, and information that will be needed if eigenbeamforming is employed. The next field is the HT-STF field, which stands for *HT-short training field*. This field is the same as the L-STF field except that different transmitters use different cyclic delays. The receiver uses this field to obtain refined time and frequency estimates. The last part of the preamble consists of two or more HT-LTF fields, each consisting of one augmented OFDM symbol, which are used by the receiver to estimate the channel matrix. The number of HT-LTF fields depends on the number of data streams that are being simultaneously transmitted by the MIMO system in a specific way that will be defined shortly.

The third type of packet structure defined in 802.11n is the so-called Greenfield packet, which can only be used if all the users in the wireless network are 802.11n-enabled. As illustrated, the main difference between the Greenfield packet and the mixed mode packet is that the Greenfield version does not have the legacy preamble, which makes the Greenfield preamble shorter and results in greater efficiency. For the purpose of our discussion, the differences between the mixed mode and Greenfield packets are not germane.

Table 11.2 summarizes the various fields in the preambles of 802.11n packets.

11.1.3 802.11n HT transmitter architecture

IEEE 802.11n supports a wide variety of waveforms and antenna configurations, ranging from SISO arrangements with no MIMO to $m \times n$ MIMO configurations with various combinations of space-time block coding and spatial multiplexing. Adding to this complexity is the fact that the transmit processing is different for the preamble than it is for the data portion of each packet. Since a full description of all the possible transmitter options is beyond the scope of this chapter, we focus, instead, only on the transmit architecture when transmitting data and only when operating in the HT mode.

11.1.3.1 Transmitter block diagram

Figure 11.3 is a block diagram of an 802.11n transmitter for the HT data field. This, of course, is a general diagram that depicts all of the functionality potentially available in an HT transmitter; however, only subsets of the diagram may be implemented for any particular transmission mode, as we see shortly.

The transmitter operates as follows. Information bits to be transmitted enter at the left and are first scrambled to reduce the adverse effect of large peak-to-average power ratios (PAPRs) when using OFDM. A packet that has a significant number of OFDM symbols with large PAPR will have a large probability of error and may never arrive at its destination because every retransmission of the packet has the same large error

Table 11.2 Description of preamble fields used in IEEE 802.11n.

Abbreviation	Field name	Description
L-STF	legacy-short training field	It is used by the receiver to determine the beginning of a packet, set the AGC, and provide initial frequency synchronization and time acquisition. The L-STF consists of two augmented OFDM symbols totaling 8 μs.
L-LTF	legacy-long training field	It is used for frequency tracking during which fine frequency correction and fine timing adjustments are made. The LTF is also used to generate the channel estimate. The L-LTF consists of two augmented OFDM symbols totaling 8 μs.
L-SIG	legacy-signal field	Tells the receiver the modulation type and coding rate, as well as the length of the data portion of the packet. The L-SIG field consists of one augmented OFDM symbol totaling 4 μs.
HT-SIG	HT-signal field	The HT-SIG field performs the same role as the L-SIG field except that it is used when operating in one of the HT modes. The HT-SIG field consists of two augmented OFDM symbols totaling 8 μs.
HT-STF	HT-short training field	The HT-STF field is the same as for 802.11a/g, except that different transmitters use different cyclic delays.
HT-LTF	HT-long training field	The HT-LTF fields are used when operating in one of the two HT modes to estimate the channel between each transmitter and each receiver. The number of HT-LTF fields (NLTF) is equal to the number of space-time streams, except when the number of streams equals 3, in which case NLTF = 4. Each HT-LTF field consists of a single augmented OFDM symbol 4 μs in duration.

probability. The scrambler reduces this problem by pseudo-randomly scrambling the incoming information bits differently on each packet transmission.

After scrambling, the bits pass through an encoder parser that operates in one of two ways depending on the bandwidth of the 802.11n device. When the bandwidth equals 20 MHz, the encoder parser simply passes the incoming bit stream to one of the two forward error correction (FEC) encoders, depending on the MCS. When the bandwidth is 40 MHz, some of the MCSs use both encoders, so in those cases the parser de-multiplexes the incoming bit stream among the two FEC encoders in a round-robin fashion. IEEE 802.11n defines two types of FEC encoding: binary convolutional coding

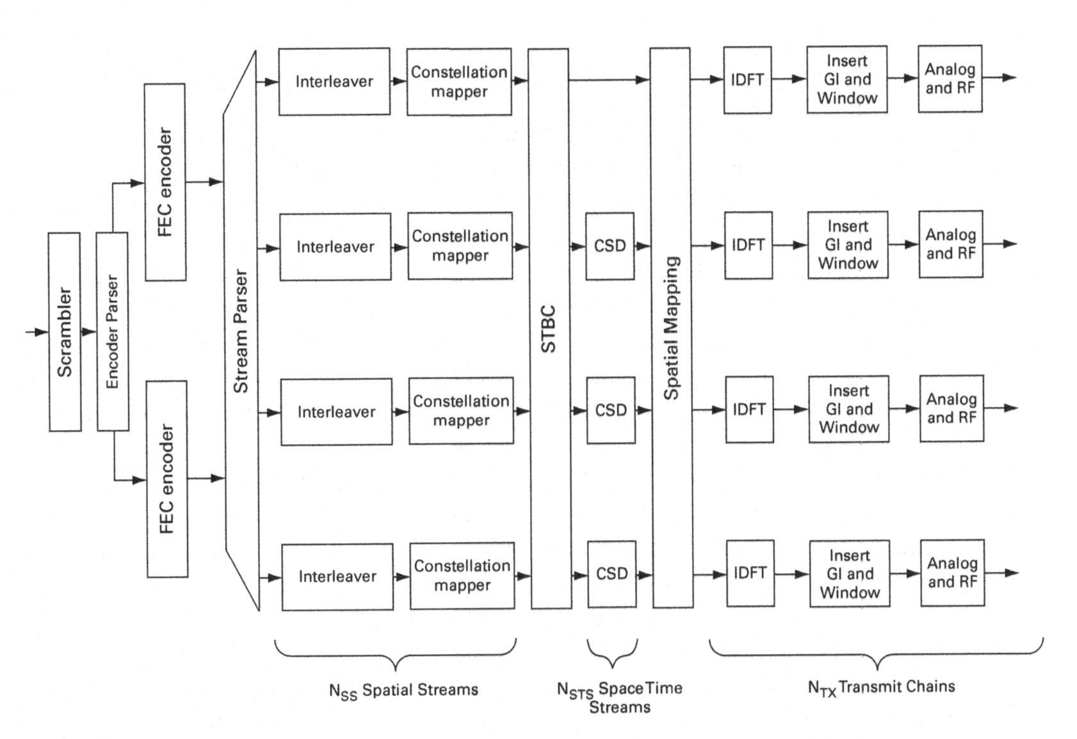

Figure 11.3 Transmitter for the HT data field of 802.11n when operating in one of the two HT modes (reproduced from [77]).

(BCC) and low density parity check coding (LDPCC). The FEC encoder block also includes puncturing that is used to achieve various code rates. Again, the specific coding used is defined by the MCS, which in turn is communicated to the receiver in the HT-SIG field of the preamble.

Following encoding, the coded bits are potentially sent to different interleaver and modulation mapping chains called *spatial streams* (SS). Up to four spatial streams may be used, which is denoted in the Standard by N_{SS}. The block that divides the coded bits into spatial streams is called the *stream parser*. Each stream is interleaved and modulated in the block labeled *constellation mapper*, which maps the sequence of bits in each spatial stream into complex baseband constellation points. Like 802.11a/g, 802.11n uses BPSK, QPSK, 16-QAM, and 64-QAM modulation, depending on the particular MCS.

The next block in the transmitter after modulation is *space-time block coding* (STBC), which applies one of four different STBCs. The number of spatial stream inputs to the STBC varies from one to three, depending on the particular type of STBC used (if any), and the number of output streams varies from two to four, and is always one larger than the number of input streams. The STBC output streams are called *space-time streams* (STS), and the number of STSs is denoted in the Standard by N_{STS}. Of course, if no space-time block coding is used, the STBC effectively disappears from the diagram and $N_{SS} = N_{STS}$. Furthermore, the restriction on N_{SS} being less than or equal to three is lifted in that case, and it would be possible to have up to four spatial streams, which is, in fact, required to achieve the full 600 MHz data rate as we discussed earlier. It should be noted that all four of the STBCs supported by 802.11n use Alamouti coding, which we discuss in greater detail shortly.

Following space-time block coding (if it is performed), the bottom three space-time streams are subjected to cyclic shifts by the cyclic shift diversity (CSD) blocks, which are needed when multiple transmit chains are employed to prevent unintentional beamforming. Unintentional beamforming is potentially problematic because it causes nulls in certain directions that can result in poor performance. The role of CSD in mitigating such nulls is described in [55] for interested readers desiring greater detail.

The next major block in the transmitter is the *spatial mapper*, which maps the space-time streams to different transmit chains, each of which feeds a separate transmit antenna. The number of transmit chains is denoted in the Standard by N_{TX}, which is equivalent to N_t in this book.

There are two types of spatial mapping defined in 802.11n: *direct mapping* and *spatial expansion*. In direct mapping, each space time stream is directly mapped to a different transmit chain (i.e., antenna). In this simple case, the spatial mapper does not perform any function other than to pass each input stream to a single output stream in a one-to-one mapping. When using direct mapping, the number of space time streams equals the number of antennas.

When spatial expansion is performed, the mapping between the space-time streams and the transmit chains is more complex. In spatial expansion, the mapping is performed by multiplying the space time streams by a matrix denoted in the 802.11n standard by Q. When spatial expansion is performed, the number of transmit chains is always equal to or greater than the number of space time streams; hence, the term *spatial expansion*. As we will see, using the Q matrix to perform spatial expansion enables the transmit architecture to implement eigenbeamforming.

After spatial mapping, each transmit chain has OFDM modulation applied to it by processing it through an IDFT block that converts a block of modulated constellation points to a time domain block of symbols. Since the IDFT block operates on blocks of data, it is necessary to perform serial-to-parallel conversion before performing IDFT processing and to apply parallel-to-serial conversion after application of the IDFT. These two steps are performed in the IDFT blocks shown in the figure. The sequence of symbols out of each IDFT block has a cyclic prefix inserted to mitigate inter-OFDM-symbol-interference, and the resulting augmented OFDM symbol is multiplied by a windowing function that smooths the edges of each OFDM symbol to decrease spectral splatter. The resulting baseband sequence of symbols in each chain is then passed to the analog and RF blocks which perform digital-to-analog conversion, mixing with a carrier frequency, filtering, and amplification before being applied to each transmit antenna.

11.1.3.2 Example MIMO implementations

Now that we have described the general 802.11n transmit architecture, we can show how this general architecture can be used to implement some of the transmit configurations we described earlier in the book.

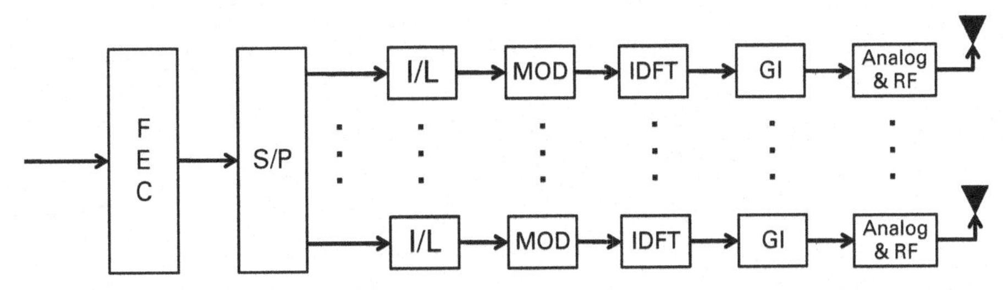

Figure 11.4 Simplified 802.11n block diagram when: a) there is only one FEC encoder; b) the stream parser is denoted as a serial-to-parallel converter; and c) the STBC and spatial mapping blocks are ignored.

(a) **BLAST**

Some versions of the BLAST architecture are supported by 802.11n. Consider, for example, if we were to make the following assumptions in Figure 11.3: 1) assume only one FEC encoder ; 2) replace the stream parser with a serial-to-parallel converter block (which is functionally what it is); 3) ignore the STBC and spatial mapping blocks, since these are optional blocks. Under these assumptions $N_{SS} = N_{STS} = N_t$ and Figure 11.3 simplifies to Figure 11.4.

Comparing this figure with Figure 8.2 shows that it is equivalent to the V-BLAST transmitter architecture (except with more of the details explicitly shown). This shows that 802.11n is capable of supporting V-BLAST.

Examination of Figure 11.3 shows that it is not possible to modify it to look exactly like Figure 8.3, even if the number of streams is restricted to two, due to the presence of the stream parser, which distributes the bits from each of the FEC blocks onto both spatial streams. Were it not for that, it would be possible to modify Figure 11.3 to mimic the H-BLAST architecture in Figure 8.3. Similarly, we see that it is not possible to configure Figure 11.3 to duplicate the functionality in Figure 8.4. We conclude that 802.11n is able to support V-BLAST, but not H-BLAST and D-BLAST.

(b) **Spatial diversity**

The block diagram in Figure 11.3 can also be simplified to implement spatial diversity through the use of space-time block coding. To do so, we make the following assumptions: 1) use only one FEC encoder as before; 2) ignore the stream parser and spatial mapper; and 3) set $N_{SS} = 1$ and $N_{STS} = N_t = 2$. Under these assumptions, Figure 11.3 simplifies to the configuration shown in Figure 11.5.

There are of course other configurations involving more than one input spatial stream and, as we will see, space-time block coding configurations that allow more than two output space-time streams. Space-time block coding in 802.11n is discussed in detail in the following section.

(c) **Eigenbeamforming**

Eigenbeamforming is also supported by 802.11n. To help see this, consider the following simplifications to Figure 11.3: 1) use only one FEC; and 2) ignore the STBC block. Under those assumptions, Figure 11.3 simplifies to the form shown in Figure 11.6, which, in turn, is functionally equivalent to the transmit side of Figure 3.2

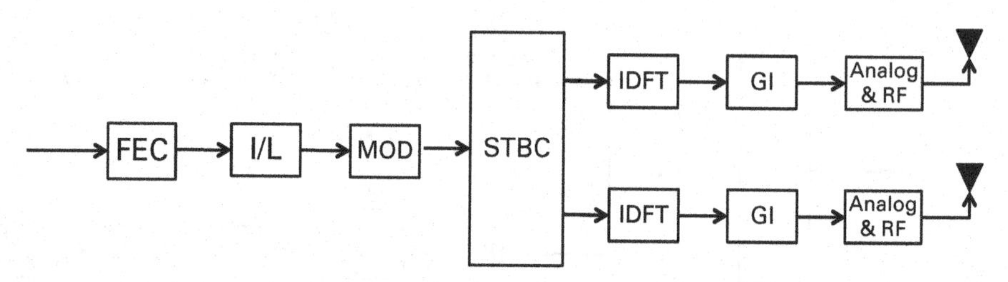

Figure 11.5 Simplified 802.11n block diagram depicting how it is possible to configure the general 802.11n transmitter configuration to implement spatial diversity using space-time block coding.

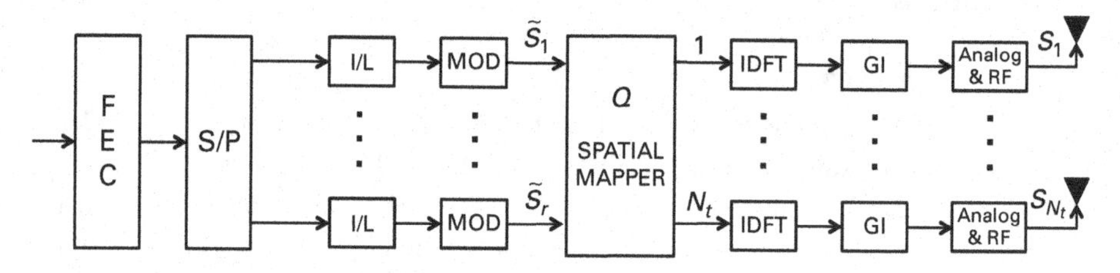

Figure 11.6 Simplified version of 802.11n block diagram depicting how it is possible to configure the general 802.11n transmitter to implement eigenbeamforming.

for eigenbeamforming with $\boldsymbol{Q} = \boldsymbol{V}$. Since eigenbeamforming requires that the number of spatial streams entering the spatial mapper, r, be less than or equal to the number of transmit antennas, the spatial mapper operates in its spatial expansion mode when conducting eigenbeamforming.

11.1.4 Space-time block coding in 802.11n

11.1.4.1 Nomenclature

IEEE 802.11n supports four different space-time block coding schemes, all of which are based on use of the Alamouti code. To understand these coding schemes, it is helpful to first understand the nomenclature that is used in the standard documentation. In 802.11n, the complex baseband modulated symbols entering the STBC block are denoted by the symbols

$$\left\{d_{k,i,n} \mid k = 0, 1, \ldots, N_{SD} - 1; i = 1, \ldots, N_{SS}; n = 0, \ldots, N_{SYM} - 1\right\}, \qquad (11.1)$$

where N_{SD} denotes the number of symbols in each OFDM symbol, N_{SS} is the number of spatial streams entering the STBC block, and N_{SYM} is the number of OFDM symbols being transmitted. Thus, $d_{k,i,n}$ denotes the kth modulation symbol in the nth OFDM symbol on the ith input stream. Similarly, the output symbols from the STBC block are denoted by $\tilde{d}_{k,i,n}$; where the subscripts have the same meaning as before except that $i = 1, \ldots, N_{STS}$. Figure 11.7, which is based on Table 20-17 from [77], defines the four STBC schemes supported by 802.11n.

An examination of this table shows that there are four distinct STBC schemes, indexed by the first column. The number of input streams to each STBC is specified

STBC number	N_{ss}	N_{STS}	i_{STS}	$\tilde{d}_{k,i,2m}$	$\tilde{d}_{k,i,2m+1}$	
1	1	2	1	$d_{k,1,2m}$	$d_{k,1,2m+2}$	Alamouti
			2	$-d^*_{k,1,2m+1}$	$d^*_{k,1,2m}$	
2	2	3	1	$d_{k,1,2m}$	$d_{k,1,2m+1}$	Alamouti
			2	$-d^*_{k,1,2m+1}$	$d^*_{k,1,2m}$	
			3	$d_{k,2,2m}$	$d_{k,2,2m+1}$	Pass through
3	2	4	1	$d_{k,1,2m}$	$d_{k,1,2m+1}$	Alamouti
			2	$-d^*_{k,1,2m+1}$	$d^*_{k,1,2m}$	
			3	$d_{k,2,2m}$	$d_{k,2,2m+1}$	Alamouti
			4	$-d^*_{k,2,2m+1}$	$d^*_{k,2,2m}$	
4	3	4	1	$d_{k,1,2m}$	$d_{k,1,2m+1}$	Alamouti
			2	$-d^*_{k,1,2m+1}$	$d^*_{k,1,2m}$	
			3	$d_{k,2,2m}$	$d_{k,2,2m+1}$	Pass through
			4	$d_{k,3,2m}$	$d_{k,3,2m+1}$	Pass through

Figure 11.7 Space-time block codes defined in IEEE 802.11n. In this figure, $k = 0, 1, \ldots, N_{SD} - 1$; $i = 1, 2, \ldots, N_{SS}$; and $m = 0, 1, \ldots, \left(\frac{N_{SYM}-1}{2}\right)$.

by the number of spatial streams, N_{SS}, and the number of output STBC streams is specified by N_{STS}, the number of space time streams. The fourth column, labeled i_{STS}, indexes the output space time streams for each code. The right side of the table shows that the STBCs in 802.11n involve only Alamouti coding, with some straight pass through connections. Because the nomenclature is a bit cumbersome, we study STBC 1 in detail in the next subsection to help clarify how to interpret the entries in Figure 11.7.

11.1.4.2 Alamouti coding in 802.11n

Interestingly, the 802.11n standard does not refer to the Alamouti code by name, but it is straightforward to show that the logic defined in Figure 11.7 is equivalent to Alamouti encoding. In Chapter 6 we defined Alamouti coding by the logic in Table 6.1, which is depicted graphically in Figure 11.8 (a). In this figure, the right-most symbol should be interpreted as the symbol that arrives or exists at the STBC block first (i.e., time increases to the left). Part (b) of this figure shows an alternative way of performing Alamouti encoding, which is evident from the fact that it has all the same properties as the earlier form. In particular, it has a code rate equal to 1 and it is an OSTBC, which follows from the fact that its rows and columns are orthogonal.

Now consider the STBC No. 1 defined in Figure 11.7. Since that code has $N_{SS} = 1$ and $N_{STS} = 2$, there is one input stream and two output streams, which are related to each other according to the rules defined in the figure. To help visualize the logic of this code, consider Figure 11.9, which depicts the first two input OFDM symbols (i.e., for $2m = 0$ and $2m + 1 = 1$). It is important to keep in mind that since the STBC block precedes the OFDM IFFT operation in the transmitter, the input and output data streams to the

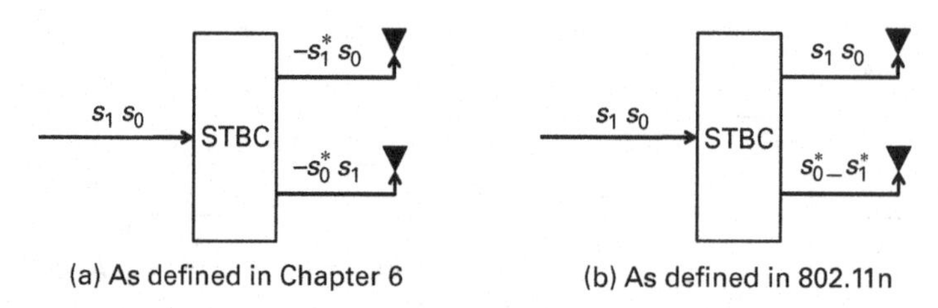

Figure 11.8 Examples of Alamouti encoding variations: (a) Corresponds to the encoding defined in Table 6.1; (b) Shows an alternative variation of Alamouti encoding used in 802.11n. (Note that the Alamouti coding in Table 6.1 indexes the symbols starting at 1 whereas the symbols in Figure 11.8 start with index 0. This chapter adopts the convention used in the IEEE 802.11n standard.)

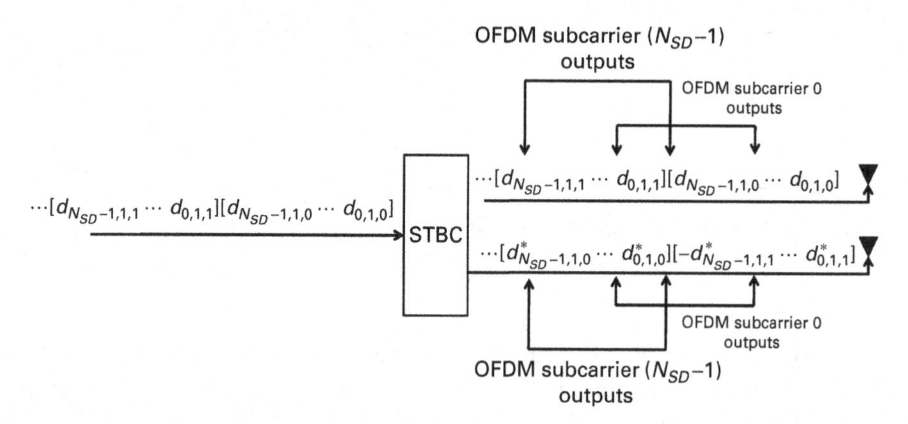

Figure 11.9 The IEEE 802.11n STBC No. 1.

STBC are in the frequency domain; thus, the outputs correspond to different frequency subcarriers. The sequence of symbols entering and exiting the STBC block is depicted by the brackets as being grouped into OFDM blocks of length N_{SD}. An examination of the figure shows that the output symbols associated with subcarrier 0 (i.e., right-most subscript equals 0) are related to the input symbols associated with subcarrier 0 in exactly the same way as the input and output symbols are in Figure 11.8 (b). In fact, this observation applies to any subcarrier, which demonstrates that STBC No. 1 is, in fact, an Alamouti code. It is clear from this discussion that 802.11n uses Alamouti coding by employing space-frequency block coding, as discussed in Section 9.6.

A careful examination of the definitions of the other STBCs in Figure 11.7 shows that they all involve Alamouti coding. Simple graphical depictions of the four STBCs in 802.11n are shown in Figure 11.10.

11.1.5 OFDM in 802.11n

As indicated in Figure 11.3, IEEE 802.11n uses OFDM modulation to mitigate the effects of frequency-selective fading, which is evident from the presence of the inverse

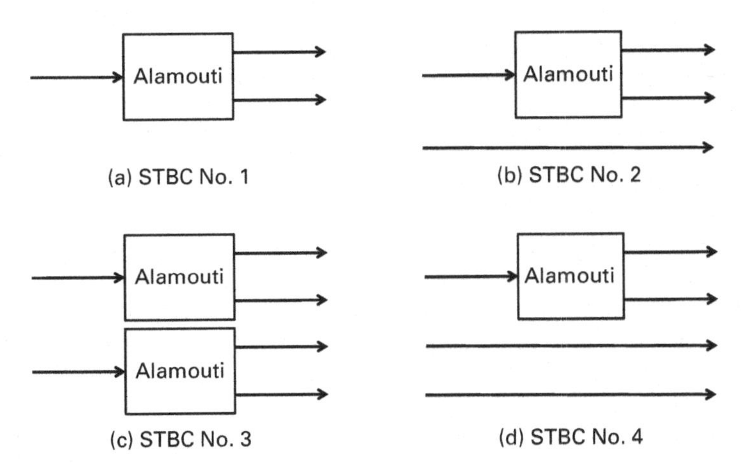

Figure 11.10 The four STBCs supported by 802.11n.

discrete Fourier transform (IDFT) blocks on each of the transmit chains. In this subsection, we describe the key 802.11n OFDM parameters and their values, and briefly describe how packets are generated.

Figure 11.2 shows that although there are three types of packets used in 802.11n, all three have a similar structure, consisting of some form of preamble that contains combinations of L-STF, L-LTF, HT-STF, and/or HT-LTF fields, followed by a SIG field, and finally a DATA field that contains the information that is being communicated. We can express the transmitted baseband signal associated with an 802.11n packet mathematically as follows:

$$s_{\mathrm{PACKET}}(t) = s_{\mathrm{PREAMBLE}}(t) + s_{\mathrm{SIG}}(t - T_{\mathrm{P}}) + s_{\mathrm{DATA}}(t - T_{\mathrm{P+S}}), \tag{11.2}$$

where T_{P} denotes the duration of the preamble and $T_{\mathrm{P+S}}$ denotes the duration of the preamble plus the SIG portion of the packet.

Each of the three terms on the right-side of Eq. 11.2 are constructed by performing an inverse Fourier transform on a set of coefficients $\{C_k\}$ that represent *training*, *pilot*, or *data* symbols, depending on the portion of the signal under consideration (i.e., s_{PREAMBLE}, s_{SIG}, or s_{DATA}). Using the nomenclature in the 802.11n Standard, each of the terms is computed using an inverse discrete Fourier transform (IDFT) as follows:

$$s(t) = w(t) \sum_{k=-N_{ST}/2}^{N_{ST}/2} C_k \exp(\mathrm{j}2\pi k\Delta_f)(t - T), \tag{11.3}$$

where $w(t)$ denotes a windowing function that is applied to each OFDM symbol to reduce spectral splatter, Δ_f denotes the OFDM subcarrier spacing, N_{ST} denotes the number of data symbols that are encoded in each OFDM symbol, and T denotes the delay relative to the start of the packet, which depends on the part of the packet under consideration.

Table 11.3 IEEE 802.11n OFDM parameter values [77].

Parameter	Non-HT 20 MHz	HT 20 MHz	HT 40 MHz
FFT length	64	64	128
N_{SD}: No. data carriers	48	52	108
N_{SP}: No. pilot carriers	4	4	6
N_{ST}: Total no. carriers	52	56	114
Δ_f: Carrier spacing	312.5 kHz	312.5 kHz	312.5 kHz
T_{DFT}: IDFT/DFT period	3.2 μs	3.2 μs	3.2 μs
T_{GI}: Guard interval	0.8 $\mu s = T_{DFT}/4$	0.8 μs	0.8 μs
T_{2GI}: Double GI	1.6 μs	1.6 μs	1.6 μs
T_{L-STF}: L-STF duration	8 $\mu s = T_{DFT}/4$	8 μs	8 μs
T_{L-LTF}: L-LTF duration	8 $\mu s = 2 \times T_{DFT} + T_{GI2}$	8 μs	8 μs
T_{SYM}: OFDM symbol duration	4 $\mu s = T_{DFT} + T_{GI}$	4 μs	4 μs
T_{L-SIG}: L-SIG field duration	4 $\mu s = T_{SYM}$	4 μs	4 μs
T_{HT-SIG}: HT-SIG field duration	N/A	8 $\mu s = 2T_{SYM}$	8 μs(1)
T_{HT-STF}: HT-STF field duration	N/A	4 μs	4 μs(1)
T_{HT-LTF}: HT-LTF field duration	N/A	4 or 8 μs(2)	4 or 8 μs(1)

(1) Not applicable in non-HT formats.
(2) 4 μs in HT mixed format, 8 μs in HT Greenfield format.

Table 11.4 Delay spread and coherence bandwidths for key 802.11n channel models.

Channel model	Description	σ_m (ns)	$\max\{T_m\}(ns)$	$B_{c,50}$ (MHz)
B	Residential (e.g., intra-room, room-to-room	15	80	13.3
D	Typical office	50	390	4.0
E	Large office (e.g., campus small hotspot)	100	730	2.0

It should be kept in mind that the IDFT defined in Eq. 11.3 is implemented using an IFFT algorithm with a length larger than N_{ST}, which is simply to say that the IFFT adds zeros (i.e., zero pads) to each group of N_{ST} symbols before processing. Two IFFT lengths are used in 802.11n: 64, when the bandwidth is 20 MHz; and 128 when operating in the 40 MHz mode. In contrast, $N_{ST} = 52$ or 56 at 20 MHz and is equal to 114 at 40 MHz. Table 11.3 lists the key OFDM parameter values used in 802.11n.

The key physical parameter that drives several of the OFDM values in Table 11.3 is the coherence bandwidth of the channel. A set of channel models were used during the development of 802.11n that specify the delay spreads, Doppler bandwidths, and multipath delay profiles for several different environments [60]. Table 11.4 lists the RMS and maximum delay spreads and the corresponding coherence bandwidths for the three most commonly used channel modes, which are designated as models B, D, and E.

To ensure that each OFDM subcarrier experiences flat fading, which is the objective of OFDM, it is necessary that $\Delta_f \leq \min\{B_c\} = 2$ MHz. Since $\Delta_f = 312.5$ kHz in 802.11n, this requirement is clearly met. The length of the IDFT/DFT period, denoted by T_{DFT} in Table 11.3, is, therefore, given by $T_{DFT} = 1/\Delta_f = 1/312.5$ kHz $= 3.2$ µs. This, of course, is the time duration of the data portion of the OFDM symbol (i.e., the portion of the symbol that encodes the data to be transmitted), which does not include the guard interval (i.e., the cyclic prefix).

In 802.11n, the guard interval is equal to either 0.4 or 0.8 µs. Based on Table 11.4, however, the guard interval should be at least equal to the maximum delay spread (i.e., 0.73 µs) to eliminate inter-OFDM-symbol interference. The 0.8 µs option achieves this objective, but the 0.4 µs value falls short when operating in the type of large spatial environment represented by Model E. The advantage of using smaller guard intervals is that their use results in less overhead and yields larger data rates. The designers of 802.11n were willing to compromise to some degree on the guard interval in order to achieve larger data rates. As we saw earlier in this chapter, the maximum supported data rate of 600 Mbps requires use of the smaller guard interval of 0.4 µs. It follows that the augmented OFDM symbol length in 802.11n, T_{AOFDM}, is, therefore, equal to $T_{DFT} + T_{GI} = 3.2 + (0.4 \text{ or } 0.8)$ µs $= 3.6$ or 4.0 µs.

11.1.6 Channel estimation

Channel estimation is performed in 802.11n using the LTF fields in the beginning of each packet. Although legacy systems also need to estimate the channel (using L-LTF fields), for the purpose of this discussion we focus on HT users and use of HT-LTFs to estimate the MIMO channel matrix when operating in the mixed and Greenfield modes. The LTF fields are generated by creating OFDM symbols from blocks of known symbols. The LTF symbol sequences are listed below for the 20 and 40 MHz modes. The HT-LTF OFDM symbols are generated by assigning each of the elements in either $C_{HT-LTF,20}$ or $C_{HT-LTF,40}$ to the C_k values in Eq.11.3, zero padding up to either 64 or 128 symbols, and then appending a cyclic prefix to the resulting IDFT signal. In practice, of course, the IDFT is implemented using an IFFT algorithm, in which case the time dimension would be discrete.

HT long training field (HT-LTF) in 20 MHz channel

$$\begin{aligned} C_{HT-LTF,20} = \{&1,1,1,1,-1,-1,1,1,-1,1,-1,1,1,1,1,1,1,-1,-1,1,1,-1,1,\\ &-1,1,1,1,1,0,1,-1,-1,1,1,-1,1,-1,1,-1,-1,-1,\\ &-1,-1,1,1,-1,-1,1,-1,1,-1,1,1,1,1,-1,-1\}. \end{aligned} \tag{11.4}$$

HT long training field (HT-LTF) in 40 MHz channel

$$\begin{aligned} C_{HT-LTF,40} = \{&1,1,-1,-1,1,1,-1,1,-1,1,1,1,1,1,1,-1,-1,1,1,-1,1,\\ &-1,1,1,1,1,1,1,-1,-1,1,1,-1,1,-1,1,-1,-1,-1,\\ &-1,-1,1,1,-1,-1,1,-1,1,-1,1,1,1,1,-1,-1,-1,1,0,\\ &0,0,-1,1,1,-1,1,1,-1,-1,1,1,-1,1,-1,1,1,1,1, \end{aligned}$$

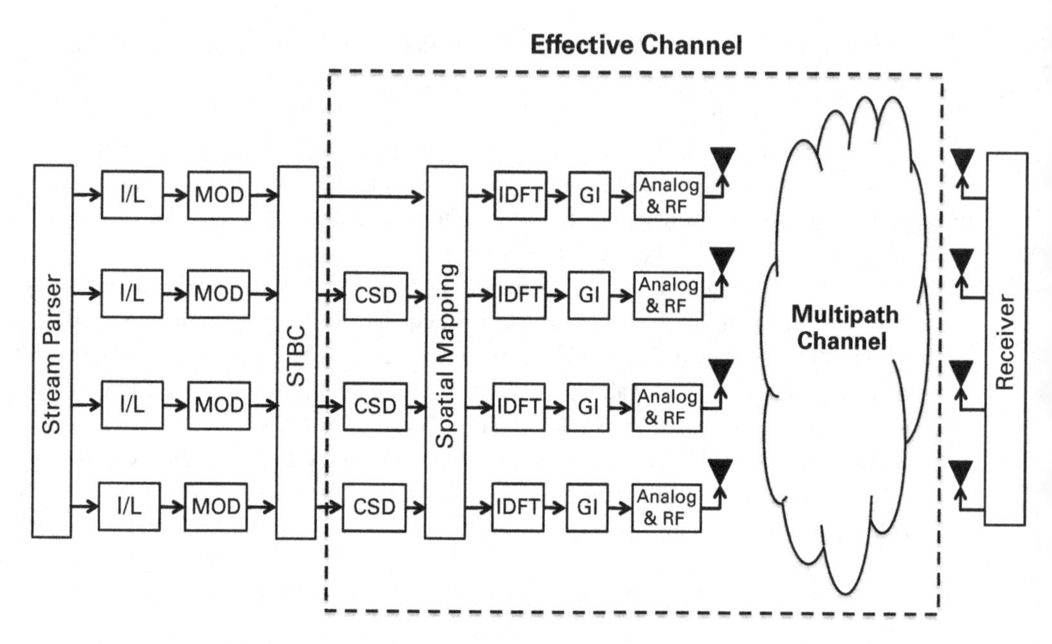

Figure 11.11 The effective channel computed during channel estimation in IEEE 802.11n.

$$1, 1, -1, -1, 1, 1, -1, 1, -1, 1, 1, 1, 1, 1, 1, -1, -1, 1, 1,$$
$$-1, 1, -1, 1, -1, -1, -1, -1, -1, 1, 1, -1, -1, 1, -1, 1, -1,$$
$$1, 1, 1, 1\}. \tag{11.5}$$

It is important to keep in mind that the term "channel matrix" in 802.11n does not refer to the naturally occurring RF multipath channel between the transmit and receive antennas as we have normally defined it throughout the book. Instead, the term *channel* in the 802.11n sense refers to the *effective channel* defined as the path between the set of modulator outputs and the outputs of the DFTs at the receiver. Figure 11.11 illustrates this point. The goal of channel estimation in 802.11n is to estimate the effective channel matrix, which is equivalent to saying that channel estimation will be performed in the frequency domain.

It should be noted that Figure 11.11 omits the STBC block that appears in Figure 11.3. The reason for this is that Figure 11.11 depicts the diagram when the HT-LTFs are being transmitted during channel estimation, whereas Figure 11.3 shows the block diagram during the data transmission portion of a packet. STBC processing is never performed when the HT-LTFs are transmitted, but it may be when data are transmitted. During channel estimation, $N_{ss} = N_{STS}$.

Our goal in this section is to explain how the effective channel matrix can be estimated using the HT-LTF fields in the preamble. To do this, we start by writing down the time-domain representation of the transmitted signal during the HT-LTF portion of the packet directly as it appears in the standard. The standard denotes the transmitted waveform during HT-LTF symbol n on transmit chain i_{TX} by $r^{n,i_{TX}}_{\text{HTLTF}}(t)$, which is given by the following expression:

$$
\begin{aligned}
r_{\mathrm{HTLTF}}^{n,i_{TX}}(t) = {} & \frac{1}{\sqrt{N_{STS} \cdot N_{\mathrm{HTLTF}}^{\mathrm{Tone}}}} \quad w_{\mathrm{HTLTF}}(t) \\
& \times \cdot \sum_{k=-N_{SR}}^{N_{SR}} \sum_{i_{STS}=1}^{N_{STS}} \left[Q_k\right]_{i_{TX},i_{STS}} [P_{\mathrm{HTLTF}}]_{i_{STS},n}\, \mathrm{HTLTF}_k \\
& \times e^{j2\pi k \Delta_F (t - T_{GI} - T_{CS}^{i_{STS}})},
\end{aligned}
\tag{11.6}
$$

where $w_{\mathrm{HTLTF}}(t)$ is a window function, $N_{\mathrm{HTLTF}}^{\mathrm{Tone}}$ is the number of non-zero elements in the LTF sequence (i.e., in either $C_{HT-LTF,20}$ or $C_{HT-LTF,40}$), $T_{CS}^{i_{STS}}$ denotes the cyclic shift values that are applied to the lower three space time streams by the CSD blocks shown in Figure 11.3, Q_k is the spatial mapping matrix, HTLTF_k is the kth element of the $C_{HT-LTF,20}$ or $C_{HT-LTF,40}$ vectors given in Eq. 11.4 or Eq. 11.5, and P_{HTLTF} is the HT-LTF mapping matrix derived from the matrix given by

$$
P = \begin{pmatrix} 1 & -1 & 1 & 1 \\ 1 & 1 & -1 & 1 \\ 1 & 1 & 1 & -1 \\ -1 & 1 & 1 & 1 \end{pmatrix}. \tag{11.7}
$$

Recall that in the mixed and Greenfield modes, there are multiple HT-LTF fields and that the exact number (N_{DLTF}) equals N_{STS} except when $N_{STS} = 3$, in which case $N_{LTF} = 4$.

The P_{HTLTF} array is an orthogonal matrix where the column dimension corresponds to the number of LTFs and the row dimension corresponds to the number of STSs. For a given MIMO implementation, P_{HTLTF} is obtained by taking the upper N_{STS} rows and left-most N_{LTF} columns of the matrix defined in Eq. 11.7.

Next, we express the transmitted signal in Eq. 11.6 in its frequency domain equivalent form, which we denote by $S_{\mathrm{HTLTF}}^{n,i_{TX}}(k)$, where k is the frequency index. To do so, recall that for a general continuous waveform $x(t)$, $0 \le t \le T$, with a discrete frequency Fourier transform $X(k)$, $0 \le k \le N-1$, and it follows that $x(t) = \sum_k X(k) \exp(\mathrm{j}2\pi k \Delta_F t)$, where Δt denotes the time between samples, and $\Delta_F = 1/T$ is the spacing between frequency samples. It follows from this that

$$
S_{\mathrm{HTLTF}}^{n,i_{TX}}(k) = \zeta \sum_{i_{STS}=1}^{N_{STS}} [Q_k]_{i_{TX},i_{STS}} [P_{\mathrm{HTLTF}}]_{i_{STS},n}\, \mathrm{HTLTF}_k e^{-\mathrm{j}2\pi k \Delta_F (T_{GI} + T_{CS}^{i_{STS}})}, \tag{11.8}
$$

where ζ is used for notational convenience to denote the coefficient in front of the double summation in Eq. 11.6.

It is mathematically convenient to express Eq. 11.8 in matrix notation. The resulting matrix version of this equation has the following form:

$$
S_{\mathrm{HTLTF}}(k) = \zeta \boldsymbol{Q}_k \boldsymbol{\Lambda} \boldsymbol{P}_{\mathrm{HTLTF}} \mathrm{HTLTF}_k, \tag{11.9}
$$

where

$$S_{\text{HTLTF}}(k) \triangleq \begin{pmatrix} S^{1,1}_{\text{HTLTF}}(k) & S^{2,1}_{\text{HTLTF}}(k) & \cdots & S^{N_{DLTF},1}_{\text{HTLTF}}(k) \\ S^{1,2}_{\text{HTLTF}}(k) & S^{2,2}_{\text{HTLTF}}(k) & \cdots & S^{N_{DLTF},2}_{\text{HTLTF}}(k) \\ \vdots & \vdots & \vdots & \vdots \\ S^{1,N_T}_{\text{HTLTF}}(k) & S^{2,N_T}_{\text{HTLTF}}(k) & \cdots & S^{N_{DLTF},N_T}_{\text{HTLTF}}(k) \end{pmatrix} \tag{11.10}$$

and

$$\mathbf{\Lambda} \triangleq \begin{pmatrix} e^{-j2\pi k\Delta_F(T_{GI}+T^1_{CS})} & 0 & \cdots & 0 \\ 0 & e^{-j2\pi k\Delta_F(T_{GI}+T^2_{CS})} & \cdots & 0 \\ \vdots & \vdots & \ddots & \vdots \\ 0 & 0 & \cdots & e^{-j2\pi k\Delta_F(T_{GI}+T^{N_{STS}}_{CS})} \end{pmatrix}. \tag{11.11}$$

In Chapter 9 we showed that when OFDM is used on a frequency-selective channel, the frequency domain of the signal is converted into a series of flat fading channels. Denoting the frequency response of the channel between the jth transmitter and the ith receiver by $\mathcal{H}_{ij}(k)$ as before, it follows that the kth frequency component of the received signal at receiver i, $r^n_i(k) = \sum_j \mathcal{H}_{ij}(k)S^{n,j}_{\text{HTLTF}}(k)+Z^n_i(k)$, where $Z^n_i(k)$ denotes the thermal noise component at receiver i at frequency index k associated with HT-LTF symbol n. We can, therefore, express the received frequency domain signal in matrix notation as follows:

$$R(k) = \boldsymbol{\mathcal{H}}(k)\boldsymbol{S}_{\text{HTLTF}}(k) + \boldsymbol{Z}(k), \tag{11.12}$$

where $[R(k)]_{in} = r^n_i(k)$ and $[\boldsymbol{\mathcal{H}}(k)]_{ij} \triangleq \mathcal{H}_{ij}(k)$. Plugging Eq. 11.9 into (11.12) yields

$$R(k) = \zeta\boldsymbol{\mathcal{H}}(k)\boldsymbol{Q}_k\mathbf{\Lambda}P_{\text{HTLTF}}\text{HTLTF}_k + Z(k). \tag{11.13}$$

Since the effective channel, $\boldsymbol{H}_{\text{eff}}(k)$, shown in Figure 11.11 includes the combined effects of the multipath channel, the spatial mapping block, and the cyclic shift delays, we can express it mathematically as

$$H_{\text{eff}}(k) = \zeta\boldsymbol{\mathcal{H}}(k)\boldsymbol{Q}_k\mathbf{\Lambda}. \tag{11.14}$$

Therefore, we can rewrite Eq. 11.13 as

$$\boldsymbol{R}(k) = \boldsymbol{H}_{\text{eff}}(k)\boldsymbol{P}_{\text{HTLTF}}\text{HTLTF}_k + \boldsymbol{Z}(k). \tag{11.15}$$

We can now use the expression we derived for the LS/ML channel estimate in Chapter 10 to write down a corresponding expression for the estimate of $\mathbf{H}_{\text{eff}}(k)$. The LS/ML estimate is obtained by replacing the terms in Eq. 10.8 as follows: $\sqrt{\rho} \rightarrow \text{HTLTF}_k$; and $\mathbf{S}_p \rightarrow \mathbf{P}_{\text{HTLFT}}$. It follows that

$$\textbf{LS/ML:} \quad \hat{\mathbf{H}}_{\text{eff}}(k) = \frac{1}{N_{DLTF}\text{HTLTF}_k}\mathbf{R}(k)\mathbf{P}^T_{\text{HTLTF}}, \tag{11.16}$$

where we have made use of the fact that $\mathbf{P}$ has orthogonal rows of length N_{DLTF} to simplify the resulting expression.

Similarly, we can write down the LMMSE estimate for $\mathbf{H}_{\text{eff}}(k)$ by making the same substitutions in Eq. 10.19. The details are left as an exercise for the reader as part of the problem set at the end of the chapter.

We conclude this section on channel estimation with some comments on the role of the matrix $\mathbf{P}_{\text{HTLTF}}$. It should be noted that the same methodology described above could be used to estimate the channel if we eliminated $\mathbf{P}_{\text{HTLTF}}$ from our equations. In that case, $\mathbf{R}(k)$ in Eq. 11.13 would be different, but Eq. 11.16 could still be used by setting $\mathbf{P}_{\text{HTLTF}} = \mathbf{P}^T_{\text{HTLTF}} = \mathbf{I}$. From this observation we might be tempted to conclude that it is not necessary to use $\mathbf{P}_{\text{HTLTF}}$ to estimate the channel, or at the very least, it might make us wonder, what is its purpose?

To understand the purpose of $\mathbf{P}_{\text{HTLTF}}$, consider the product $\mathbf{P}_{\text{HTLTF}}\text{HTLTF}_k$ in the first term of Eq. 11.13. This product has dimension $[N_{STS} \times N_{LTF}]$, and it defines the N_{STS} input signals to the effective channel (see Figure 11.3) at the kth frequency index in N_{LTF} consecutive HT-LTF OFDM symbols. The important thing to notice is that eliminating matrix $\mathbf{P}_{\text{HTLTF}}$ from our analysis is equivalent to setting $\mathbf{P}_{\text{HTLTF}} = \mathbf{I}$, which means that $\mathbf{P}_{\text{HTLTF}}\text{HTLTF}_k$ would be equal to $\text{HTLTF}_k\mathbf{I}$, implying that at any instant of time, only one of the inputs to the effective channel would be non-zero. One of the key goals in the design of 802.11n was to ensure that the total power is the same during all parts of the packet so that the AGC gain can be computed once during the preamble and a single amplifying gain can be used for the entire packet. If $\mathbf{P}_{\text{HTLTF}} = \mathbf{I}$, the power would be less during the LTF portion of the packet than during the data portion, violating this goal. Employing an orthogonal matrix for $\mathbf{P}_{\text{HTLTF}}$ with equal element magnitudes ensures that all inputs to the effective channel have equal power.

11.1.7 Modulation and coding schemes in 802.11n

IEEE 802.11n defines a large number of modulation, coding and data rate combinations for use in its HT mode. There are a total of 77 different modulation and coding schemes (MCSs), and of those, 33 define modes in which the same modulation and coding are used on all spatial streams. The remaining 44 MCSs, called *unequal modes*, define different modulation and coding combinations on each stream.

Each of these MCSs has four data rates associated with it, depending on whether the OFDM guard interval is short (400 ns) or long (800 ns) and whether the bandwidth is 20 or 40 MHz. The lowest data rate, which is associated with MCS 32 and $T_{GI} = 800$ ns, is equal to 6 Mbps, and the highest data rate, 600 Mbps, is defined by MCS 31 operating with a short guard interval.

Although there are a total of 77 MCSs defined in the standard, not all need to be supported in order for a device to be 802.11n compliant. Only MCSs 0–7 are mandatory for non-AP devices and MCSs 0–15 are mandatory for AP devices. Appendix F lists the MCS parameters for the non-unequal modes for 802.11n. The interested reader can find all the MCS listed in [77].

11.2 LTE

11.2.1 Overview and history

LTE, which stands for *long-term evolution*, is the name of a technology standard for cellular communications that is the next step beyond 3G technology. The LTE standard was developed by the 3rd Generation Partnership Project (3GPP) [52], which is a collaboration of telecommunications associations that define standards and specifications for wireless communications. Each major new update to the 3GPP standard is called a release, and 3GPP Release 8, which was finalized in December 2009, defines the original version of LTE. The first large-scale deployments of LTE systems began in late 2010. As of this writing, the most recent finalized version of the LTE standard is Release 9.

LTE was marketed by cellular service providers as the first 4G wireless technology; however, that is not technically correct. The International Telecommunications Union-Radio communications sector (ITU-R) defines 4G as a wireless system that provides peak data rates of 100 Mbps for high mobility communications (such as from trains and cars) and 1 Gbps for low mobility communication (such as pedestrians and stationary users) [62]. As we will see, however, the maximum downlink (from base station to mobile user) data rate that is supported by LTE is 300 Mbps. Nevertheless, LTE is often advertised as a 4G service. In recognition of the fact that LTE does not meet true 4G requirements, LTE is sometimes referred to as a 3.9G technology.

3GPP Release 10, which was frozen in April 2011, defines an upgrade to LTE called LTE-advanced, which will achieve true ITU-R 4G performance. The first deployments of LTE-advanced technology are expected to occur in 2013, and large-scale deployments are anticipated to occur over the period from 2013 to 2015. In this chapter, we focus on LTE, however, because LTE and LTE-advanced use similar signaling and MIMO schemes, and so the concepts we discuss are generally applicable to LTE-advanced as well.

LTE (and LTE-advanced) have grown out of the 2G and 3G cellular technologies called GSM and UMTS, which were based on TDMA and CDMA, respectively. The physical layer details of LTE are much different than these predecessors, but some of the networking concepts and terminology of the earlier GSM and UMTS technology exists in LTE. Our focus in this chapter is on the physical layer details of LTE in general, and on MIMO aspects in particular. As we will learn, LTE MIMO supports transmit diversity, receive diversity, spatial multiplexing such as BLAST, eigenbeamforming, and MU-MIMO. In the rest of the book, we focus on SU-MIMO.

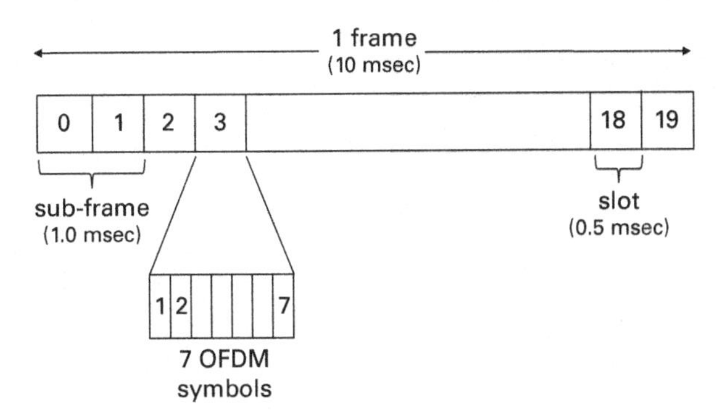

Figure 11.12 LTE frame structure.

11.2.2 LTE waveform structure

The LTE/LTE-advanced systems are based on the use of OFDMA on the downlink (DL) and on single-carrier FDMA (SC-FDMA) on the uplink (UL). In LTE (as well as in UMTS and LTE-advanced), the base station is called the eNB and the mobile user is called the UE. Although LTE is not packet-based like 802.11n, it uses a blocked timing structure that divides time into *frames*, which are 10 ms in length, *sub-frames*, which are 1.0 ms long, *slots*, which are 0.5 ms in duration, and OFDM symbols. There are either seven or six OFDM symbols per slot, depending on if the cyclic prefix is short (seven) or long (six). Figure 11.12 illustrates the LTE frame structure, where for illustration purposes, short cyclic prefixes are assumed.

Since LTE uses OFDMA on the downlink, the downlink signal consists of a series of OFDM symbols. Figure 11.13 shows the downlink time-frequency structure of the transmitted waveform, which is called the *downlink resource grid* in the LTE standard. The vertical dimension represents frequency and the horizontal dimension corresponds to time. Each column, therefore, corresponds to an OFDM symbol. The smallest component of the DL resource structure is called a resource element (RE), which is a single time-frequency cell having dimensions of 1 augmented OFDM symbol duration (T_{AOFDM}) in the time dimension and $1/T_{\mathrm{OFDM}}$ Hz in the frequency dimension.[1] Users request and are assigned time-frequency resources in units called *resource blocks* (RBs), which consist of 12 subcarriers-by-7 OFDM symbols, as illustrated in the figure. Users are allocated RBs on a dynamic basis based on their need.

Although the uplink uses SC-FDMA, which is different than OFDMA, the UL still has a transmission time-frequency structure that is similar to the one in Figure 11.13. The dimensions of the resource elements on the UL are the same as those on the DL, and the definition of a resource block is the same in both directions. On the uplink, UEs are allocated RBs sufficient in number to support each UE's UL data rate requirement. On the UL, the RBs assigned to each UE are contiguous; on the DL, the RBs allocated to each UE do not have to be contiguous.

[1] Recall from Chapter 9 that the term *augmented OFDM symbol* refers to the OFDM IDFT plus the cyclic prefix; thus, $T_{\mathrm{AOFDM}} = T_{\mathrm{OFDM}} + T_{CP}$; where, T_{CP} is the duration of the cyclic prefix.

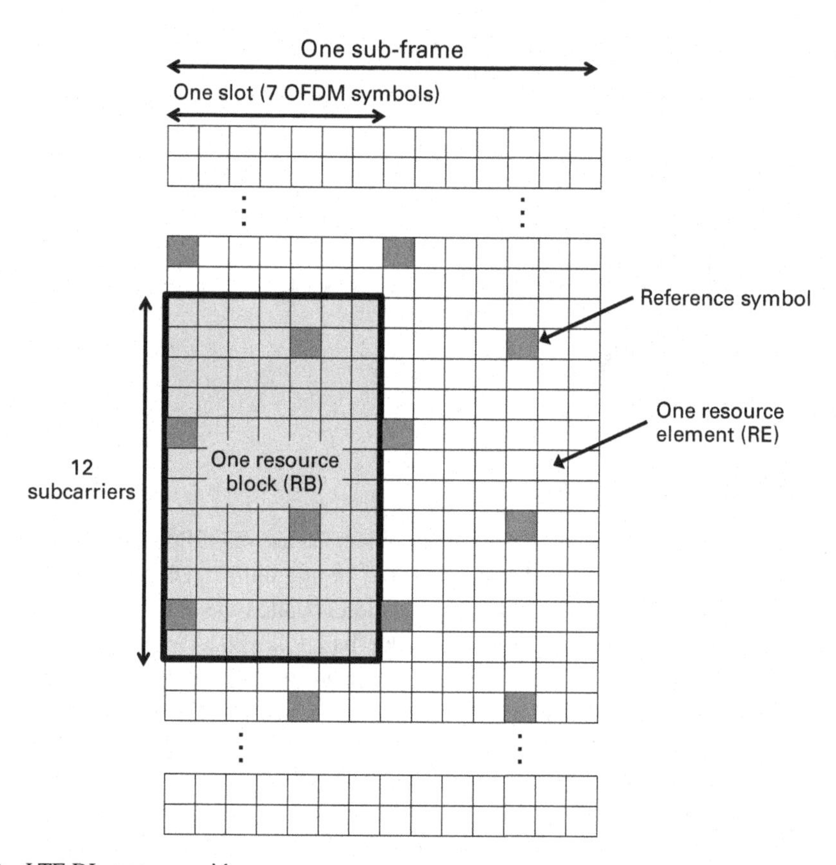

Figure 11.13 LTE DL resource grid.

Data that are to be transmitted over the LTE physical layer, arrive in blocks called *transmission blocks* (TB) from the data link medium access control (MAC) layer. The symbols in each TB are mapped into RBs in the transmission time-frequency space. As we see in the next section, up to two TBs can be transmitted simultaneously on the DL, whereas only one TB can be transmitted at a time from each UE on the UL.

The OFDMA structure used in LTE is based on the requirement that the subcarrier spacing be equal to 15 kHz. In addition, three different cyclic prefix values are used to combat multipath delay spread. For normal cellular data, the cyclic prefix is 4.69 μs. The first OFDM symbol in each sub-frame, however, is used for control, so to provide additional protection against ISI, the first OFDM symbol in each sub-frame has a slightly longer cyclic prefix equal to 5.2 μs. Both of these values refer to the short prefix. LTE also supports broadcast services, which have a cyclic prefix equal to 16.67 μs. The longer prefix on broadcast services is required because of the longer distances that can be involved in broadcasting, resulting in potentially larger delay spreads. Since the subcarrier spacing is 15 kHz, it follows that $T_{\text{OFDM}} = 1/15 \times 10^3 = 66.666\ \mu\text{s}$. The augmented LTE OFDM symbol duration is, therefore, equal to either $66.666 + 5.2 = 71.8666\ \mu\text{s}$ for the control OFDM symbol, $66.666 + 4.69 = 71.35666\ \mu\text{s}$ for the non-control symbols, or $66.666 + 16.67 = 83.33666\ \mu\text{s}$ for symbols that use the long prefix.

Table 11.5 LTE DL OFDM parameters.

Bandwidth (MHz)	1.4	3	5	10	15	20
FFT size	128	256	512	1024	1536	2048
Occupied subcarriers	76	151	301	601	901	1201
Guard subcarriers	52	105	211	423	635	847
Number of resource blocks	6	12	25	50	75	100
Occupied bandwidth (MHz)	1.140	2.265	4.515	9.015	13.515	18.015
OFDM symbols/sub-frame	7/6 (short/long CP)					
CP length (short CP) (μs)	5.2 (first symbol) / 4.69 (six following symbols)					
CP length (long CP) (μs)	16.67					

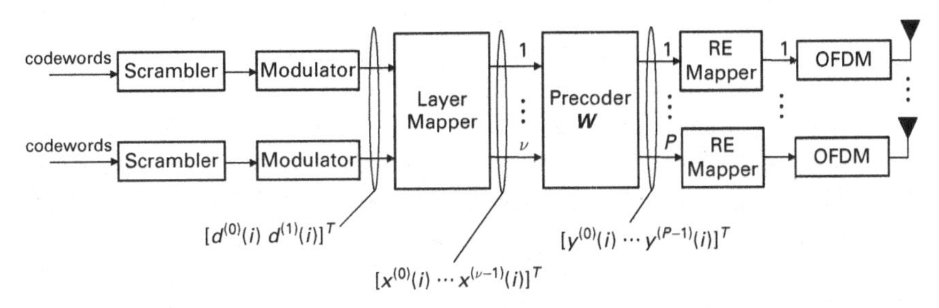

Figure 11.14 LTE DL transmitter.

Table 11.5 lists the key OFDM parameters used in LTE. This table shows that LTE is designed to operate using six different channel bandwidths and that many of the OFDM parameters vary with the bandwidth. It should be noted that the number of occupied subcarriers is significantly less than the FFT length for each bandwidth. Only occupied subcarriers carry data; the remainder are zeros that are used to provide spectral guard bands.

11.2.3 LTE transmitter block diagrams

The transmit block diagram used at the eNB for DL transmission differs from the transmit block diagram used by the UE for UL transmission. We consider each separately below.

11.2.3.1 DL block diagram

Figure 11.14 shows a block diagram of the transmitter for the downlink path. The transmitter operates on one or two TBs at a time, which enter at the left of the diagram as input to the scrambler blocks (denoted by the arrows labeled “codewords”). The data in each TB are assumed to have undergone some form of forward error correction before entering the scrambler, so we refer to each data block that enters the scrambler as a codeword. The LTE scrambler block prevents long strings of 0s or 1s, which, in turn: a) makes the transmitted bit stream random, a characteristic normally required by the

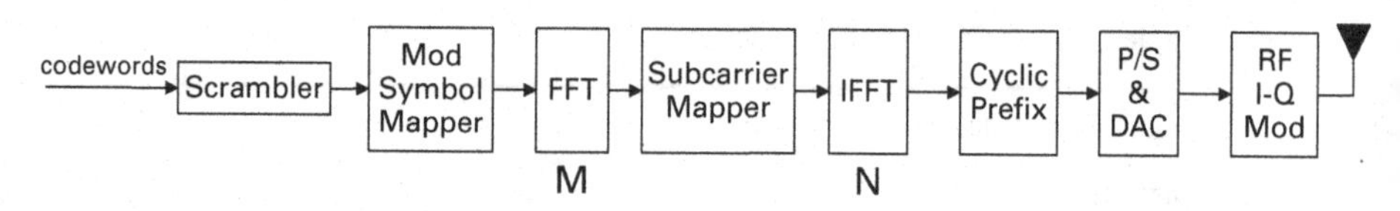

Figure 11.15 LTE UL transmitter.

decoder at the receiver; and b) prevents spectral spikes (i.e., whitens the transmitted spectrum) that would otherwise occur due to periodicity in the transmitted waveform, resulting in potential violation of spectral mask requirements.

Following the scrambler, the data bits are modulated using either BPSK or QPSK for signaling, or using QPSK, 16QAM or 64QAM for data. The modulation symbols are then fed to the *layer mapper*, which accepts either one or two input streams and maps those streams to ν spatial streams, also called *layers*. In LTE, $\nu \leq 4$. Following the layer mapper, the spatial streams are processed by the *precoder* block, which maps the layers to P antenna ports, where $P \leq 4$. The precoding block is where space-time block coding and eigenbeamforming are performed. LTE only defines two types of STBCs for achieving spatial diversity. In general, the precoder block can be represented by a matrix, $\mathbf{W}$, that premultiplies the input. In this sense, the LTE precoder block is analogous to the 802.11n spatial mapper block.

The P output streams from the precoder are then applied to RE mapper blocks, which map the symbols on each antenna port to blocks of subcarriers that are allocated to the UE being communicated with. The final step before the RF chain (which is not shown) involves performing OFDM processing, which consists of a serial-to-parallel conversion block, an IFFT, a cyclic extension, and then parallel-to-serial conversion. The symbol nomenclature shown below the block diagram is the same as that used in the LTE standard, which we use in subsequent discussions.

11.2.3.2 UL block diagram

Figure 11.15 is a block diagram of the UL transmitter used by an LTE UE. In LTE, the UL uses SC-FDMA instead of OFDMA. The main difference between SC-FDMA and OFDMA is that SC-FDMA has an FFT block in addition to the IFFT block. As illustrated, codewords arrive from the MAC layer, just as they do in the eNB transmitter, before being scrambled and modulated. After modulation, the symbols are processed by an M-point FFT block, which generates M frequency domain samples. Those M samples are then applied to an N-point IFFT, where $N \geq M$. Since the length of the FFT is generally less than the IFFT length, it is necessary to decide which of the N IFFT inputs should receive the M FFT outputs. That is the role of the *subcarrier mapper*.

In general, there are two ways to map the FFT subcarriers: *localized mapping* or *distributed mapping*. In localized mapping, the FFT output samples are mapped to a contiguous set of M IFFT input subcarriers; in distributed mapping, the FFT output samples are mapped to non-contiguous IFFT inputs. Figure 11.16 illustrates these two types of mappings by showing the frequencies to which the FFT outputs are mapped. The different shades correspond to different UEs. Each UE maps the output of its FFT

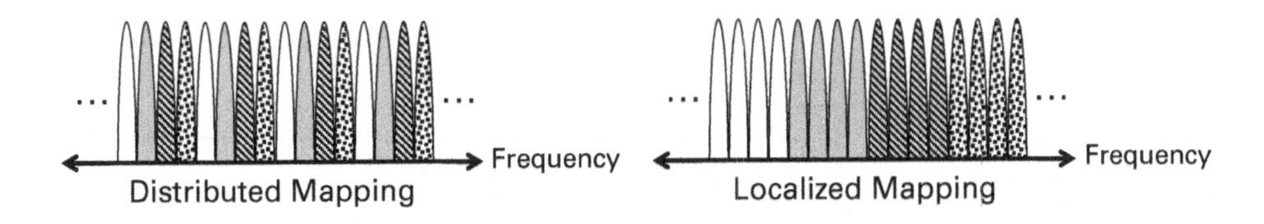

Figure 11.16 SC-FDMA carrier mapping options.

to its assigned subcarrier frequencies and fills in the rest of the IFFT inputs with zeros. In practice, localized mapping is primarily used in LTE.

SC-FDMA is used on the UL channel of LTE because it has lower PAPR than OFDMA. As we discussed earlier, when a signal has a large PAPR, it is necessary to operate at large power amplifier back off levels to avoid operating in the non-linear region of the PA, which results in low power efficiency. Since the UE typically is power constrained, it is important for its PA to operate with high power efficiency, which means that its transmitted signal should have a low PAPR. Berardinelli *et al.* [13] show that the PAPR of SC-FDMA is on average about 2 dB less than the PAPR of OFDMA. Readers desiring to learn more about the technical details and the historical development of SC-FDMA are referred to [17].

It is evident from Figure 11.15 that LTE only supports single-antenna transmissions by the UE on the uplink.[2] This means the only type of MIMO processing that is possible on the uplink is receive diversity combining at the base station. STBC, eigenbeamforming, and single-user spatial multiplexing are precluded.

11.2.4 DL transmit diversity

LTE achieves transmit diversity through the use of SFBC on the DL. Two types of SFBCs are defined in the LTE standard for use with either two or four transmit antennas. When operating in the transmit diversity mode, only one stream feeds the layer mapper in Figure 11.14 and the layer mapper generates either $\nu = 2$ or $\nu = 4$ output layers, depending on which of the two SFBCs is implemented. When transmit diversity is employed, the UE receiver can use one, two, three, or four receive antennas; however, the details of the decoding are not specified in the LTE standard. We describe each of the SFBCs separately below.

SFBC No. 1 The first code is based on a straightforward implementation of the Alamouti code for two transmit antennas, which assumes that $\nu = P = 2$ (i.e., $N_t = 2$). In this case, the layer mapper operates as a simple serial-to-parallel converter and the precoder performs Alamouti encoding on the two inputs from the mapper using the same

[2] It should be noted that although the LTE standard does not support multiple antennas at the UE when transmitting, there is nothing to preclude the UE from using any number of antennas when receiving.

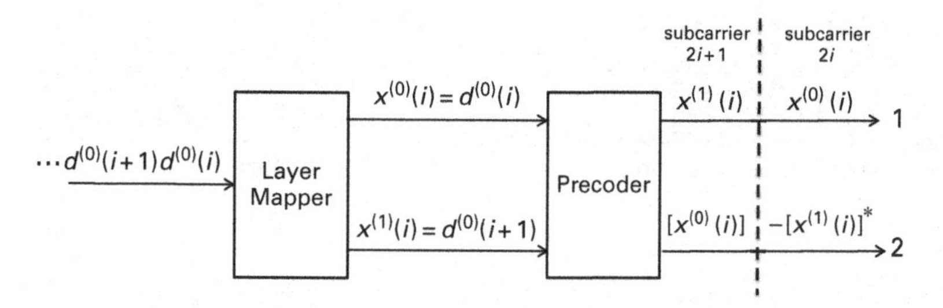

Figure 11.17 LTE DL transmit diversity mapping scheme for two antennas.

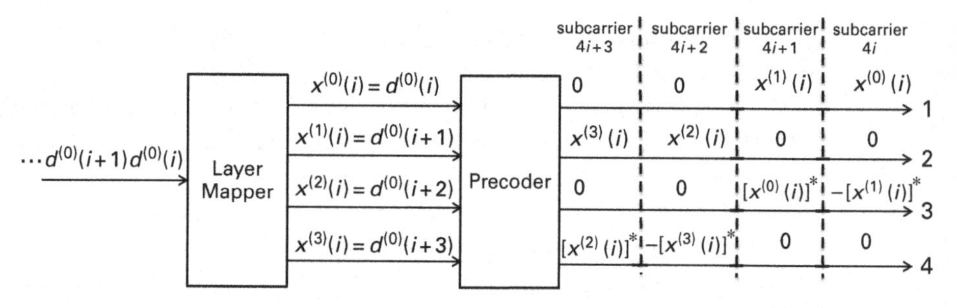

Figure 11.18 LTE DL transmit diversity mapping scheme for four antennas.

encoding depicted in Figure 11.8(b) for 802.11n. Figure 11.17 illustrates the operation of the layer mapper and precoder for SFBC 1.

The precoder block processes each pair of input symbols exactly the same way that the Alamouti encoder does in 802.11n. However, since the precoder output feeds OFDM blocks (see Figure 11.14), the output symbols from the precoder are interpreted as being in the frequency domain; thus, transmit diversity is implemented in LTE using Alamouti space-frequency block coding instead of space-time block coding. Mathematically, however, they are identical.

SFBC No. 2 The second code used in LTE for transmit diversity is designed to operate with four transmit antennas. The layer mapper and precoder operations for this second code are depicted in Figure 11.18. As illustrated, the layer mapper samples the serial input symbols by mapping them in round-robin fashion to each of the four output layers. The precoder then performs space-frequency block coding on each set of four input symbols, resulting in the 4-by-4 space-frequency block code shown at the precoder output. Examination of this part of the figure reveals that it consists of two separate space-frequency Alamouti codes. One of the Alamouti codes involves the input symbols $x^{(0)}(i)$ and $x^{(1)}(i)$, which are encoded on antenna ports 1 and 3 in subcarriers $4i$ and $4i+1$; the other code involves symbols $x^{(2)}(i)$ and $x^{(3)}(i)$, which are encoded on antenna ports 2 and 4 in subcarriers $4i+2$ and $4i+3$, where $i = 0, 1, \ldots$

In general, if the OFDM encoding involves N-point IFFTs, then for SFBC 2, the precoder needs to generate $N/4$ space-frequency blocks of encoded symbols to fill the IFFTs, since each block consists of four subcarrier symbols. We see from this that

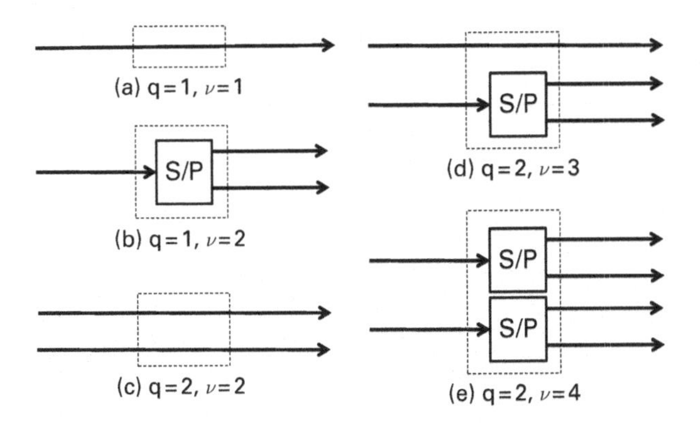

Figure 11.19 LTE layer mapper configurations for spatial multiplexing. The dashed boxes denote the layer mapper block. The figure shows the effective functionality in the layer mapper block for each configuration.

synchronization in time and frequency at the receiver is critical in order for the receiver to correctly decode each Alamouti space-frequency block. Assuming that such synchronization occurs, then, in principle, the receiver would be able to perform maximum likelihood decoding on each encoded block using Eq. 6.30.

11.2.5 Spatial multiplexing

LTE supports multiple kinds of spatial multiplexing using two, three, or four transmit antennas. The LTE standard defines five different layer mapper configurations for spatial multiplexing for various combinations of the numbers of input streams and output layers. Figure 11.19 graphically depicts the five different types of layer mappers defined in the standard. In this figure, q denotes the number of input streams to the layer mapper, which can be either one or two, and ν denotes the number of output layers, which can be one, two, three, or four. The blocks labeled "S/P" denote serial-to-parallel converters, which take the stream of input symbols and map them in an alternating fashion onto their two outputs. This figure shows that the layer mapper block in Figure 11.14 consists of a combination of either simple pass-through or serial-to-parallel conversion operations. It should be noted that configuration (d) is unique because it reads symbols from the bottom input stream twice as fast as it does from the top. This enables different code rates to be used on the two inputs.

The operation of the precoder depends on the type of spatial multiplexing that is being used. For example, to implement BLAST-type spatial multiplexing, all the precoder needs to do is perform a pass-through on each of its inputs (i.e., do nothing). In that case, $\mathbf{W} = \mathbf{I}$. The layer mapper would, in turn, perform one of the five operations depicted in Figure 11.19, depending on the number of transmit antennas (i.e., spatial streams).

LTE also supports eigenbeamforming. Recall that in eigenbeamforming the transmitter is assumed to have knowledge of the channel matrix, $\mathbf{H}$, and can therefore perform a singular value decomposition on $\mathbf{H}$ resulting in $\mathbf{H} = \mathbf{UDV}^H$. Since eigenbeamforming involves premultiplication of the data by the matrix $\mathbf{V}$, it follows that eigenbeamforming

Table 11.6 Precoding codebook for two transmit antennas (based on Table 6.3.4.2.3-1 of [50]).

Codebook index	Number of layers ν	
	1	2
0	$\frac{1}{\sqrt{2}}\begin{pmatrix}1\\1\end{pmatrix}$	$\frac{1}{\sqrt{2}}\begin{pmatrix}1 & 0\\0 & 1\end{pmatrix}$
1	$\frac{1}{\sqrt{2}}\begin{pmatrix}1\\-1\end{pmatrix}$	$\frac{1}{\sqrt{2}}\begin{pmatrix}1 & 1\\1 & -1\end{pmatrix}$
2	$\frac{1}{\sqrt{2}}\begin{pmatrix}1\\j\end{pmatrix}$	$\frac{1}{\sqrt{2}}\begin{pmatrix}1 & 1\\j & -j\end{pmatrix}$
3	$\frac{1}{\sqrt{2}}\begin{pmatrix}1\\-j\end{pmatrix}$	-

can be implemented by setting the precoder matrix $\mathbf{W} = \mathbf{V}$. Although this is the way eigenbeamforming is performed in theory, LTE implements eigenbeamforming differently. To reduce the amount of data that needs to be transmitted from the UE back to the eNB, LTE does not send the channel matrix back to the base station (i.e., the transmitter). Instead, the UE estimates $\mathbf{W}$ by choosing from a discrete list of predefined choices in a lookup table known as a *codebook*. Only the code index number is transmitted back to the eNB, which is used by the eNB to determine which precoding matrix to use.

LTE only supports eigenbeamforming with two or four transmit antennas. For $N_t = 2$, four precoding matrices exist for $\nu = 1$ and three exist for $\nu = 2$. Table 11.6 lists the allowable precoding matrices when there are two transmit antennas. In general, the precoding matrices are dimensioned $[P \times \nu]$. A similar table exists for $N_t = 4$. In the case of four antennas, however, there is a total of 16 precoding matrices for each value of $\nu = 1, 2, 3,$ and 4. The interested reader can find a listing of those matrices in Table 6.3.4.2.3-2 of [50].

11.2.6 LTE data rates

Since one of the key benefits of using MIMO is its ability to enhance data rates, we conclude our overview of LTE with a discussion of the peak data rates that are achievable with LTE, the role of MIMO in achieving those rates, and a description of how the peak data rates are computed. First, we present an approximation for the peak data rates, which provides an intuitive understanding of why the values are what they are. Following that, we explain how to compute the exact data rates using the LTE standard.

In general, the peak DL and UL data rates depend on three parameters: a) the channel bandwidth, which can be either 1.4, 3, 5, 10, 15, or 20 MHz; b) the number of antennas at the eNB and UE; and c) the *UE category*. The UE category is a number between 1 and

5 defined in [52] that specifies the performance capabilities of the UE. UE category 5 denotes the category of UE equipment with the most capabilities. For example, category 5 UEs are capable of having up to four receive antennas and employing 64QAM modulation, the highest order modulation supported by LTE. Categories 2–4 only support two receive antennas and can only support 16QAM on the UL. Category 1 UEs are only capable of using one receive antenna.

The maximum DL LTE data rate is often quoted to be 300 Mbps (e.g., see [49]). To understand how this value is obtained, we begin by assuming that the bandwidth, B, is 20 MHz and compute the number of RBs (i.e., chunks of 12 OFDM subcarriers $\times$ 7 OFDM symbol periods) that fall within that bandwidth over a period of one time slot (i.e., 7 OFDMA symbol periods). In general, the usable bandwidth (i.e., the "occupied bandwidth" in Table 11.5) is 90% of the channel bandwidth in order to provide spectral guard protection. It follows that for the 20 MHz band, the usable bandwidth $B_u = 0.9B = 18$ MHz. Since the OFDMA subcarrier spacing in LTE is fixed at 15 kHz, it follows that the number of resource blocks in a slot is $N_{RB,20} = 18 \times 10^6/(12 \times 15 \times 10^3) = 100$, which is equivalent to $100 \times 12 \times 7 = 8400$ REs. We denote this number by $N_{RE,20}$. Since each RE can hold a modulation symbol, then the total number of bits in a 20 MHz $\times$ 0.5 ms frequency-time block is $N_{b,20} = \log_2(M)N_{RE,20}$, where M denotes the modulation order. Not all of these bits, however, are available for data transmission. In general, LTE uses about 25% of the REs for overhead associated with either: a) reference symbols; b) control channel symbols; c) synchronization symbols; and d) forward error correction. Since the highest order modulation in LTE is 64QAM, $\log_2(M) = 6$. Combining all of this together, we conclude that the maximum data rate over one antenna is approximately $(1 - \eta_{OH}) \times N_{b,20}/0.5 \times 10^{-3} = (1 - \eta_{OH}) \times \log_2(M) \times N_{RE,20}/0.5 \times 10^{-3} = 0.75 \times 6 \times 8.4 \times 10^3/0.5 \times 10^{-3} = 75.6$ Mbps, where η_{OH} denotes the overhead fraction. If 4×4 MIMO is used, then we need to multiply this value by 4 to get a total maximum DL data rate of $75.6 \times 4 = 302.4$ Mbps.

It follows from the preceding discussion that a simple approximation for the peak DL data rate as a function of bandwidth and number of antennas can be expressed as

$$R_{\max} = 75.6N\left(\frac{B_{\text{MHz}}}{20}\right) \quad \text{Mbps,} \tag{11.17}$$

where N denotes the number of antennas at the transmitter and receiver (i.e., we assume $N \times N$ MIMO) and B_{MHz} is the channel bandwidth in units of MHz.[3]

The results above are approximations based on assuming $\eta_{OH} = 0.25$. Precise peak values can be obtained using Tables 7.1.7.1-2 and 7.1.7.2-1 in [51]. These tables specify the *transport block size* (TBS) as a function of bandwidth. The TBS is the number of data bits that are transmitted in each sub-frame (with a duration 10^{-3} seconds) taking

[3] It is important to keep in mind that the approximation in Eq. 11.17 assumes that the error correction code rate is nearly one, which results in the 25% overhead value that is implicitly assumed. Lower code rates would result in commensurately lower data rate values.

Table 11.7 Transport block sizes as a function of bandwidth for LTE DL.

B (MHz)	N_{RB}	TBS
1.4	7	5160
3	15	11 064
5	25	18 336
10	50	36 696
15	75	55 056
20	100	75 376

Table 11.8 Peak LTE DL data rates (in Mbps) for all combinations of bandwidths and antenna numbers. In this table, we assume $N \times N$ MIMO.

	N			
B (MHz)	1	2	3	4
1.4	5.16	10.32	15.48	20.64
3	11.064	22.128	33.192	44.256
5	18.336	36.672	55.008	73.344
10	36.696	73.392	110.088	146.784
15	55.056	110.112	165.168	220.224
20	75.376	150.752	226.128	301.504

Table 11.9 Peak LTE DL and UL data rates (in Mbps) for different UE categories. Peak values in this table are from Tables 4.1-1 and 4.1-2 in [52].

UE category	Peak DR		Max. DL mod	Max. UL mod	Max. layers for DL SM
	DL	UL			
1	10.296	5.160	64QAM	16QAM	1
2	51.024	25.456	64QAM	16QAM	2
3	102.048	51.024	64QAM	16QAM	2
4	150.752	51.024	64QAM	16QAM	2
5	301.504	75.376	64QAM	64QAM	4

into account the actual overhead. Therefore, the actual DL peak data rate is given by

$$R_{\max} = \frac{\text{TBS}(BW)}{10^{-3}} N_t \tag{11.18}$$

where the TBS values are listed in Table 11.7. Table 11.8 lists the peak DL data rates for all combinations of bandwidths and antenna numbers using Eq. 11.18. A comparison of the predictions using Eq. 11.17 with the values in Table 11.8 shows that the approximation is quite accurate.

It should be noted that the peak data rates in Table 11.8 assume a category 5 UE with maximum capability. Lower numbered UE categories have lower peak data rates. Table 11.9 lists the peak LTE data rates on both the UL and DL paths, assuming full MIMO capabilities, for all five UE categories. The right-most column specifies the maximum number of data layers that are supported for spatial multiplexing on the DL. The smaller UL rates are due primarily to the generally lower order modulation used by the UE on the UL and the fact that multiple streams are not supported on the uplink.

Problems

11.1 Compute the sampling rates in 802.11n for the 20 MHz and 40 MHz modes. How do the sampling rates compare to the bandwidths?

11.2 Assume that an 802.11n transmitter is configured to achieve transmit diversity using Alamouti coding. Write down the expression for the HT-LTF mapping matrix for this system.

11.3 (a) Prove Eq. 11.16. Do this by replacing $\sqrt{\rho}$ in Eq. 10.8 with HTLFT_k and $\mathbf{S}_p$ with $\mathbf{P}_{\text{HTLTF}}$ and using the fact that $\mathbf{P}$ has orthogonal rows. (b) Show that $\hat{\mathbf{H}}_{\text{eff}}(k) = H_{\text{eff}}(k) + \frac{\mathbf{N}(k)\mathbf{P}^T}{N_{DLTF}\text{HTLTF}_k}$.

11.4 Table 11.5 lists the number of occupied carriers for each LTE channel bandwidth. Show that the numbers of occupied subcarriers match the channel bandwidths exactly.

Appendices

Appendix A MIMO system equation normalization

An examination of the MIMO literature shows that different authors express the MIMO system equation in various forms, which can be confusing to students new to the MIMO field. The seeming inconsistencies among the different forms of the system equation are due to different assumptions various authors make about how to normalize the parameters in the system equation. When those normalizations are taken into account, one finds that all the forms are equivalent. The purpose of this appendix is to explain the issue of normalization in general and to prove why the normalizations in Eqs. 3.2–3.5 are implied by the form of the system equation in Eq. 3.1.

We begin by expressing the MIMO system equation in the following general form:

$$\mathbf{r} = \alpha \mathbf{H}\mathbf{s} + \mathbf{z}, \tag{A.1}$$

where α is a general scalar coefficient, $\mathbf{s}$ is a vector of transmitted baseband symbols, $\mathbf{H}$ is the channel matrix, $\mathbf{z}$ is a vector of baseband noise voltages at each of the receivers, and $\mathbf{r}$ is a vector of received signals.[1] Next, we compute the covariance matrix of the received signal vector as follows:

$$\begin{aligned}
\mathbf{R}_{rr} &\triangleq \mathbb{E}\left\{\mathbf{r}\mathbf{r}^H\right\} \\
&= \mathbb{E}\left\{(\alpha\mathbf{H}\mathbf{s} + \mathbf{z})(\alpha\mathbf{H}\mathbf{s} + \mathbf{z})^H\right\} \\
&= \mathbb{E}\left\{\alpha^2\mathbf{H}\mathbf{s}\mathbf{s}^H\mathbf{H}^H + \alpha\mathbf{H}\mathbf{s}\mathbf{z}^H + \alpha\mathbf{z}\mathbf{s}^H\mathbf{H}^H + \mathbf{z}\mathbf{z}^H\right\} \\
&= \alpha^2\mathbb{E}\left\{\mathbf{H}\mathbf{s}\mathbf{s}^H\mathbf{H}^H\right\} + \mathbb{E}\left\{\mathbf{z}\mathbf{z}^H\right\} \\
&= \alpha^2\mathbb{E}\left\{\mathbf{H}\,\mathbb{E}\left\{\mathbf{s}\mathbf{s}^H\right\}\mathbf{H}^H\right\} + \mathbb{E}\left\{\mathbf{z}\mathbf{z}^H\right\} \\
&= \alpha^2\mathbb{E}\left\{\mathbf{H}\,\mathbf{R}_{ss}\mathbf{H}^H\right\} + \sigma_z^2\mathbf{I}_{N_r} \\
&= \alpha^2\sigma_s^2\mathbb{E}\left\{\mathbf{H}\mathbf{H}^H\right\} + \sigma_z^2\mathbf{I}_{N_r},
\end{aligned} \tag{A.2}$$

where we assume that $\mathbf{H}$ and $\mathbf{s}$ are independent, that the transmitted signals have equal power from each transmit antenna and are uncorrelated, σ_s^2 denotes the variance of each transmitted signal, and σ_z^2 is the variance of the thermal noise at each receive antenna.

[1] It should be noted that although we call $\mathbf{H}$ the "channel matrix," there is a distinction between it and the true "physical channel matrix" used in Eq. 2.15. In general, when the system equation is written with a coefficient like that shown in Eq. 2.15, the product $\alpha\mathbf{H}$ is equal to the physical channel matrix and $\mathbf{H}$ is a "normalized channel matrix." For simplicity, however, we will not make that distinction except to note it here.

The total received power from all transmitters summed over all receivers, $P_{r,\text{tot}}$, is equal to the sum of the diagonal terms of $\mathbf{R}_{rr}$, which, in turn, is equal to the trace of $\mathbf{R}_{rr}$. Therefore, we can write

$$\begin{aligned} P_{r,\text{tot}} &= \text{Tr}\left[\mathbf{R}_{rr}\right] \\ &= \alpha^2\sigma_s^2 \text{Tr}\left[\mathbb{E}\left\{\mathbf{H}\mathbf{H}^H\right\}\right] + \sigma_z^2 \text{Tr}\left[\mathbf{I}_{N_r}\right] \\ &= \alpha^2\sigma_s^2 \mathbb{E}\left\{\text{Tr}\left[\mathbf{H}\mathbf{H}^H\right]\right\} + N_r\sigma_z^2 \\ &= \alpha^2\sigma_s^2 \mathbb{E}\left\{\|\mathbf{H}\|_F^2\right\} + N_r\sigma_z^2. \end{aligned} \tag{A.3}$$

Since the received powers at the receiver antennas are assumed to be equal to each other, it follows that the received power at *each* receive antenna, P_r, is

$$\begin{aligned} P_r &= \frac{P_{r,\text{tot}}}{N_r} \\ &= \frac{\alpha^2\sigma_s^2 \mathbb{E}\left\{\|\mathbf{H}\|_F^2\right\}}{N_r} + \sigma_z^2. \end{aligned} \tag{A.4}$$

The first term is the power associated with the desired signal and the second term is just the variance of the noise. The signal-to-noise ratio at each receiver, ρ, is, therefore, equal to

$$\rho = \frac{\alpha^2\sigma_s^2 \mathbb{E}\left\{\|\mathbf{H}\|_F^2\right\}}{N_r\sigma_z^2}. \tag{A.5}$$

This equation gives the most general form of the relationship between the variances of the signal and noise, the Frobenius norm of the channel matrix, the signal-to-noise ratio at the receiver, the coefficient, α, and the number of receivers. The convention in this book is to make $\alpha = \sqrt{\rho}$. Under that assumption, Eq. A.5 simplifies to the following form:

$$1 = \frac{\sigma_s^2 \mathbb{E}\left\{\|\mathbf{H}\|_F^2\right\}}{N_r\sigma_z^2}. \tag{A.6}$$

Any assumptions we make about the variances of the signal, the noise, and the statistics of the elements of the normalized channel matrix $\mathbf{H}$ that satisfy Eq. A.6 are consistent with the form of the system equation used in this book. This shows that there are actually an infinite number of ways to choose combinations of σ_z^2, σ_s^2, and the statistics of the normalized channel matrix that satisfy Eq. A.6. Collectively, the set of assumptions about these three parameters is what we mean by the system equation normalization. In particular, the normalization defined by Eqs. 3.2–3.5 is one normalization that satisfies Eq. A.6. To see this, consider the fact that since our normalization assumes that $\mathbb{E}\left\{|h_{ij}|^2\right\} = 1$ (assuming that $\mathbf{H}$ is random), it follows that $\mathbb{E}\left\{\|\mathbf{H}\|_F^2\right\} = N_rN_t$. Clearly, when this is the case, Eq. A.6 is satisfied when $\sigma_z^2 = 1$ and $\sigma_s^2 = 1/N_t$. If, on the other hand, $\mathbf{H}$ is deterministic, the criterion is satisfied when $\|\mathbf{H}\|_F^2 = N_tN_r$, which justifies Eq. 3.3.

Appendix B Proof of theorem 5.2

This appendix proves Theorem 5.2. The proof involves two halves: the first half proves that if $\mathbf{H} = \mathbf{R}_r^{1/2}\mathbf{G}\mathbf{R}_t^{H/2}$, then $\mathbf{R} = \mathbf{R}_r \otimes \mathbf{R}_t^T$, and the second half proves the converse. Both halves are required to prove the *if and only if* aspect of the theorem.

To prove the first half, we compute the elements of $\mathbf{R}$ using the definition of $\mathbf{R}$ in Eq. 5.6 and assume that $\mathbf{H} = \mathbf{R}_r^{1/2}\mathbf{G}\mathbf{R}_t^{H/2}$. The elements of $\mathbf{R}$ have the form $\mathbb{E}\left\{H_{ij}H_{mn}^*\right\}$, which can be expressed as follows:

$$
\begin{aligned}
\mathbb{E}\left\{H_{ij}H_{mn}^*\right\} &= \mathbb{E}\left\{\sum_{l,k,u,v} R_{r_{i,l}}^{1/2} G_{l,k} R_{t_{k,j}}^{H/2} R_{r_{n,u}}^{1/2^*} G_{u,v}^* R_{t_{v,m}}^{H/2^*}\right\} \\
&= \sum_{l,k,u,v} R_{r_{i,l}}^{1/2} R_{t_{k,j}}^{H/2} \mathbb{E}\left\{G_{l,k}G_{u,v}^*\right\} R_{r_{n,u}}^{1/2^*} R_{t_{v,m}}^{H/2^*} \\
&= \sum_{l,k,u,v} R_{r_{i,l}}^{1/2} R_{t_{k,j}}^{H/2} \delta_{l,u}\delta_{k,v} R_{r_{n,u}}^{1/2^*} R_{t_{v,m}}^{H/2^*} \\
&= \sum_{l,k} R_{r_{i,l}}^{1/2} R_{t_{k,j}}^{H/2} R_{r_{n,l}}^{1/2^*} R_{t_{k,m}}^{H/2^*} \\
&= \sum_{l} R_{r_{i,l}}^{1/2} R_{r_{n,l}}^{1/2^*} \sum_{k} R_{t_{k,j}}^{H/2} R_{t_{m,k}}^{1/2} \\
&= (\mathbf{R}_r^{1/2}\mathbf{R}_r^{H/2})_{i,n} (\mathbf{R}_t^{1/2}\mathbf{R}_t^{H/2})_{m,j} \\
&= \mathbf{R}_{r_{i,n}} \mathbf{R}_{t_{m,j}},
\end{aligned} \tag{B.1}
$$

where we have used the assumption that $\mathbf{G}$ is spatially white (i.e., $\mathbb{E}\left\{\mathbf{G}_{i,j}\mathbf{G}_{m,n}\right\} = \delta_{i,m}\delta_{j,n}$).

The right side of Eq. B.1 is equivalent to $\mathbf{R}_r \otimes \mathbf{R}_t^T$. To show that this is true, we consider a specific example where $\mathbf{H}$ is dimensioned 3×4. We do this for simplicity; however, the result is self-evidently true for any dimension of $\mathbf{H}$. Substituting each of the terms in this equation with elements of $\mathbf{R}_r$ and $\mathbf{R_t}$ using Eq. B.1 results in the following:

$$
\mathbf{R} = \begin{pmatrix}
\mathbf{R}_{r_{1,1}}\mathbf{R}_{t_{1,1}} & \mathbf{R}_{r_{1,2}}\mathbf{R}_{t_{1,1}} & \mathbf{R}_{r_{1,3}}\mathbf{R}_{t_{1,1}} & \mathbf{R}_{r_{1,4}}\mathbf{R}_{t_{2,1}} & \mathbf{R}_{r_{1,5}}\mathbf{R}_{t_{2,1}} & \mathbf{R}_{r_{1,6}}\mathbf{R}_{t_{2,1}} \\
\mathbf{R}_{r_{2,1}}\mathbf{R}_{t_{1,1}} & \mathbf{R}_{r_{2,2}}\mathbf{R}_{t_{1,1}} & \mathbf{R}_{r_{2,3}}\mathbf{R}_{t_{1,1}} & \mathbf{R}_{r_{2,4}}\mathbf{R}_{t_{2,1}} & \mathbf{R}_{r_{2,5}}\mathbf{R}_{t_{2,1}} & \mathbf{R}_{r_{2,6}}\mathbf{R}_{t_{2,1}} \\
\mathbf{R}_{r_{1,1}}\mathbf{R}_{t_{1,1}} & \mathbf{R}_{r_{1,2}}\mathbf{R}_{t_{1,1}} & \mathbf{R}_{r_{1,3}}\mathbf{R}_{t_{1,1}} & \mathbf{R}_{r_{1,4}}\mathbf{R}_{t_{2,1}} & \mathbf{R}_{r_{1,5}}\mathbf{R}_{t_{2,1}} & \mathbf{R}_{r_{1,6}}\mathbf{R}_{t_{2,1}} \\
\mathbf{R}_{r_{1,1}}\mathbf{R}_{t_{1,2}} & \mathbf{R}_{r_{1,2}}\mathbf{R}_{t_{1,2}} & \mathbf{R}_{r_{1,3}}\mathbf{R}_{t_{1,2}} & \mathbf{R}_{r_{1,1}}\mathbf{R}_{t_{2,2}} & \mathbf{R}_{r_{1,2}}\mathbf{R}_{t_{2,2}} & \mathbf{R}_{r_{1,3}}\mathbf{R}_{t_{2,1}} \\
\mathbf{R}_{r_{2,1}}\mathbf{R}_{t_{1,2}} & \mathbf{R}_{r_{2,2}}\mathbf{R}_{t_{1,2}} & \mathbf{R}_{r_{2,3}}\mathbf{R}_{t_{1,2}} & \mathbf{R}_{r_{2,1}}\mathbf{R}_{t_{2,2}} & \mathbf{R}_{r_{2,2}}\mathbf{R}_{t_{2,2}} & \mathbf{R}_{r_{2,3}}\mathbf{R}_{t_{2,2}} \\
\mathbf{R}_{r_{3,1}}\mathbf{R}_{t_{1,2}} & \mathbf{R}_{r_{3,2}}\mathbf{R}_{t_{1,2}} & \mathbf{R}_{r_{3,3}}\mathbf{R}_{t_{1,2}} & \mathbf{R}_{r_{3,1}}\mathbf{R}_{t_{2,2}} & \mathbf{R}_{r_{3,2}}\mathbf{R}_{t_{2,2}} & \mathbf{R}_{r_{3,3}}\mathbf{R}_{t_{2,2}}
\end{pmatrix}. \tag{B.2}
$$

Examination of Eq. B.2 shows that it is equivalent to $\mathbf{R}_r \otimes \mathbf{R}_t^T$ when

$$\mathbf{R}_r = \mathbb{E}\left\{\begin{pmatrix} h_{i,k}h^*_{i,k} & h_{i,k}h^*_{2,k} & h_{1,k}h^*_{3,k} \\ h_{2,k}h^*_{1,k} & h_{2,k}h^*_{2,k} & h_{2,k}h^*_{3,k} \\ h_{3,k}h^*_{1,k} & h_{3,k}h^*_{2,k} & h_{3,k}h^*_{3,k} \end{pmatrix}\right\} \tag{B.3}$$

and

$$\mathbf{R}_t = \mathbb{E}\left\{\begin{pmatrix} h_{k,1}h^*_{k,1} & h_{k,2}h^*_{k,1} \\ h_{k,1}h^*_{k,2} & h_{k,2}h^*_{k,2} \end{pmatrix}\right\}, \tag{B.4}$$

since $\mathbf{R}_r$ and $\mathbf{R}_t$ are assumed to be independent of k. This proves the first half of the theorem for the specific case of a 3×2 channel matrix. Similar methods could be used to prove it for any dimension of $\mathbf{H}$.

We now consider the second half of the proof, which states that if $\mathbf{R} = \mathbf{R}_r \otimes \mathbf{R}_t^T$, it follows that $\mathbf{H}$ can be written in the form $\mathbf{H} = \mathbf{R}_r^{1/2}\mathbf{G}\mathbf{R}_t^{H/2}$ where $\mathbf{G}$ is spatially white. Since it is clearly always possible to express the channel matrix in the form $\mathbf{H} = \mathbf{R}_r^{1/2}\mathbf{G}\mathbf{R}_t^{H/2}$ simply by defining $\mathbf{G} = \mathbf{R}_r^{-1/2}\mathbf{H}\mathbf{R}_t^{-H/2}$, the only requirement for the second half of the proof is that we show that if $\mathbf{R} = \mathbf{R}_r \otimes \mathbf{R}_t^T$ and $\mathbf{H} = \mathbf{R}_r^{1/2}\mathbf{G}\mathbf{R}_t^{H/2}$, then $\mathbf{G}$ is spatially white. This is what we now show.

We start by assuming that $\mathbf{G} = \mathbf{R}_r^{-1/2}\mathbf{H}\mathbf{R}_t^{-H/2}$ and examine the elements of the covariance matrix of $\mathbf{G}$ to see if $\mathbf{G}$ is spatially white, which occurs if all the elements of the covariance matrix are zero except the diagonal elements. The elements of the covariance matrix of $\mathbf{G}$ are expressed as follows:

$$\begin{aligned}
\mathbb{E}\left\{G_{i,j}G^*_{n,m}\right\} &= \mathbb{E}\left\{\sum_{l,k,u,v} R^{-1/2}_{r_{i,l}} H_{l,k} R^{-H/2}_{t_{k,j}} R^{-1/2^*}_{r_{n,u}} H^*_{u,v} R^{-H/2^*}_{t_{v,m}}\right\} \\
&= \sum_{l,k,u,v} R^{-1/2}_{r_{i,l}} R^{-H/2}_{t_{k,j}} R^{-1/2^*}_{r_{n,u}} R^{-H/2^*}_{t_{v,m}} \mathbb{E}\left\{H_{l,k}H^*_{u,v}\right\} \\
&= \sum_{l,k,u,v} R^{-1/2}_{r_{i,l}} R^{-H/2}_{t_{k,j}} R^{-1/2^*}_{r_{n,u}} R^{-H/2^*}_{t_{v,m}} R_{r_{l,u}} R_{t_{v,k}} \\
&= \sum_{l,u} R^{-1/2}_{r_{i,l}} R_{r_{l,u}} R^{-1/2^*}_{r_{n,u}} \sum_{k,v} R^{-H/2}_{t_{k,j}} R_{t_{v,k}} R^{-H/2^*}_{t_{v,m}} \\
&= \sum_{l,u} R^{-1/2}_{r_{i,l}} \left[\sum_q R^{1/2}_{r_{l,q}} R^{H/2}_{r_{q,u}}\right] R^{-1/2^*}_{r_{n,u}} \\
&\quad \times \sum_{k,v} R^{-H/2}_{t_{k,j}} \left[\sum_{q'} R^{1/2}_{t_{v,q'}} R^{H/2}_{t_{q',k}}\right] R^{-H^*/2}_{t_{v,m}}
\end{aligned}$$

$$= \sum_{u,q} R_{r_{q,u}}^{H/2} \left[\sum_{l} R_{r_{l,q}}^{1/2} R_{r_{i,l}}^{-1/2} \right] R_{r_{n,u}}^{-1/2^*}$$

$$\times \sum_{v,q'} R_{t_{v,q'}}^{1/2} \left[\sum_{k'} R_{t_{q',k}}^{H/2} R_{t_{k,j}}^{-H/2} \right] R_{t_{v,m}}^{-H^*/2}. \tag{B.5}$$

Next, note that

$$\sum_{l} R_{r_{l,q}}^{1/2} R_{r_{i,l}}^{-1/2} = \sum_{l} R_{r_{i,l}}^{-1/2} R_{r_{l,q}}^{1/2}$$

$$= \left[\mathbf{R}_r^{-1/2} \mathbf{R}_r^{1/2} \right]_{i,q}$$

$$= [\mathbf{I}]_{i,q} = \delta_{i,q}. \tag{B.6}$$

Similarly,

$$\sum_{k} R_{t_{q',k}}^{H/2} R_{t_{k,j}}^{-H/2} = \left[\mathbf{R}_t^{H/2} \mathbf{H}_t^{-H/2} \right]_{q'j}$$

$$= [\mathbf{I}]_{q',j} = \delta_{q',j}. \tag{B.7}$$

Using these two relationships in Eq. B.5 results in the following:

$$\mathbb{E}\left\{ G_{i,j} G_{n,m}^* \right\} = \sum_{u,q} R_{r_{q,u}}^{H/2} \delta_{i,q} R_{r_{n,u}}^{-1/2^*} \sum_{v,q'} R_{t_{v,q'}}^{1/2} \delta_{q',j} R_{t_{v,m}}^{-H^*/2}$$

$$= \sum_{u} R_{r_{i,u}}^{H/2} R_{r_{n,u}}^{-1/2^*} \sum_{v} R_{t_{v,j}}^{1/2} R_{t_{v,m}}^{-H^*/2}$$

$$= \sum_{u} R_{r_{i,u}}^{H/2} R_{r_{u,n}}^{-H/2} \sum_{v} R_{t_{v,j}}^{1/2} R_{t_{m,v}}^{-1/2}$$

$$= \sum_{u} R_{r_{i,u}}^{H/2} R_{r_{u,n}}^{-H/2} \sum_{v} R_{t_{m,v}}^{-1/2} R_{t_{v,j}}^{1/2}$$

$$= [\mathbf{I}]_{i,n} [\mathbf{I}]_{m,j}, \tag{B.8}$$

which means that the covariance of $\mathbf{G}$ is spatially white and completes the proof of Theorem 5.2.

Appendix C Derivation of Eq. 7.9

This appendix derives the general expression for the maximum likelihood (ML) estimate of a transmitted space-time codeword **S** based on the received space-time codeword **R**, where **S** and **R** are related as follows:

$$\mathbf{R} = \sqrt{\rho}\mathbf{HS} + \mathbf{Z}. \tag{C.1}$$

As usual, ρ denotes the signal-to-noise ratio, **H** is the narrowband channel matrix, and **Z** is the noise matrix, which is assumed to consist of iid (independent and identically distributed) circularly symmetric complex Gaussian random variable elements, each with variance σ_n. In order to make this derivation as general as possible, we replace the $\sqrt{\rho}$ coefficient with an arbitrary real scalar that we denote by α. Thus, we seek the ML estimate of **S** where

$$\mathbf{R} = \alpha\mathbf{HS} + \mathbf{Z}. \tag{C.2}$$

For given values of α, **H**, and **S**, it follows that **R** also consists of iid Gaussian random variables, which we can characterize in terms of a multivariate Gaussian probability density function (pdf) $f_{\mathbf{R}}(r_{11}, r_{1,2}, \ldots, r_{N_r,p})$, where we are assuming that **R** is dimensioned $N_r \times p$. Since the elements of **R** are identically distributed and independent, we can write the multivariate pdf as

$$f_{\mathbf{R}}(r_{11}, r_{1,2}, \ldots, r_{N_r,p}) = \prod_{i=1}^{N_r}\prod_{j=1}^{p} f_R(r_{i,j}), \tag{C.3}$$

where $f_R(r_{i,j})$ denotes the Gaussian distribution function of a single element of **R**. In general, we can write the (i,j)th element of **R** as

$$[\mathbf{R}]_{i,j} = \alpha\,[\mathbf{HS}]_{i,j} + [\mathbf{Z}]_{i,j}\,. \tag{C.4}$$

It follows that $[\mathbf{R}]_{i,j} \sim \mathcal{N}(\alpha\,[\mathbf{HS}]_{i,j}\,, \sigma_z^2)$. Thus, the pdf of $r_{i,j} \triangleq [\mathbf{R}]_{i,j}$ is given by

$$f_R(r_{i,j}|\alpha, \mathbf{H}, \mathbf{S}) = \frac{1}{\sqrt{2\pi\sigma_z^2}} e^{-|r_{i,j} - [\alpha\mathbf{HS}]_{i,j}|^2/2\sigma_z^2}, \tag{C.5}$$

where we explicitly indicate that this pdf is conditioned on α, **H**, and **S** having specific values. It follows from Eq. C.3 that the multivariate pdf for the elements of **R** is given by

$$f_{\mathbf{R}}(r_{11}, r_{1,2}, \ldots, r_{N_r,p}|\alpha, \mathbf{H}, \mathbf{S})$$
$$= (2\pi\sigma_z^2)^{N_r p/2} \times \exp\left(-\sum_{i=1}^{N_r}\sum_{j=1}^{p} |r_{i,j} - [\alpha\mathbf{HS}]_{i,j}|^2/2\sigma_z^2\right). \quad \text{(C.6)}$$

Since the ML estimate of $\mathbf{S}$, which we denote by $\hat{\mathbf{S}}$, is given by

$$\hat{\mathbf{S}} = \arg\max_{\mathbf{S}} f_{\mathbf{R}}(r_{11}, r_{1,2}, \ldots, r_{N_r,p}|\alpha, \mathbf{H}, \mathbf{S}), \quad \text{(C.7)}$$

it follows from Eq. C.6 that

$$\hat{\mathbf{S}} = \arg\min_{\mathbf{S}} \sum_{i=1}^{N_r}\sum_{j=1}^{p} |r_{i,j} - [\alpha\mathbf{HS}]_{i,j}|^2. \quad \text{(C.8)}$$

This is clearly equivalent to

$$\hat{\mathbf{S}} = \arg\min_{\mathbf{S}} \|\mathbf{R} - \alpha\mathbf{HS}\|_F^2, \quad \text{(C.9)}$$

which completes the proof for $\alpha = \sqrt{\rho}$.

Appendix D Maximum likelihood decoding rules for selected OSTBCs

This appendix provides the decoding formulas for the $\mathcal{G}_2$, $\mathcal{G}_3$, $\mathcal{G}_4$, $\mathcal{H}_3$, and $\mathcal{H}_4$ OSTBCs. These formulas are based on results presented in [72] using the nomenclature and dimensioning conventions used in this book rather than those in the paper.

The decoder for $\mathcal{G}_2$ (i.e., the Alamouti code) is as follows:

$$\hat{s}_1 = \arg\min_{\{\mathbf{S}\}} \left\{ \left| \sum_{i=1}^{N_r} \left(r_i(1) h_{i,1}^* + r_i^*(2) h_{i,2} \right) - s_1 \right|^2 + \left(\sum_{i=1}^{N_r} \sum_{j=1}^{2} |h_{i,j}|^2 - 1 \right) |s_1|^2 \right\} \tag{D.1}$$

and

$$\hat{s}_2 = \arg\min_{\{\mathbf{S}\}} \left\{ \left| \sum_{i=1}^{N_r} \left(r_i(1) h_{i,2}^* - r_i^*(2) h_{i,1} \right) - s_2 \right|^2 + \left(\sum_{i=1}^{N_r} \sum_{j=1}^{2} |h_{i,j}|^2 - 1 \right) |s_2|^2 \right\}. \tag{D.2}$$

The decoder for $\mathcal{G}_3$ is as follows:

$$\hat{s}_1 = \arg\min_{\{s_1\}} \left\{ \left| \left[\sum_{i=1}^{N_r} \left(r_i(1) h_{i,1}^* + r_i(2) h_{i,2}^* + r_i(3) h_{i,3}^* + (r_i(5))^* h_{i,1} + (r_i(6))^* h_{i,2} + (r_i(7))^* h_{i,3} \right) \right] - s_1 \right|^2 + \left(2 \sum_{i=1}^{N_r} \sum_{j=1}^{3} |h_{i,j}|^2 - 1 \right) |s_1|^2 \right\}, \tag{D.3}$$

$$\hat{s}_2 = \arg\min_{\{s_2\}} \left\{ \left| \left[\sum_{i=1}^{N_r} \left(r_i(1) h_{i,2}^* - r_i(2) h_{i,1}^* + r_i(4) h_{i,3}^* + (r_i(5))^* h_{i,2} - (r_i(6))^* h_{i,1} + (r_i(8))^* h_{i,3} \right) \right] - s_2 \right|^2 + \left(2 \sum_{i=1}^{N_r} \sum_{j=1}^{3} |h_{i,j}|^2 - 1 \right) |s_2|^2 \right\}, \tag{D.4}$$

$$\hat{s}_3 = \arg\min_{\{s_3\}} \left\{ \left| \left[\sum_{i=1}^{N_r} \left(r_i(1)h_{i,3}^* - r_i(3)h_{i,1}^* - r_i(4)h_{i,2}^* + (r_i(5))^* h_{i,3} - (r_i(7))^* h_{i,1} - (r_i(8))^* h_{i,2} \right) \right] - s_3 \right|^2 + \left(2\sum_{i=1}^{N_r}\sum_{j=1}^{3} \left|h_{i,j}\right|^2 - 1 \right) |s_3|^2 \right\}, \tag{D.5}$$

$$\hat{s}_4 = \arg\min_{\{s_4\}} \left\{ \left| \left[\sum_{i=1}^{N_r} \left(-r_i(2)h_{i,3}^* + r_i(3)h_{i,2}^* - r_i(4)h_{i,1}^* - (r_i(6))^* h_{i,3} + (r_i(7))^* h_{i,2} - (r_i(8))^* h_{i,1} \right) \right] - s_4 \right|^2 + \left(2\sum_{i=1}^{N_r}\sum_{j=1}^{3} \left|h_{i,j}\right|^2 - 1 \right) |s_4|^2 \right\}. \tag{D.6}$$

The decoder for $\mathcal{G}_4$ is as follows:

$$\hat{s}_1 = \arg\min_{\{\mathbf{S}\}} \left\{ \left| \sum_{i=1}^{N_r} \left(r_i(1)h_{i,1}^* + r_i^*(2)h_{i,2}^* + r_i(3)h_{i,3}^* + r_i(4)h_{i,4}^* + r_i^*(5)h_{i,1}^* + r_i(6)^* h_{i,2} + r_i^*(7)h_{i,3} + r_i^*(8)h_{i,4} \right) - s_1 \right|^2 + \left(2\sum_{i=1}^{N_r}\sum_{j=1}^{4} |h_{i,j}|^2 - 1 \right) |s_1|^2 \right\}, \tag{D.7}$$

$$\hat{s}_2 = \arg\min_{\{\mathbf{S}\}} \left\{ \left| \sum_{i=1}^{N_r} \left(r_i(1)h_{i,1}^* - r_i^*(2)h_{i,1}^* - r_i(3)h_{i,4}^* + r_i(4)h_{i,3}^* + r_i^*(5)h_{i,2} - r_i^*(6)h_{i,1} - r_i^*(7)h_{i,4} + r_i^*(8)h_{i,3} \right) - s_2 \right|^2 + \left(2\sum_{i=1}^{N_r}\sum_{j=1}^{4} |h_{i,j}|^2 - 1 \right) |s_2|^2 \right\}, \tag{D.8}$$

$$\hat{s}_3 = \arg\min_{\{\mathbf{S}\}} \left\{ \left| \sum_{i=1}^{N_r} \left(r_i(1)h_{i,3}^* + r_i(2)h_{i,4}^* - r_i(3)h_{i,1}^* - r_i(4)h_{i,2}^* + r_i(5)^* h_{i,3}^* + r_i^*(6)h_{i,4} - r_i^*(7)h_{i,1} - r_i^*(8)h_{i,2} \right) - s_3 \right|^2 + \left(2\sum_{i=1}^{N_r}\sum_{j=1}^{4} |h_{i,j}|^2 - 1 \right) |s_3|^2 \right\}, \tag{D.9}$$

$$\hat{s}_4 = \arg\min_{\{\mathbf{S}\}} \left\{ \left| \sum_{i=1}^{N_r} \left(r_i(1)h_{i,4}^* - r_i(2)h_{i,3}^* + r_i(3)h_{i,2}^* - r_i(4)h_{i,1}^* + r_i^*(5)h_{i,4} - r_i^*(6)h_{i,3} + r_i^*(7)h_{i,2} - r_i^*(8)h_{i,1} \right) - s_4 \right|^2 + \left(2\sum_{i=1}^{N_r}\sum_{j=1}^{4} |h_{i,j}|^2 - 1 \right) |s_4|^2 \right\}. \tag{D.10}$$

The decoder for $\mathcal{H}_3$ is as follows:

$$\hat{s}_1 = \arg\min_{\{\mathbf{S}\}} \left\{ \left| \sum_{i=1}^{N_r} \left(r_i(1)h_{i,2}^* + r_i^*(2)h_{i,2} + \frac{(r_i(4) - r_i(3))\, h_{i,3}^*}{2} - \frac{(r_i(3) + r_i(4))^*\, h_{i,3}}{2} \right) - s_1 \right|^2 + \left(\sum_{i=1}^{N_r} \sum_{j=1}^{3} |h_{i,j}|^2 - 1 \right) |s_1|^2 \right. \tag{D.11}$$

$$\hat{s}_2 = \arg\min_{\{\mathbf{S}\}} \left\{ \left| \sum_{i=1}^{N_r} \left(r_i(1)h_{i,2}^* - r_i^*(2)h_{i,1} + \frac{(r_i(4) + r_i(3))\, h_{i,3}^*}{2} + \frac{(-r_i(3) + r_i(4))^*\, h_{i,3}}{2} \right) - s_2 \right|^2 + \left(\sum_{i=1}^{N_r} \sum_{j=1}^{3} |h_{i,j}|^2 - 1 \right) |s_2|^2, \right. \tag{D.12}$$

$$\hat{s}_3 = \arg\min_{\{\mathbf{S}\}} \left\{ \left| \sum_{i=1}^{N_r} \left(\frac{(r_i(1) + r_i(2))\, h_{i,3}^*}{\sqrt{2}} + \frac{r_i(3)^* \left(h_{i,1} + h_{i,2}\right)}{\sqrt{2}} + \frac{r_i(4)^* \left(h_{i,1} - h_{i,2}\right)}{\sqrt{2}} \right) - s_3 \right|^2 + \left(\sum_{i=1}^{N_r} \sum_{j=1}^{3} |h_{i,j}|^2 - 1 \right) |s_3|^2. \right. \tag{D.13}$$

The decoder for $\mathcal{H}_4$ is as follows:

$$\hat{s}_1 = \arg\min_{\{\mathbf{S}\}} \left\{ \left| \sum_{i=1}^{N_r} \left(r_i(1)h_{i,1}^* + r_i^*(2)h_{i,2} + \frac{(r_i(4) - r_i(3)) \left(h_{i,3}^* - h_{i,4}^*\right)}{2} - \frac{(r_i(3) + r_i(4))^* \left(h_{i,3} + h_{i,4}\right)}{2} \right) - s_1 \right|^2 + \left(\sum_{i=1}^{N_r} \sum_{j=1}^{4} |h_{i,j}|^2 - 1 \right) |s_1|^2, \right. \tag{D.14}$$

$$\hat{s}_2 = \arg\min_{\{\mathbf{S}\}} \left\{ \left| \sum_{i=1}^{N_r} \left(r_i(1)h_{i,2}^* - r_i^*(2)h_{i,1} + \frac{(r_i(4) + r_i(3)) \left(h_{i,3}^* - h_{i,4}^*\right)}{2} - \frac{(-r_i(3) + r_i(4))^* \left(h_{i,3} + h_{i,4}\right)}{2} \right) - s_2 \right|^2 + \left(\sum_{i=1}^{N_r} \sum_{j=1}^{4} |h_{i,j}|^2 - 1 \right) |s_2|^2, \right. \tag{D.15}$$

$$\hat{s}_3 = \arg\min_{\{\mathbf{S}\}} \left\{ \left| \sum_{i=1}^{N_r} \left(\frac{(r_i(1) + r_i(2))\, h_{i,3}^*}{\sqrt{2}} + \frac{(r_i(1) - r_i(2))\, h_{i,4}^*}{\sqrt{2}} + \frac{r_i(3)^* \left(h_{i,1} - h_{i,2}\right)}{\sqrt{2}} + \frac{r_i(4)^* \left(h_{i,1} - h_{i,2}\right)}{\sqrt{2}} \right) - s_3 \right|^2 + \left(\sum_{i=1}^{N_r} \sum_{j=1}^{4} |h_{i,j}|^2 - 1 \right) |s_3|^2. \right. \tag{D.16}$$

Appendix E Derivation of Eq. 8.68

This appendix derives the expression for $\tilde{\mathbf{r}}_{N_t-1}^{(N_t-1)}(k)$ in Eq. 8.68. To do so, recall that $\tilde{\mathbf{r}}_{N_t-1}^{(N_t-1)}(k)$ is defined as the kth element of the $(N_t - 1)$th row of $\tilde{\mathbf{R}}^{(N_t-1)}$ defined in Eq. 8.67, where $\tilde{\mathbf{R}}^{(N_t-1)}$ is defined in Eq. 8.66. It, therefore, follows that

$$\begin{aligned}\tilde{\mathbf{R}}^{(N_t-1)} &= \mathbf{W}_o\left(\sqrt{\rho}\mathbf{HS} + \mathbf{Z} - \sqrt{\rho}\mathbf{P}_{N_t}\right)\\ &= \sqrt{\rho}\,\left(\mathbf{W}_o\mathbf{HS} - \mathbf{W}_o\mathbf{P}_{N_t}\right) + \tilde{\mathbf{Z}},\end{aligned} \tag{E.1}$$

where $\tilde{\mathbf{Z}} \triangleq \mathbf{W}_o\mathbf{Z}$. Next, express the **HS** term in expanded matrix form as follows:

$$\begin{aligned}\mathbf{HS} &= \begin{pmatrix} h_{1,1} & \dots & h_{1,N_t} \\ \vdots & \ddots & \vdots \\ h_{N_r,1} & \dots & h_{N_r,N_t}\end{pmatrix}\begin{pmatrix} s_1(1) & \dots & s_1(p) \\ \vdots & \ddots & \vdots \\ s_{N_t}(1) & \dots & s_{N_t}(p)\end{pmatrix}\\ &= \begin{pmatrix} \sum_{n=1}^{N_t} h_{1,n}s_n(1) & \dots & \sum_{n-1}^{N_t} h_{1,n}s_n(p) \\ \vdots & \ddots & \vdots \\ \sum_{n=1}^{N_t} h_{N_r,n}s_n(1) & \dots & \sum_{n-1}^{N_t} h_{N_r,n}s_n(p)\end{pmatrix}.\end{aligned} \tag{E.2}$$

Therefore,

$$\begin{aligned}\mathbf{W}_o\mathbf{HS} = &\begin{pmatrix} w_{1,1} & \dots & w_{1,N_r} \\ \vdots & \ddots & \vdots \\ w_{N_t,1} & \dots & w_{N_t,N_r}\end{pmatrix}\\ &\times\begin{pmatrix} \sum_{n=1}^{N_t} h_{1,n}s_n(1) & \dots & \sum_{n-1}^{N_t} h_{1,n}s_n(p) \\ \vdots & \ddots & \vdots \\ \sum_{n=1}^{N_t} h_{N_r,n}s_n(1) & \dots & \sum_{n-1}^{N_t} h_{N_r,n}s_n(p)\end{pmatrix}.\end{aligned} \tag{E.3}$$

It follows that the kth element of the $(N_t - 1)$th row of $\mathbf{W}_o\mathbf{HS}$ is given by

$$\begin{aligned}[\mathbf{W}_o\mathbf{HS}]_{N_t-1,k} = w_{N_t-1,1}\sum_{n=1}^{N_t} h_{1,n}s_n(k) + w_{N_t-1,2}\sum_{n=1}^{N_t} h_{2,n}s_n(k) + \dots\\ + w_{N_t-1,N_r}\sum_{n=1}^{N_t} h_{N_r,n}s_n(k)\end{aligned}$$

$$= \sum_{m=1}^{N_r} w_{N_t-1,m} \sum_{n=1}^{N_t} h_{m,n} s_n(k). \tag{E.4}$$

The other term in Eq. E.1 that we need to consider is $\mathbf{W}_o \mathbf{P}_{N_t}$, which we can express in expanded form as follows:

$$\mathbf{W}_o \mathbf{P}_{N_t} = \begin{pmatrix} w_{1,1} & \dots & w_{1,N_r} \\ \vdots & \ddots & \vdots \\ w_{N_t,1} & \dots & w_{N_t,N_r} \end{pmatrix} \begin{pmatrix} h_{1,j}\hat{s}_j(1) & \dots & h_{1,j}\hat{s}_j(p) \\ \vdots & \ddots & \vdots \\ h_{N_r,j}\hat{s}_j(1) & \dots & h_{N_r,j}\hat{s}_j(p) \end{pmatrix}. \tag{E.5}$$

It follows that the kth element of the $(N_t - 1)$th row of $\mathbf{W}_o \mathbf{P}_{N_t}$ is equal to

$$\left[\mathbf{W}_o \mathbf{P}_{N_t}\right]_{N_t-1,k} = \sum_{m=1}^{N_r} w_{N_t-1,m} h_{m,N_t} \hat{s}_{N_t}(k). \tag{E.6}$$

Substituting Eqs. E.4 and E.6 into Eq. E.1 results in the following:

$$\begin{aligned}
\tilde{\mathbf{r}}_{N_t-1}^{(N_t-1)}(k) &\triangleq \left[\tilde{\mathbf{R}}^{(N_t-1)}\right]_{N_t-1,k} \\
&= \sqrt{\rho}\,[\mathbf{W}_o \mathbf{H}\mathbf{S}]_{N_t-1,k} - \sqrt{\rho}\left[\mathbf{W}_o \mathbf{P}_{N_t}\right]_{N_t-1,k} + \tilde{z}_{N_t-1}(k) \\
&= \sqrt{\rho} \sum_{m=1}^{N_r} w_{N_t-1,m} \sum_{n=1}^{N_t} h_{m,n} s_n(k) - \sqrt{\rho} \sum_{m=1}^{N_r} w_{N_t-1,m} h_{m,N_t} \hat{s}_{N_t}(k) \\
&\quad + \tilde{z}_{N_t-1}(k) \\
&= \sqrt{\rho} \sum_{m=1}^{N_r} w_{N_t-1,m} \left[\sum_{n=1}^{N_t} h_{m,n} s_n(k) - h_{m,N_t} \hat{s}_{N_t}(k)\right] + \tilde{z}_{N_t-1}(k) \\
&= \sqrt{\rho} \sum_{m=1}^{N_r} w_{N_t-1,m} \Bigg[h_{m,N_t-1} s_{N_t-1}(k) + \sum_{n=1}^{N_t-2} h_{m,n} s_n(k) \\
&\quad + h_{m,N_t}\left(s_{N_t}(k) - \hat{s}_{N_t}(k)\right)\Bigg] + \tilde{z}_{N_t-1}(k) \\
&= \sqrt{\rho} \left(\sum_{m=1}^{N_r} w_{N_t-1,m} h_{m,N_t-1}\right) s_{N_t-1}(k) \\
&\quad + \sqrt{\rho} \sum_{m=1}^{N_r} w_{N_t-1,m} \left[\sum_{n=1}^{N_t-2} h_{m,n} s_n(k) + h_{m,N_t}\left(s_{N_t}(k) - \hat{s}_{N_t}(k)\right)\right] \\
&\quad + \tilde{z}_{N_t-1}(k), \qquad k = 1, \dots, p,
\end{aligned} \tag{E.7}$$

which is equivalent to Eq. 8.68, thereby completing the proof.

Appendix F Parameters for the non-unequal HT modulation and coding schemes in IEEE 802.11n

This appendix lists the key parameters associated with the non-unequal modulation and coding schemes (MCSs) in IEEE 802.11n. This refers to those modes where the same modulation is used on all the input spatial streams. These tables are derived from [77]. In the following tables, r denotes the rate of the forward error correction code and GI denotes the guard interval of the OFDM symbol.

Table F.1 *20 MHz* MCS parameters for *one* spatial stream.

			Data rate (Mb/s)	
MCS index	Modulation	r	800 ns GI	400 ns GI
0	BPSK	1/2	6.5	7.2
1	QPSK	1/2	13.0	14.4
2	QPSK	3/4	19.5	21.7
3	16-QAM	1/2	26.0	28.9
4	16-QAM	3/4	39.0	43.3
5	64-QAM	2/3	52.0	57.8
6	64-QAM	3/4	58.5	65.0
7	64-QAM	5/6	65.0	72.2

Table F.2 *20 MHz* MCS parameters for *two* spatial streams.

			Data rate (Mb/s)	
MCS index	Modulation	r	800 ns GI	400 ns GI
8	BPSK	1/2	13.0	14.4
9	QPSK	1/2	26.0	28.9
10	QPSK	3/4	39.0	43.3
11	16-QAM	1/2	52.0	57.8
12	16-QAM	3/4	78.0	86.7
13	64-QAM	2/3	104.0	115.6
14	64-QAM	3/4	117.0	130.0
15	64-QAM	5/6	130.0	144.4

Table F.3 *20 MHz* MCS parameters for *three* spatial streams.

MCS index	Modulation	r	Data rate (Mb/s)	
			800 ns GI	400 ns GI
16	BPSK	1/2	19.5	21.7
17	QPSK	1/2	39.0	43.3
18	QPSK	3/4	58.5	65.0
19	16-QAM	1/2	78.0	86.7
20	16-QAM	3/4	117.0	130.0
21	64-QAM	2/3	156.0	173.3
22	64-QAM	3/4	175.5	195.0
23	64-QAM	5/6	195.0	216.7

Table F.4 *20 MHz* MCS parameters for *four* spatial streams.

MCS index	Modulation	r	Data rate (Mb/s)	
			800 ns GI	400 ns GI
24	BPSK	1/2	26.0	28.9
25	QPSK	1/2	52.0	57.8
26	QPSK	3/4	78.0	86.7
27	16-QAM	1/2	104.0	115.6
28	16-QAM	3/4	156.0	173.3
29	64-QAM	2/3	208.0	231.1
30	64-QAM	3/4	234.0	260.0
31	64-QAM	5/6	260.0	288.9

Table F.5 *40 MHz* MCS parameters for *one* spatial stream.

MCS index	Modulation	r	Data rate (Mb/s)	
			800 ns GI	400 ns GI
0	BPSK	1/2	13.5	15.0
1	QPSK	1/2	27.0	30.0
2	QPSK	3/4	40.5	45.0
3	16-QAM	1/2	54.0	60.0
4	16-QAM	3/4	81.0	90.0
5	64-QAM	2/3	108.0	120.0
6	64-QAM	3/4	121.5	135.0
7	64-QAM	5/6	135.0	150.0

Table F.6 *40 MHz* MCS parameters for *two* spatial streams.

MCS index	Modulation	r	Data rate (Mb/s)	
			800 ns GI	400 ns GI
8	BPSK	1/2	27.0	30.0
9	QPSK	1/2	54.0	60.0
10	QPSK	3/4	81.0	90.0
11	16-QAM	1/2	108.0	120.0
12	16-QAM	3/4	162.0	180.0
13	64-QAM	2/3	216.0	240.0
14	64-QAM	3/4	243.0	270.0
15	64-QAM	5/6	270.0	300.0

Table F.7 *40 MHz* MCS parameters for *three* spatial streams.

MCS index	Modulation	r	Data rate (Mb/s)	
			800 ns GI	400 ns GI
16	BPSK	1/2	40.5	45.0
17	QPSK	1/2	81.0	90.0
18	QPSK	3/4	121.5	135.0
19	16-QAM	1/2	162.0	180.0
20	16-QAM	3/4	243.0	270.0
21	64-QAM	2/3	324.0	360.0
22	64-QAM	3/4	364.5	405.0
23	64-QAM	5/6	405.0	450.0

Table F.8 *40 MHz* MCS parameters for *four* spatial streams.

MCS index	Modulation	r	Data rate (Mb/s)	
			800 ns GI	400 ns GI
24	BPSK	1/2	54.0	60.0
25	QPSK	1/2	108.0	120.0
26	QPSK	3/4	162.0	180.0
27	16-QAM	1/2	216.0	240.0
28	16-QAM	3/4	324.0	360.0
29	64-QAM	2/3	432.0	480.0
30	64-QAM	3/4	486.0	540.0
31	64-QAM	5/6	540.0	600.0
32	BPSK	1/2	6.0	6.7

References

[1] MIMO discussion summary. Technical Report 3GPP TSG R1-02-0181, 3rd Generation Partnership Project (3GPP), January 2002.

[2] R. Adams. Simplified baseband diversity combiner. *IRE Transactions on Communications Systems,* **8**(*4*):247–249, December 1960.

[3] Jeffrey G. Andrews, Arunabha Ghosh, and Rias Muhamed. *Fundamentals of WiMAX: Understanding Broadband Wireless Networking,* Prentice Hall, 2007.

[4] A. Alexiou and M. Haardt. Smart antenna technologies for future wireless systems: trends and challenges. *IEEE Communications Magazine,* **42**(*9*):90–97, September 2004.

[5] M.S. Akram. *Pilot-Based Channel Estimation in OFDM Systems.* Master's thesis, Technical University of Denmark, 2007.

[6] S.M. Alamouti. A simple transmit diversity technique for wireless communications. *IEEE Journal on Selected Areas in Communications,* **16**(*8*):1451–1458, October 1998.

[7] Luis Alvarez. Biography. `http://www.nobelprize.org/nobel_prizes/physics/laureates/1968/alvarez.html`, 1968. [Online; accessed 24 September 2012].

[8] D. Asztely. On antenna arrays in mobile communication systems: Fast fading and GSM base station receiver algorithms. Technical Report IR-S3-SB-9611, Royal Institute of Technology, Stockholm, Sweden, March 1996.

[9] M. Baghaie, P.A. Martin, and D.P. Taylor. Grouped multilevel space-time trellis codes. *IEEE Communications Letters,* **14**(*3*):232–234, March 2010.

[10] M. Biguesh and A.B. Gershman. Training-based MIMO channel estimation: a study of estimator tradeoffs and optimal training signals. *IEEE Transactions on Signal Processing,* **54**(*3*):884–893, March 2006.

[11] H.H. Beverage and H.O. Peterson. Diversity receiving system of R.C.A. communications, Inc., for radiotelegraphy. *Proceedings of the Institute of Radio Engineers,* **19**(*4*):529–561, April 1931.

[12] Karl Braun. Electrical oscillations and wireless telegraphy, Nobel Lecture, 11 December, 1909 (pp. 239–240). `http://nobelprize.org/nobel_prizes/physics/laureates/1909/braun-lecture.pdf`, 1909. [Online; accessed 24 September 2012].

[13] G. Berardinelli, L.A. Ruiz de Temino, S. Frattasi, *et al.* OFDMA vs. SC-FDMA: performance comparison in local area IMT-A scenarios. *IEEE Wireless Communications,* **15**(*5*):64–72, October 2008.

[14] D.G. Brennan. Linear diversity combining techniques. *Proceedings of the IRE,* **47**(*6*):1075–1102, June 1959.

[15] S. Coleri, M. Ergen, A. Puri, and A. Bahai. Channel estimation techniques based on pilot arrangement in OFDM systems. *IEEE Transactions on Broadcasting,* **48**(*3*):223–229, September 2002.

[16] J.W. Craig. A new, simple and exact result for calculating the probability of error for two-dimensional signal constellations. In *Military Communications Conference, 1991. MILCOM '91, Conference Record, Military Communications in a Changing World,* IEEE, pp. 571–575, vol.2, Nov. 1991.

[17] C. Ciochina and H. Sari. A review of OFDMA and single-carrier FDMA. In *2010 European Wireless Conference (EW),* pp. 706–710, April 2010.

[18] Thomas M. Cover and Joy A. Thomas. *Elements of Information Theory*, 2nd edition (Wiley Series in *Telecommunications and Signal Processing*). Wiley-Interscience, 2006.

[19] D. Devasirvatham. Time delay spread and signal level measurements of 850 MHz radio waves in building environments. *IEEE Transactions on Antennas and Propagation,* **34**(*11*):1300–1305, Nov. 1986.

[20] P.F. Driessen and G.J. Foschini. On the capacity formula for multiple input-multiple output wireless channels: a geometric interpretation. 1999 IEEE International Conference on, *Communications, ICC '99.* **3**:1603–1607, 1999.

[21] Tolga M. Duman and Ali Ghrayeb. *Coding for MIMO Communication Systems.* Wiley, 2007.

[22] R.B. Ertel, P. Cardieri, K.W. Sowerby, T.S. Rappaport, and J.H. Reed. Overview of spatial channel models for antenna array communication systems. *IEEE Personal Communications,* **5**(*1*):10–22, February 1998.

[23] P. Elias, A. Feinstein, and C. Shannon. A note on the maximum flow through a network. *IRE Transactions on Information Theory,* **2**(*4*):117–119, December 1956.

[24] H. El Gamal, A.R.Jr. Hammons, Youjian Liu, M.P. Fitz, and O.Y. Takeshita. On the design of space-time and space-frequency codes for MIMO frequency-selective fading channels. *IEEE Transactions on Information Theory,* **49**(*9*):2277–2292, September 2003.

[25] Ericsson, Nokia, Motorola, and Rohde & Schwarz. Proposal for LTE channel models. Technical Report 3GPP TSG R1-02-0181, 3GPP TSG-RAN Working Group 4 (Radio), meeting 43, Kobe, Japan, May 2007.

[26] V. Erceg. IEEE P802.11 Wireless LANs, TGn channel models. doc.: IEEE 802.11-03/940r4, May 2004.

[27] M.M. Errasti. *Effects of Channel Estimation and Implementation on the Performance of MIMO Wireless Systems.* PhD thesis, Mondragon University, 2008.

[28] G.J. Foschini, D. Chizhik, M.J. Gans, C. Papadias, and R.A. Valenzuela. Analysis and performance of some basic space-time architectures. *IEEE Journal on Selected Areas in Communications,* **21**(*3*), April 2003.

[29] F.R. Farrokhi, A. Lozano, G.J. Foschini, and R.A. Valenzuela. Spectral efficiency of FDMA/TDMA wireless systems with transmit and receive antenna arrays. *IEEE Transactions on Wireless Communications,* **1**(*4*):591–599, October 2002.

[30] G.J. Foschini. Layered space-time architecture for wireless communication in a fading environment when using multi-element antennas. *Bell Labs Technical Journal*, pp. 41–59, 1996.

[31] G.J. Foschini. V-BLAST: an architecture for realizing very high data rates over the rich-scattering wireless channel. International symposium on *Signals, systems, and electronics,* ISSSE, pp. 295–300, 1998.

[32] H.T. Friis. A note on a simple transmission formula. *Proceedings of IRE,* **34**:254–256, 1946.

[33] M.J. Gans. A power-spectral theory of propagation in the mobile-radio environment. *IEEE Transactions on Vehicular Technology,* **21**(*1*):27–38, February 1972.

[34] D. Gesbert, H. Bolcskei, D.A. Gore, and A.J. Paulraj. Outdoor MIMO wireless channels: models and performance prediction. *IEEE Transactions on Communications*, **50**(*12*):1926–1934, December 2002.

[35] R.W. Heath, Jr. H. Bolcskei, and A.J. Paulraj. Space-time signaling and frame theory. In Proceedings. (ICASSP '01). 2001 IEEE International Conference on Acoustics, Speech, and Signal Processing, volume 4, pp. 2445–2448, 2001.

[36] A. Hjorungnes, D. Gesbert, and D.P. Palomar. Unified theory of complex-valued matrix differentiation. In ICASSP 2007. IEEE International Conference on Acoustics, Speech and Signal Processing, volume 3, pages III–345 to III–348, April 2007.

[37] B. Hassibi and B.M. Hochwald. How much training is needed in multiple-antenna wireless links? *IEEE Transactions on Information Theory*, **49**(*4*):951–963, April 2003.

[38] Roger A. Horn and Charles R. Johnson. *Matrix Analysis*. Cambridge University Press, 2nd edition, 2012.

[39] J.R. Hampton, N.M. Merheb, W.L. Lain, *et al.* Urban propagation measurements for ground based communication in the military UHF band. *IEEE Transactions on Antennas and Propagation*, **54**(*2*):644–654, February 2006.

[40] R.W. Heath, Jr. and A.J. Paulraj. Linear dispersion codes for MIMO systems based on frame theory. *IEEE Transactions on Signal Processing*, **50**(*10*):2429–2441, October 2002.

[41] Meng-Han Hsieh and Che-Ho Wei. Channel estimation for OFDM systems based on comb-type pilot arrangement in frequency selective fading channels. *IEEE Transactions on Consumer Electronics*, **44**(*1*):217–225, February 1998.

[42] Guidelines for evaluation of radio transmission technologies for IMT-2000. Technical Report ITU-2 M.1225, ITU-R, 1997.

[43] Hamid Jafarkhani. *Space-Time Coding: Theory and Practice*. Cambridge University Press, 2010.

[44] William C. Jakes, editor. *Microwave Mobile Communications*. Wiley-IEEE Press, 2nd edition, 1994.

[45] Yi Jiang, M.K. Varanasi, and Jian Li. Performance analysis of ZF and MMSE equalizers for MIMO systems: An in-depth study of the high SNR regime. *IEEE Transactions on Information Theory*, **57**(*4*):2008–2026, April 2011.

[46] J.P. Kermoal, L. Schumacher, K.I. Pedersen, P.E. Mogensen, and F. Frederiksen. A stochastic MIMO radio channel model with experimental validation. *IEEE Journal on Selected Areas in Communications*, **20**(*6*):1211–1226, August 2002.

[47] S. Loyka and F. Gagnon. Performance analysis of the V-BLAST algorithm: an analytical approach. *IEEE Transactions on Wireless Communications*, **3**(*4*):1326–1337, July 2004.

[48] X. Li, H. Huang, G.J. Foschini, and R.A. Valenzuela. Effects of iterative detection and decoding on the performance of BLAST. In *Global Telecommunications Conference, 2000. GLOBECOM '00.* IEEE, volume 2, 2000.

[49] 3GPP LTE main web site. `http://www.3gpp.org/Technologies/Keywords-Acronyms/LTE`.

[50] 3GPP; Technical Specification Group Radio Access Network; Evolved Universal Terrestrial Radio Access (E-UTRA); Physical channels and modulation (Release 9), TS 36.211 v9.1.0, September 2010.

[51] 3GPP; Technical Specification Group Radio Access Network; Evolved Universal Terrestrial Radio Access (E-UTRA); Physical layer procedures (Release 9), TS 36.213 v9.3.0, September 2010.

[52] 3GPP; Technical Specification Group Radio Access Network Evolved Universal Terrestrial Radio Access (E-UTRA); User Equipment (UE) radio access capabilities (Release 9), TS 36.306 v9.6.0, December 2011.

[53] Hung-Quoc Lai, B. Zannetti, T. Chin, *et al.* Measurements of multiple-input multiple-output (MIMO) performance under army operational conditions. In *Military Communications Conference, 2010 – MILCOM 2010*, pp. 2119–2124, 31, 3 November 2010.

[54] Carl D. Meyer. *Matrix Analysis and Applied Linear Algebra Book and Solutions Manual.* SIAM: Society for Industrial and Applied Mathematics, 2001.

[55] S. Plass, A. Dammann, and S. Sand. An overview of cyclic delay diversity and its applications. In *68th, Vehicular Technology Conference, 2008. VTC 2008-Fall.* IEEE pp. 1–5, September 2008.

[56] R. Penrose. A generalized inverse for matrices. In *Proc. Cambridge Philos. Soc*, volume 51, page C655. Cambridge University Press, 1955.

[57] Arogyaswami Paulraj, Rohit Nabar, and Dhananjay Gore. *Introduction to Space-Time Wireless Communications.* Cambridge University Press, 2008.

[58] Athanasios Papoulis and S. Unnikrishna Pillai. *Probability, Random Variables and Stochastic Processes.* McGraw-Hill Europe, 4th edition, 2002.

[59] John Proakis and Masoud Salehi. *Digital Communications,* 5th edition. McGraw-Hill Science/Engineering/Math, 2007.

[60] Eldad Perahia and Robert Stacey. *Next Generation Wireless LANs: Throughput, Robustness, and Reliability in 802.11n.* Cambridge University Press, 2008.

[61] Theodore S. Rappaport. *Wireless Communications: Principles and Practice*, 2nd edition. Prentice Hall, 2002.

[62] M. Rumney. IMT-advanced 4G wireless takes shape in an olympic year. Technical Report 59989-9793EN, Agilent Technologies, September 2008.

[63] M. Sandell and O. Edfors. A comparative study of pilot-based channel estimators for wireless OFDM. *Lulea Univ. of Technol., Lulea, Sweden, Res. Rep. TULEA*, 1996.

[64] R. Schober, W.H. Gerstacker, and L.H.-J. Lampe. Performance analysis and design of STBCS for frequency-selective fading channels. *IEEE Transactions on Wireless Communications,* **3**(*3*):734–744, May 2004.

[65] C.E. Shannon. A mathematical theory of communication. *Bell System Technical Journal,* **27**:379–423, July 1948.

[66] C. E. Shannon. A mathematical theory of communication. *Bell System Technical Journal,* **27**:623–656, October 1948.

[67] Hyundong Shin and Jae Hong Lee. Closed-form formulas for ergodic capacity of MIMO Rayleigh fading channels. In *Communications, 2003. ICC '03.* IEEE International Conference, volume 5, pp. 2996–3000, vol.5, May 2003.

[68] N. Seshadri, C.E.W Sundberg, and Weerackody V. Advanced technologies for modulation, error correction, channel equalization, and diversity. *ATT Technical Journal*, 1993.

[69] R. Stridh, Kai Yu, B. Ottersten, and P. Karlsson. MIMO channel capacity and modeling issues on a measured indoor radio channel at 5.8 ghz. *IEEE Transactions on Wireless Communications,* **4**(*3*):895–903, May 2005.

[70] Emre Telatar. Capacity of multi-antenna gaussian channels. *European Transactions on Telecommunications*, **10**(*6*):585–595, 1999.

[71] V. Tarokh, H. Jafarkhani, and A.R. Calderbank. Space-time block codes from orthogonal designs. *IEEE Transactions on Information Theory,* **45**(*5*):1456–1467, July 1999.

[72] V. Tarokh, H. Jafarkhani, and A.R. Calderbank. Space-time block coding for wireless communications: performance results. *IEEE Journal on Selected Areas in Communications,* **17**(*3*):451–460, March 1999.

[73] V. Tarokh, N. Seshadri, and A.R. Calderbank. Space-time codes for high data rate wireless communication: performance criterion and code construction. *IEEE Transactions on Information Theory,* **44**(2):744–765, March 1998.

[74] R. van Nee and R. Prasad. *OFDM for Wireless Multimedia Communications (Artech House Universal Personal Communications).* Artech House Publishers, 1999.

[75] P.W. Wolniansky, G.J. Foschini, G.D. Golden, and R.A. Valenzuela. V-BLAST: an architecture for realizing very high data rates over the rich-scattering wireless channel. In *Signals, Systems, and Electronics, 1998. ISSSE 98.* URSI International Symposium, pp. 295–300, 1998.

[76] C-X. Wang, X. Hong, H. Wu, and W. Xu. Spatial temporal correlation properties of the 3GPP spatial channel model and the Kronecker MIMO channel model. *EURASIP Journal on Wireless Communiations,* **21**(*1*):27–38, 2007.

[77] IEEE Std 802.11n-2009, Part 11: *Wireless LAN Medium Access Control (MAC) and Physical Layer (PHY) Specifications; Amendment 5: Enhancements for Higher Throughput,* October 2009.

[78] J.H. Winters. Switched diversity with feedback for DPSK mobile radio systems. *IEEE Transactions on Vehicular Technology,* **32**(*1*):134–150, February 1983.

[79] J.H. Winters. The diversity gain of transmit diversity in wireless systems with Rayleigh fading. In *Communications, 1994. ICC '94, SUPERCOMM/ICC '94, Conference Record, "Serving Humanity Through Communications."* IEEE International Conference, pp. 1121–1125, vol.2, May 1994.

[80] J.H. Winters. Smart antennas for wireless systems. *Personal Communications, IEEE,* **5**(*1*):23–27, February 1998.

[81] A. Wittneben. Basestation modulation diversity for digital simulcast. In *Vehicular Technology Conference, 1991. Gateway to the Future Technology in Motion., 41st IEEE,* pp. 848–853, May 1991.

[82] A. Wittneben. A new bandwidth efficient transmit antenna modulation diversity scheme for linear digital modulation. In *Communications, 1993. ICC 93. Geneva. Technical Program, Conference Record,* IEEE International Conference, volume 3, pp. 1630–1634, vol.3, May 1993.

[83] John M. Wozencraft and Irwin Mark Jacobs. *Principles of Communication Engineering.* Waveland Pr Inc., 1990.

[84] Leang Yeh. Simple methods for designing troposcatter circuits. *IRE Transactions on Communications Systems,* **8**(*3*):193–198, September 1960.

[85] Zheng Zhang, T.M. Duman, and E.M. Kurtas. Achievable information rates and coding for MIMO systems over ISI channels and frequency-selective fading channels. *IEEE Transactions on Communications,* **52**(*10*):1698–1710, October 2004.

[86] W. Zhang, X.-G. Xia, and P.C. Ching. High-rate full-diversity space–time–frequency codes for broadband MIMO block-fading channels. *IEEE Transactions on Communications,* **55**(*1*):25–34, January 2007.

[87] E. Viterbo and J. Boutros. A universal lattice code decoder for fading channels. *IEEE Transactions on Information Theory,* **45**(*5*):1639–1642, July 1999.

[88] Hu Jun and T.M. Duman. Graph-based detector for BLAST architecture. *In Communications 2007,* IEEE International Conference, pp. 1018–1023, June 2007.

[89] A. Al Rustamani and B. R. Vojcic. A new approach to greedy multiuser detection. *IEEE Transactions on Communications,* **50**(*8*):1326–1336, August 2002.

[90] M. Sellathurai and S. Haykin. Turbo-BLAST for wireless communications: theory and experiments. *IEEE Transactions on Signal Processing,* **50**(*10*):2538–2546, October 2002.

[91] D. Wubben, R. Bohnke, V. Kuhn, and K.-D. Kammeyer. MMSE extension of V-BLAST based on sorted QR decomposition. *IEEE 58th Vehicular Technology Conference 2003,* **1**:508–512, October 2003.

[92] V. Tarokh, A. Naguib, N. Seshadri, and A.R. Calderbank. Combined array processing and space-time coding. *IEEE Transactions on Information Theory,* **45**(*4*):1121–1128, May 1999.

Index

Printed in the United States
by Baker & Taylor Publisher Services